P9-BZJ-421

BATTLE of BRITAIN ILLUSTRATED

Changing of the technology guard—a Gladiator flies over a new Chain Home radar site.

BATTLE of BRITAIN ILLUSTRATED

Paul Jacobs
Robert Lightsey

McGraw-Hill
New York San Francisco Washington, D.C. Auckland Bogotá
Caracas Lisbon London Madrid Mexico City Milan
Montreal New Delhi San Juan Singapore
Sydney Tokyo Toronto

The McGraw-Hill Companies

Cataloging-in-Publication Data is on file with the Library of Congress.

1 2 3 4 5 6 7 8 9 0 DOC/DOC 8 7 6 5 4 3 2

ISBN 0-07-138545-2

The sponsoring editor for this book was Shelley Carr and the production supervisor was Pamela Pelton. This book was set in Centaur MT by North Market Street Graphics.

Printed and bound by RR Donnelley.

This book was printed on acid-free paper.

To Paul Jacobs Sr.: He did something really brave. He landed on Omaha Beach with the Rangers on D-Day and was eventually wounded five times. From him I learned respect for those who wear the uniform of their country and go in harm's way.

To Dolly Jacobs: From her I learned patience and compassion. She has listened to a lot of war stories; many, not for the first time.

To Susan Jacobs: A quilt of many facets; my best friend

—Paul Jacobs

To Susan Lightsey: My gentle critic, unfailing supporter, friend, and wife.

To Scott and Kristin: Who have given us such pride and pleasure.

—Bob Lightsey

CONTENTS

FOREWORD

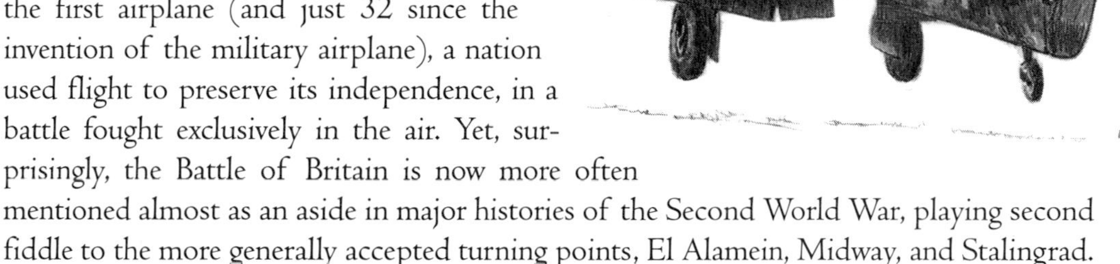

The Battle of Britain must be counted among history's great and decisive battles. Less than 40 years after the invention of the first airplane (and just 32 since the invention of the military airplane), a nation used flight to preserve its independence, in a battle fought exclusively in the air. Yet, surprisingly, the Battle of Britain is now more often mentioned almost as an aside in major histories of the Second World War, playing second fiddle to the more generally accepted turning points, El Alamein, Midway, and Stalingrad.

This is unfortunate, for the Battle of Britain was of seminal importance to the Allied victory in the war. The battle enabled the survival of Great Britain as an independent nation and, as such, as a base from which to project air power against the Continent and, ultimately, a seaborne invasion in 1944. From 1940 onward, Germany and German-occupied Europe came under increasingly fierce air attacks that targeted Nazi industry, transportation, and military targets. As a result, the Third Reich was not free to pursue its wartime objectives without diverting many resources, forces, and personnel to confront this aerial "second front." Acquisition strategies reflected this, as Germany's weapons acquisition increasingly began to stress defensive, not offensive weapons (antiaircraft artillery, production of antiaircraft shells and fuses, heavily armed day and night fighters, radars, etc.). Production of more offensive weapons—bombers, tanks, and submarines, for example—was de-emphasized. Further, even as German production increased to meet wartime needs, the rate of increase was far less than German planners desired because of the tremendous toll Allied air attacks were taking of German production resources, capabilities, and forces.

What impact did this have on the war itself? By 1943, according to Nazi medical records, Allied air attacks had become the primary cause of enemy casualties on the various fighting fronts. This continued through the end of the war. In 1944, when Dwight Eisenhower led the Allied invasion of Europe, he remarked to his son, "Without air supremacy I wouldn't be here." That in itself was a remarkable tribute to the accomplishments of the Allied airmen who, operating from bases in Great Britain, had carried the fight into the heartland of Hitler's odious regime. But perhaps the most telling comment of all came from Hitler's own propaganda chief, Joseph Goebbels, in late March 1945. Writing in his diary just weeks away from committing suicide, Goebbels recounted a conversation he had with Hitler himself, "Again and again we return to the starting point of our conversation. *Our whole military predicament is due to enemy air superiority.*"

The beginning of the Nazi collapse can be traced to one seminal event: the Battle of Britain. In 1940, in three months, the men and women of the Royal Air Force destroyed the Luftwaffe over southern England, dispersed the German navy's fleet of invasion barges and craft along the Channel coast, and threw off any timetables and plans that Hitler and his senior military leadership had for the total subjugation of Europe. Then, until joined by the United States and the other Allied air forces, they carried the fight back to the Third Reich. It was a victory—at tremendous cost—that contributed mightily to the saving of freedom. Had the battle been lost, the likely outcome would have been a long and draining war from afar as the United States sought to preserve its own freedom against a Germany essentially free of external threats and free as well to pursue a total war against the Soviet Union. The entire future of the war would have been in doubt, including the greatest (and gravest) issue of all, the race to develop and field the first atomic weapons.

The Battle of Britain is now over 60 years old, and it is critically important that it not go the way of the famous poem of the Battle of Blenheim, where little Peterkin asks what the battle was about, only to be informed vaguely "'Twas a famous victory." Thanks to Paul Jacobs, Robert Lightsey, and this remarkable book, the story of the battle is presented in an exciting and memorable fashion, certain to please those who are already aware of it and to inform those who are seeking information about it.

Dr. Richard P. Hallion
The U.S. Air Force Historian

INTRODUCTION

On September 15, 1940, shortly before three o'clock in the afternoon, Pilot Officer Paddy Stephenson's Hurricane collided with a Dornier Do 17 light bomber over Appledore, Kent. Stephenson, slightly injured, bailed out safely. A few hours later, Unteroffizier Heinrich Kopperschlager of 1 Staffel, JG 53, landed his Me 109 at Etaples, France. He had completed his last sortie of the day. Although neither pilot realized it at the time, Germany had just lost the Battle of Britain.

Somewhere in the middle of those events, you take off as a nervous 20-year-old leading a flight of fighters. Turning out of the traffic pattern, your wingmen join you in a fluid four-ship formation. You proceed toward the combat area. Your mission today is to engage enemy aircraft and destroy them so that the friendly bombers that will soon follow behind you can attack the factories that are producing airplanes for the other side. In the distance you spot a flight of enemy aircraft. Remembering what the 25-year-old "old-timers" in the squadron taught you, you climb and position your flight so that, when you attack, you will dive at them with the sun at your back. . . .

This relatively uncomplicated operational scenario symbolizes all of the elements that make up this book. For this situation to take place, there must be a coming together of interrelated factors that include policy, strategy, technology, and tactics. First a political decision must be made that commits the nation to a prescribed course of action that leads to war. High-level strategies must be developed to decide how to employ the nation's resources to support the national policies. In the scenario described, the decision to use bombers to attack another nation's manufacturing facilities represents a strategic decision. Furthermore, the technologies needed to conduct warfare are the products of decisions made about how to allocate and direct the resources of the nation involved. The fighters and bomber aircraft mentioned are the results of technology development programs selected to support the chosen strategies. Technology development includes not only the airplanes themselves, but also the communications, sensors, and weapons that have made an interception and attack possible. And, finally, tactics must be developed so that the pilots of the aircraft understand how to conduct the operation assigned. These tactics include the development of the flying formations used to assemble the fighters and bombers as well as the maneuvers used to position the fighters for an attack on an enemy formation. When all these factors are considered, even a seemingly uncomplicated event can be quite complex.

This book addresses these intertwined factors by looking at the Battle of Britain from four perspectives—historical, technical, tactical, and, as the title suggests, visual. This is accomplished by organizing the book around four timelines. The first timeline is a summary of the political events that led from the last years of World War I through October 1940, when the Battle of Britain ended. The second illustrates the major technical developments that evolved over the same period. The third timeline shows the evolution of tactics and strategy during the same period. In most cases, a battle outcome happened because of tactics used, which were dictated by technology available, which was the result of a political decision. This interrelationship makes up the fourth, how-and-why, timeline. It also explains why the fourth timeline runs backward, if you will.

Of the four perspectives mentioned, the fourth is that of the aviation artist. Artwork has always been an effective way to make events and relationships come alive for the reader. The Battle of Britain stands as a singular event, not only for military historians, tacticians, and technologists, but for aviation artists as well. It has come to epitomize the romance of air warfare. The Hurricane, Spitfire, and Me 109 have a special hold on all aviation enthusiasts. The Spitfire, with its classic, graceful lines, might be the most popular aviation art subject ever. Artists' fascination with this battle lies partly in the fact that this was a period of transition in aircraft design. This one battle saw the use of biplanes, monoplanes, stressed skin and fabric, fighters, bombers, and single- and multiengine airplanes. To tell this story, 33 different aircraft have been illustrated. We have used the visual perspective to present the Battle of Britain with a fresh approach.

The chapters are presented in three levels of detail. First, a timeline drawing provides a condensed, visual summary of the battle from a specific perspective. A summary article then follows the timelines. These summaries provide a top-level overview of the events shown in the timeline drawing. For the chapters on technology and tactics, a third level of detail is provided. Key events on those timeline drawings are numbered. The numbers direct the reader to an accompanying drawing and illustration in the chapter that provides more information about the significance of that particular subject. In addition, a few good stories have been added throughout.

The reader will find that some of the information referred to in one timeline summary also appears in others. It is the intertwining of history, technology, and tactics that makes this necessary. You might think of the architecture of a house that can be described from many different perspectives—structural, electrical, heating and ventilation, to name but a few. A description that addresses only a single perspective is necessarily incomplete. This need for multiple perspectives is equally true for history. As we describe the Battle of Britain from the tactical and strategic perspective, for example, we will refer to events already addressed in the earlier political or technical summaries; however, we hope to give the reader a more complete picture of the importance of the event by approaching it from a different perspective.

In addition to overlaps and similarities depicted in the first three timelines, there are also distinct differences between the way events in the three areas evolve. For example, the political events that define history often tend to develop in a fairly even and measured pace. Technical developments, on the other hand, tend to be characterized by periods of relatively low visibility base building, which are then followed by periods of explosive development as new designs are synthesized from the technologies developed earlier. Defense technologies tend to see their most rapid evolution during periods of preparation for conflict, and they are usually the product of many minds in collaboration. Tactics, on the other hand, tend to evolve most rapidly during periods of war, when survival is at stake and lives are on the line. Furthermore, tactical developments are often attributed to a single, innovative individual. In addition, good tactics tend to endure for years; air forces all over the world are still using the formation tactics developed by the German ace, Moelders, during the Spanish civil war. Technology, to the contrary, never lasts; what is good today is soon replaced by what is better. Look for these differences, as well as the similarities and interrelationships, among the events that define the Battle of Britain.

In the deep recesses of your imagination you can hear the faint, unsynchronized drone of the He 111's twin engines in the distance. You feel the dry thickening in your throat as the Tannoy crackles. Come join us as the shout of "Red flight scramble!" reverberates across the grass airstrip. Pull on your leather flying helmet as you run toward your Spitfire and relive the Battle of Britain through this illustrated account.

BATTLE
of
BRITAIN
ILLUSTRATED

CHAPTER 1

THE HISTORICAL TIMELINE

TIMELINE OVERVIEW

The Battle of Britain, fought during a few months in the summer of 1940, changed the course of history. It pitted a relatively few people—primarily the officers and men of the Royal Air Force's Fighter Command—against a superior German air force, fighting huge odds until their reserves, both machines and men, were exhausted. This tiny force, flying aircraft that are still recalled for their beauty and capability today, halted the German juggernaut when the entire world thought that task impossible. They showed that a marriage of technology and human resolve produces a formidably strong force. The battle that they won at huge personal cost provided the time, the resolve, and the physical stage needed to build the Allied coalition and launch the counterattack that finally won this greatest of human conflicts. Western history would be a different story today had the Battle of Britain been recorded as a German victory.

There is little new under the sun. What seems so novel today can often be traced to the events and efforts of yesterday. We are, at times, tempted to recall World War I as a time of trench warfare and its associated horrors and to think that aerial warfare and strategic bombing were inventions of the later twentieth century. That would be a mistake. Much of what took place during the Battle of Britain had roots in the events that took place during and after World War I. It was during World War I that the Germans introduced the concept of total warfare—warfare where cities and civilians are considered legitimate targets for military action.

Imagine yourself in England—London—in 1915. Your country is engaged in the war, having sent an expeditionary force to fight with the Allies in France against the German Army. But life goes on in England. Even though you are close to the war in terms of

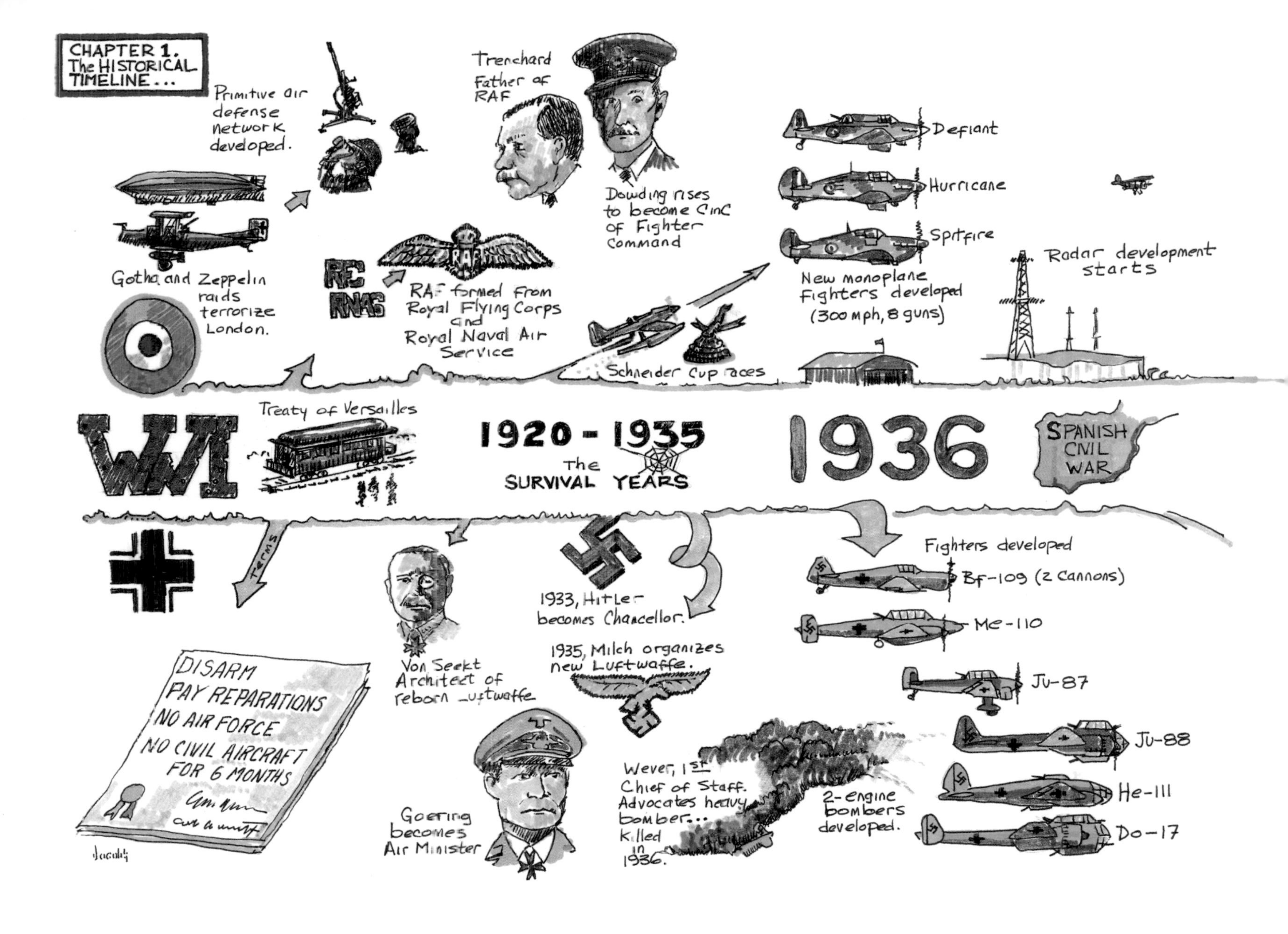

CHAPTER 1. The HISTORICAL TIMELINE...
Primitive air defense network developed.
Trenchard Father of RAF
Dowding rises to become CinC of Fighter Command
Defiant
Hurricane
Spitfire
Gotha and Zeppelin raids terrorize London.
RFC
RNAS
RAF
RAF formed from Royal Flying Corps and Royal Naval Air Service
New monoplane Fighters developed (300 mph, 8 guns)
Radar development starts
Schneider Cup races
Treaty of Versailles
WWI
1920 - 1935
The SURVIVAL YEARS
1936
SPANISH CIVIL WAR
Terms
Fighters developed
Bf-109 (2 cannons)
Me-110
1933, Hitler becomes Chancellor.
1935, Milch organizes new Luftwaffe.
Von Seekt Architect of reborn Luftwaffe
Ju-87
Ju-88
He-111
Do-17
DISARM
PAY REPARATIONS
NO AIR FORCE
NO CIVIL AIRCRAFT FOR 6 MONTHS
Wever, 1st Chief of Staff. Advocates heavy bomber... Killed in 1936.
2-engine bombers developed.
Goering becomes Air Minister
Jacoby

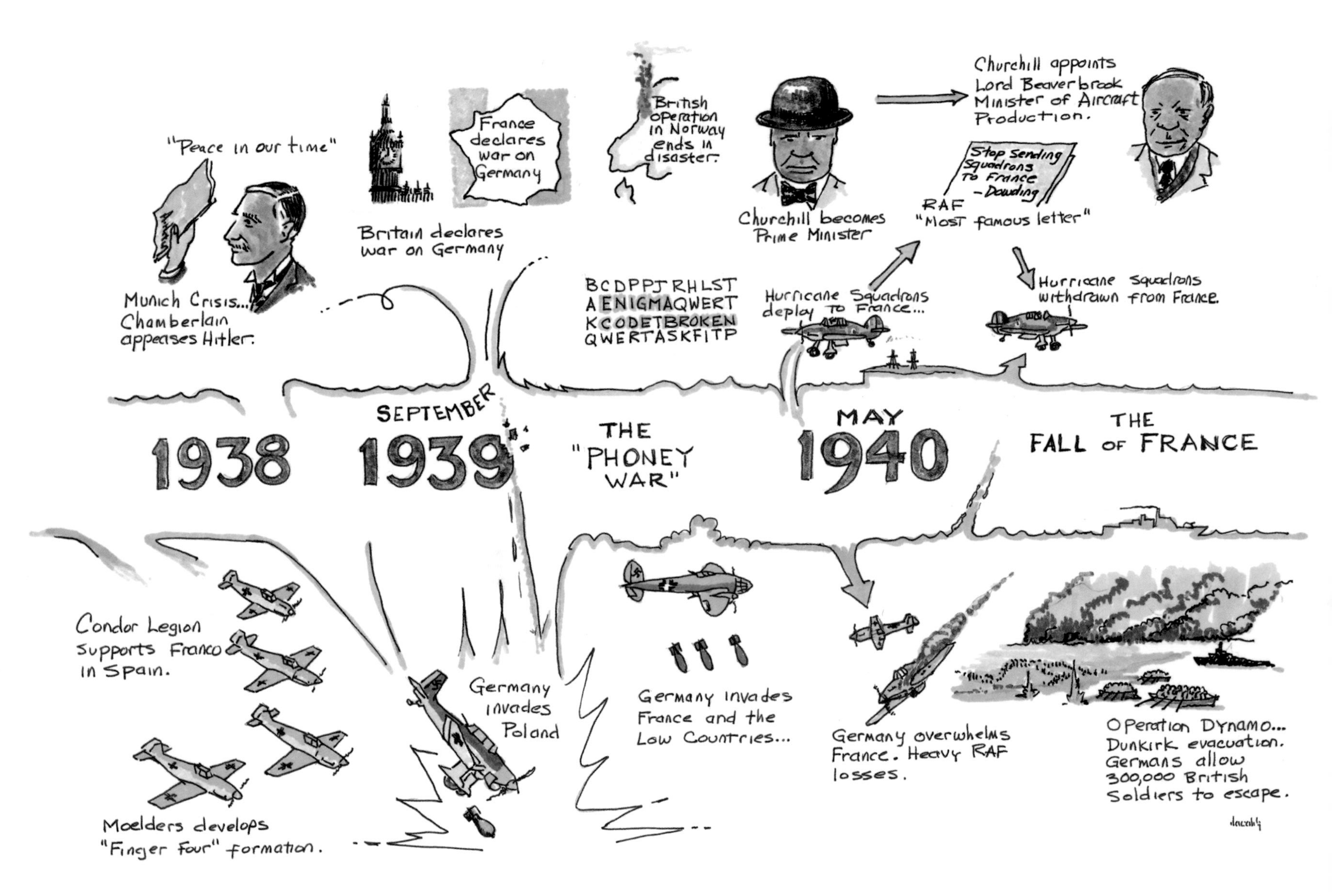

"Peace in our time"
Munich Crisis... Chamberlain appeases Hitler.
Britain declares war on Germany
France declares war on Germany
British operation in Norway ends in disaster.
Churchill becomes Prime Minister
Churchill appoints Lord Beaverbrook Minister of Aircraft Production.
Stop sending Squadrons to France – Dowding
RAF "Most famous letter"
BCDPPJRHLST
AENIGMAQWERT
KCODETBROKEN
QWERTASKFITP
Hurricane Squadrons deploy to France...
Hurricane squadrons withdrawn from France.
1938
SEPTEMBER
1939
THE "PHONEY WAR"
MAY
1940
THE FALL OF FRANCE
Condor Legion supports Franco in Spain.
Moelders develops "Finger Four" formation.
Germany invades Poland
Germany invades France and the Low Countries...
Germany overwhelms France. Heavy RAF losses.
Operation Dynamo... Dunkirk evacuation. Germans allow 300,000 British Soldiers to escape.

June 4th
Churchill's "We will fight them on the beaches" speech.
June 18th
Churchill gives "their finest hour" speech. Coins term "Battle of Britain."
British destroy French Fleet
Britain prepares for Invasion
13 Group SAUL
12 Group LEIGH-MALLORY
10 Group BRAND
11 Group PARK
RAF Order of Battle
British run convoys at night only
28 July In battle of aces, Malan forces Moelders to crash land
Ambassador Joe Kennedy claims Britain will lose. Evacuates Embassy staff.
Britain rejects pecce offer
First day score: RAF 45 Luft. 13
FRANCE SURRENDERS
JUNE 1940
(same rail car from WWI)
10 JULY OFFICIAL START BATTLE OF BRITAIN
The "CHANNEL PHASE"
AUGUST 1940
13 AUGUST ADLERTAG
Luftwaffe stands down to recover from Battle of France
Convoys attacked in Channel
First 100-plane dogfight
16 July
DIRECTIVE 16
"SEALION" Prepare For Invasion
1 AUGUST
DIRECTIVE 17
"OVERPOWER RAF SOONEST" Prepare for invasion by Sep 15
?
?
Initial confusion. Then 1500 sorties in 24 hours
Luftwaffe ORDER OF BATTLE for "KANALKAMPF"
Luftflotte 5 Stumpf
Luftflotte 2 Kesselring
Luftflotte 3 Sperrle
Hitler "Peace offer" speech in Reichstag
Goering orders
"Adlerangriff" (Eagle attack)
Jacobi

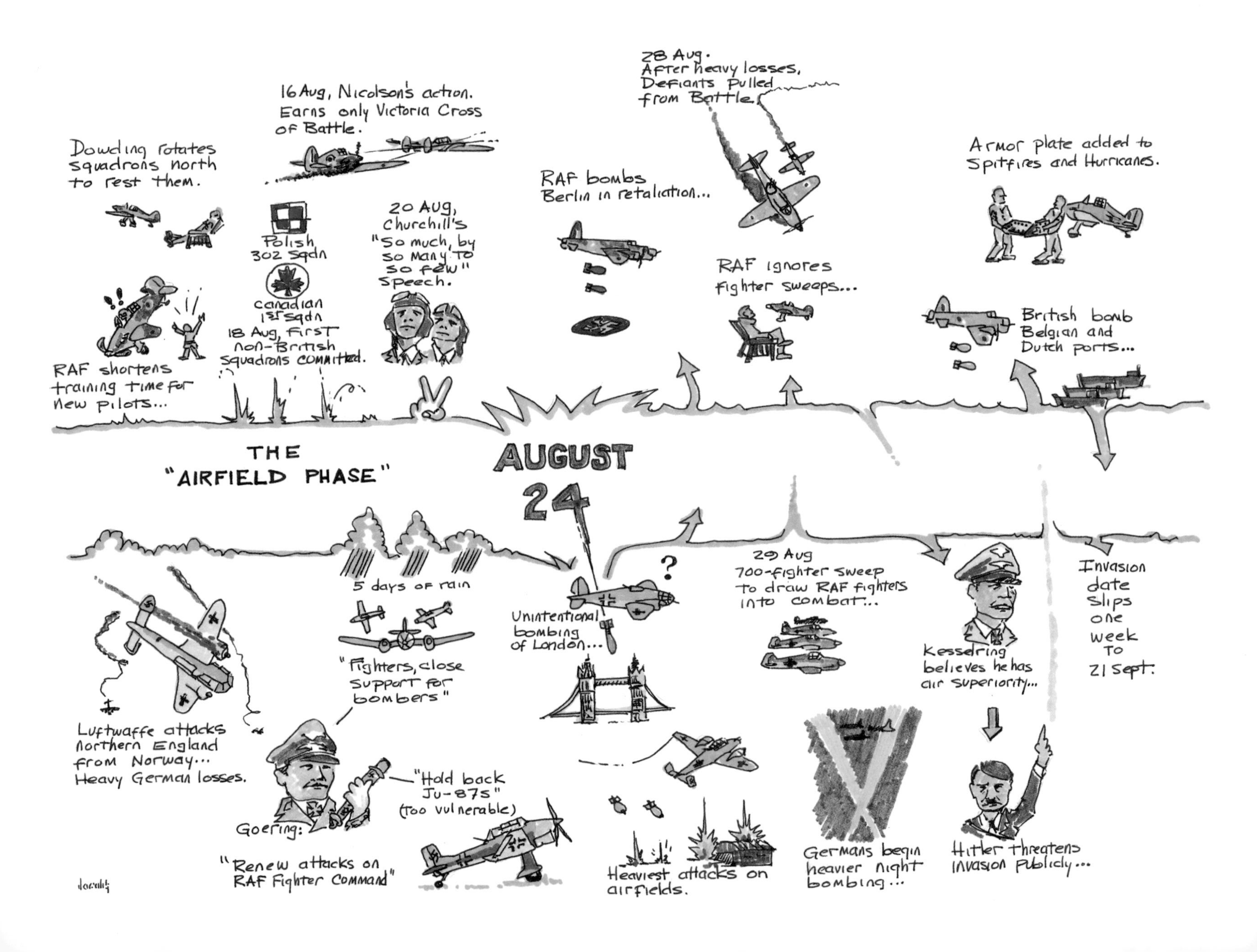

Dowding rotates squadrons north to rest them.
16 Aug, Nicolson's action. Earns only Victoria Cross of Battle.
Polish 302 Sqdn
Canadian 1st Sqdn
18 Aug, First non-British squadrons committed.
20 Aug, Churchill's "So much, by so many to so few" speech.
RAF shortens training time for new pilots...
RAF bombs Berlin in retaliation...
28 Aug. After heavy losses, Defiants pulled from Battle.
RAF ignores fighter sweeps...
Armor plate added to Spitfires and Hurricanes.
British bomb Belgian and Dutch ports...
THE "AIRFIELD PHASE"
AUGUST 24
5 days of rain
"Fighters, close support for bombers"
Unintentional bombing of London...
?
29 Aug 700-fighter sweep to draw RAF fighters into combat...
Kesselring believes he has air superiority...
Invasion date slips one week to 21 Sept.
Luftwaffe attacks northern England from Norway... Heavy German losses.
"Hold back Ju-87s" (Too vulnerable)
Goering:
"Renew attacks on RAF Fighter Command"
Heaviest attacks on airfields.
Germans begin heavier night bombing...
Hitler threatens invasion publicly...

31 Aug
RAF single
worst day.
39 aircraft
lost.

British issue
invasion alert

"Cromwell"
Highest alert
status.

Airfields and
Fighter Command recover

Airfield Phase ends
Score: RAF 670
Luft. 400

Park to Churchill:
"there are no
reserves."

24 RAF
Squadrons
committed

Aircraft factories
hit

Night bombing continues...

SEPT
1940

THE
"LONDON
PHASE"

**BATTLE
OF
BRITAIN
DAY
15 sept**

OCTOBER
1940

Luftwaffe switches
targets. Begins bombing
London...

7 Sept.
1000-plane
raid.

2 large raids
over London and
Southeast coast.
Largest ever.

Rain 5 days

17 Sept.
Hitler postpones
invasion, but
Preparations
continue.

30 Sept.
Last day of
large daylight raids

Battle of Britain
Day score:
RAF 60
Luft. 26

German fleet
dispersed

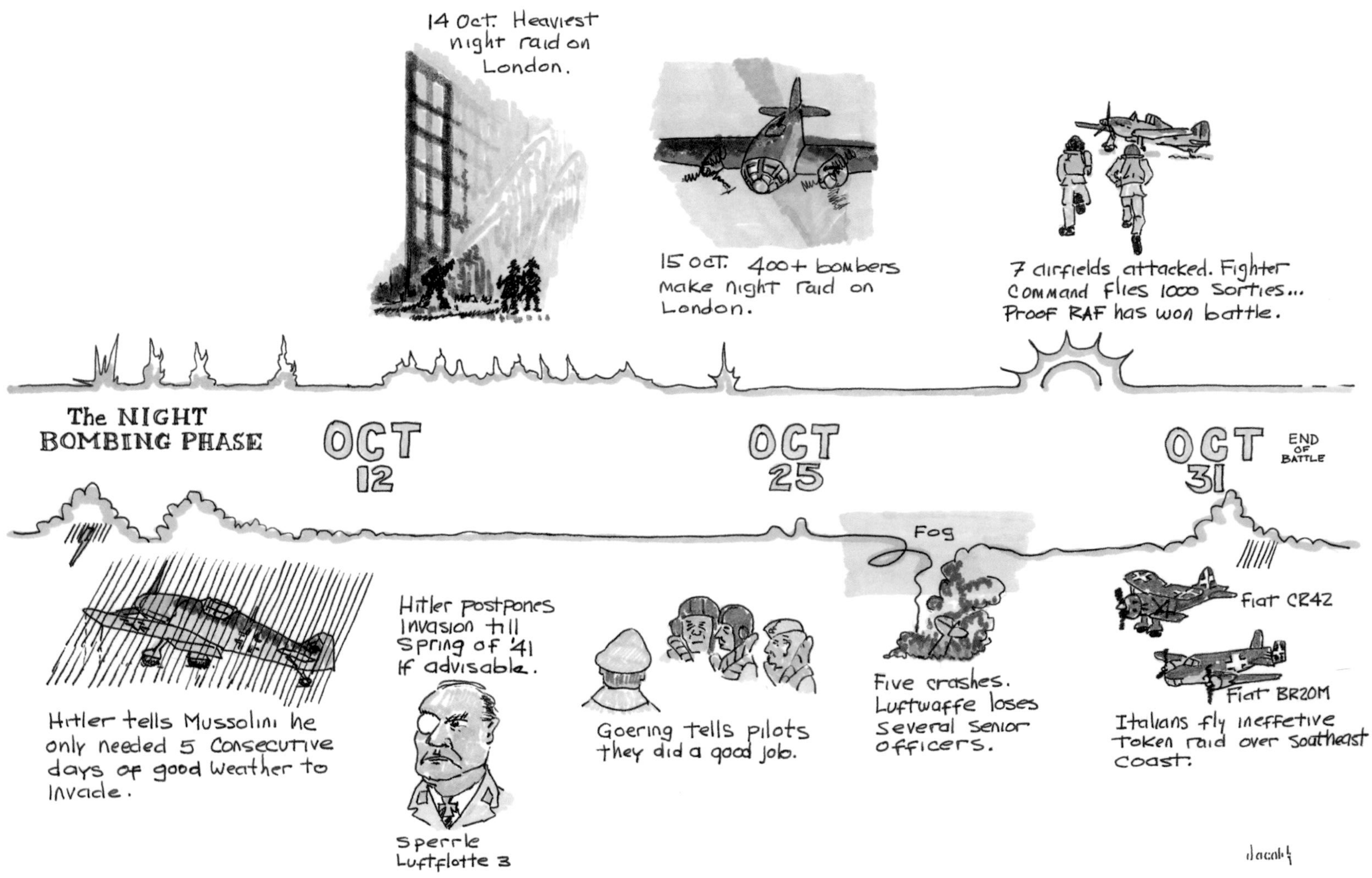

The NIGHT BOMBING PHASE

OCT 12

14 Oct. Heaviest night raid on London.

Hitler tells Mussolini he only needed 5 consecutive days of good weather to invade.

Hitler postpones invasion till Spring of '41 if advisable.

Sperrle Luftflotte 3

15 Oct. 400+ bombers make night raid on London.

Goering tells pilots they did a good job.

OCT 25

Fog

Five crashes. Luftwaffe loses several senior officers.

7 airfields attacked. Fighter Command flies 1000 sorties... Proof RAF has won battle.

OCT 31

END OF BATTLE

Fiat CR42

Fiat BR20M

Italians fly ineffetive token raid over southeast coast.

Jacobs

Moelders Meets Malan *On July 28, each side's top gun met near Dover. England's most famous ace, Sailor Malan, led his 74 Squadron to intercept approaching bombers. At the last minute, the bombers turned away. Malan's Spitfires then met the four escorting squadrons of Me 109s led by the equally famous Werner Moelders, inventor of the "finger four" formation. Moelders attempted to turn onto Malan's tail, but Malan was quick enough to preemptively turn into his foe and rake the Me 109 with gunfire from nose to tail, wounding Moelders in the leg. The German managed to disengage and nurse his ailing aircraft back across the Channel, where he crash-landed at Wissant.*

Nicolson's VC *The only Victoria Cross won by a pilot during the battle was awarded to Flight Lieutenant James Nicolson. On August 16, he attacked a group of Me 110s near Southampton. On the run-in, he was attacked by Me 109s. His Hurricane was hit by cannon fire and burst into flames. He began to bail out, but a 110 overshot and ended up in front of him. He resumed his position and fired at the 110, sending it crashing into the sea. He then bailed out, suffering from severe burns. His final brush with death came during the parachute descent. Locals on the ground saw his and his wingman's parachutes and assumed they were part of an enemy parachute assault. The locals began firing at the two airmen. Nicolson survived, but his wingman did not.*

miles, it is a distant affair on a sunny day in London. Then you notice a dirigible—a huge airship cruising gracefully, carried by the breeze, silent in its approach. You watch, fascinated by the balloonlike craft, at once ponderous and graceful. As it nears the city, you notice something falling from the dirigible. And then the explosions start—violent, each one progressively nearer to where you stand. And it begins to dawn on you that war is no longer a remote thing happening somewhere "over there." War has come to Britain.

The Germans initially used Zeppelin airships to bomb Britain. As Britain improved her defensive tactics, the vulnerabilities of the airships rendered them less and less effective. When it became apparent that a greater level of destruction was required than was achievable by the airships, Germany turned to the Gotha bomber aircraft as the primary attack weapon. In May 1917, 23 Gothas attacked the town of Folkestone, killing 95 people. This was followed in June by an attack on London, which killed 162 (Mason 1969). A few weeks later, a raid was conducted against the village of Harwich. Eighty-three British fighter-interceptors were launched; not only did they not stop the raid, none even saw the bombers. Then, in July, another large attack was mounted against London. Outraged as much as injured, the British people demanded that the government take action. To better anticipate and intercept these raids and to provide the population with advance warning, the British developed a version of an early warning system. Over time, a well-organized control system was developed. The British established control zones with a structured communication hierarchy. They designed a layered defense system around London consisting of a gun belt, interceptor aircraft, and artillery. This was augmented by a system of arrow pointers to direct the fighters toward the positions of attacking bombers and controllers who tracked the progress of attacking forces on plotting tables. The British, in building this system, had recognized a critically important fact in air defense—as important as interceptor aircraft are, they are impotent without a system to communicate and control them. This rudimentary early warning and control system provided the foundation upon which was built the control system that would enable the RAF to operate with such efficiency in 1940.

At the time, the air forces of Britain were divided into two arms, the Royal Flying Corps, which was the air arm of the army, and the Royal Naval Flying Service. The inclination on the part of each service was to use its air force to accomplish unique tasks and missions assigned to the service branch. The army preferred to use air forces to provide reconnaissance for ground forces in France; the navy preferred to use their air forces to provide coastal defense. The demands for improved defense finally led to the conclusion that an air arm independent of army and navy control was needed. In April 1918 an independent air arm, the Royal Air Force, was created by merging the two service-controlled air branches. Major General Hugh Trenchard, commander of the Royal Flying Corps, was selected to be the first chief of staff of the new RAF.

World War I finally ground to a halt in late 1918. In November, the Germans surrendered, and in June 1919 the Allies and the Germans met at Versailles to sign the peace treaty that the Allies expected would put an end to all war. The treaty was laden with restrictions intended to ensure that the events of World War I would never be repeated, but like many regulatory documents written to prevent past problems, unintended consequences intervened. Rather than ensuring that war would never occur again, the onerous restrictions, in fact, guaranteed that war would certainly occur again—between the same countries and much sooner than anyone might have anticipated. The treaty prohibited military aviation in Germany; it prohibited submarines; and it required that Germany pay reparations to the conquering Allies. The Germans were required to give all aircraft engines to the Allies, and the terms of the treaty further stipulated that Germany could possess no civil aircraft for at least six months. This last clause proved to be a mistake. It was a loophole that enabled the Germans to begin early development of aircraft which later were easily converted into bombers and transports. Germany was left with a token army—having been allowed to retain a force of 100,000, commanded by General Hans von Seeckt. The treaty was to become a constant reminder of the humiliation and deprivation that a proud nation suffered. The Germans were given five days to agree to the terms dictated; they signed the treaty on June 28, 1919. Finding ways to avoid compliance with the terms of the treaty was to become an obsession for the core of the new German leadership that would emerge and lead Germany into World War II and the Battle of Britain.

With the war at last over, the British moved quickly to disband the armed forces and to turn their economy to peacetime production. Faced with both budgetary constraints and political infighting, Trenchard battled continuously to keep the young RAF from being absorbed back into the two traditional armed forces. Eighteen months after the armistice was signed, the RAF had been reduced from 188 operational squadrons to a total of 25.

In 1921 an Italian general and theorist, Giulio Douhet, wrote a book asserting that air power would be the determinant of future conflicts between nations. General Trenchard followed this shortly with a recommendation that the RAF take over the traditional homeland defense role that had heretofore been assigned to the navy, saying that air power had now reduced the value of battleships as weapons in the context of the defense of Britain. This served only to heighten the animosity of the Admiralty toward the upstart RAF. A long and arduous political battle ensued that pitted the RAF against the Admiralty and often involved the civilian government. Finally, in August 1923, the prime minister decided in favor of retaining an independent air force. One might argue that winning the Battle of Britain began on that day. Until he retired in 1929, General Trenchard, with a single-minded determination that recognized no insurmountable barriers, proceeded to build the RAF as a technically capable, highly motivated professional armed service, worthy of the confidence and trust of the British people.

General von Seeckt, meanwhile, was encouraging the regeneration of German military forces. Resurrection of a viable air force was a theme that resonated with the frustrated Germans. The problem was, of course, that the Versailles treaty explicitly prohibited that kind of activity; consequently, the campaign had to be undertaken in secret. In 1922 Germany entered into a treaty with Russia to provide technical know-how in exchange for German access to secret military bases in Russia. A base at Lipetsk became a training site where hundreds of German pilots were eventually trained, unknown to the Allies. In 1922, even as the British aircraft industry contracted, the Allies agreed to allow the Germans to once again build civil aircraft. This allowed Junkers, Dornier, and Heinkel—mainstays of German aircraft production—to begin producing in factories located inside Germany (they had already been producing secretly in factories outside the country). Soon after, glider clubs proliferated throughout Germany, becoming a training ground for future combat pilots. In 1926, the national airline, Lufthansa, was founded. Flying in all weather conditions and to locations throughout Europe and the rest of the world, Lufthansa became a veritable test bed for concepts associated with flying, air traffic control, and the employment of large aircraft—all of which would be important as Germany rebuilt her military capability.

Toward the end of the 1920s, Europe and the rest of the world fell into a deep economic depression. During this period, Hermann Goering, who had taken over the squadron led by Manfred von Richthofen, the famous Red Baron, upon his death in World War I, was rising in the German political establishment. He became a deputy in the Nazi party, elected in 1928 with Adolf Hitler's support. In January 1932 the Nazi party took control of the Reichstag, and Goering was elected president. In 1933 Hitler became the chancellor of the Third Reich, and with that, war in Europe became inevitable. As the government began the transition to a militaristic footing, Hitler selected Goering as his new air minister. His deputy was Erhard Milch, a brilliant organizer and the man most responsible for the organization and development of Lufthansa. Using the skills and experience gained in building Lufthansa, Milch would become the principal architect and organizer of the new German Luftwaffe.

As the Germans continued to secretly build their air forces, the British were also beginning to emerge from their postwar hibernation and recognize that all the signs pointed to a serious imbalance in air forces between themselves and their old rival. Britain became active in aircraft design and development. In 1931 a British Supermarine S.6B aircraft won the Schneider Trophy race for the third time. This highly prestigious trophy was awarded to the fastest floatplane designs, and by winning it three years running Britain claimed permanent ownership, a tribute to her emerging designs. Three years later, the S.6B design led to a new aircraft and power plant, the Supermarine Spitfire and the Merlin engine.

In 1934 a young captain of the RAF briefed the Air Ministry on his calculations, concluding that the average pilot would need eight guns firing at the rate of 1000 rounds per minute to down a bomber. This led to the development of a new fighter aircraft specification (F.5/34) which required that designers develop an eight-gun fighter with retractable landing gear and a closed cockpit. Britain was moving into the modern age in fighter technology. In 1934 the government announced that the RAF would be expanded by 41 squadrons. Between 1935 and 1937 the inventory of combat aircraft doubled from 1020 to 2031, although most were outdated designs harking back to World War I technologies (Mason 1969). The real challenge became the design, development, and production of the newer Spitfire and Hurricane fighter aircraft that were emerging in response to Specification F.5/34 and its updated version, F.10/35. The Hurricane, which was employed in more numbers than any other fighter during the Battle of Britain, was a follow-on design to the Hawker Fury biplane. The design was almost complete in 1935, although the final design was altered to meet the requirements of the new specification. The Spitfire, on the other hand, although based on earlier designs for the Supermarine S6.B, was a more revolutionary development than was the Hurricane. Not all designs were as successful as the Spitfire and the Hurricane. The Gloster Gladiator entered service in 1937. A biplane, it was hopelessly behind the times technologically and was largely ineffectual against the German monoplanes. Similarly, the Boulton-Paul Defiant entered service in 1939 after production delays. This unfortunate design not only performed poorly in dogfights, but was almost impossible to escape from once hit. It was finally employed as a night fighter, but was not capable of performing that mission well either.

As aircraft design and development proceeded at breakneck speed, another development was under way that would rival the single-winged fighter airplane as the most important technical achievement of the period. Robert Watson-Watt, a scientist, demonstrated to the Air Ministry in 1935 that he could detect and track an aircraft using reflected radio waves. Within nine months, the Air Council would recommend that a chain of linked transmitting and receiving stations be constructed along Britain's east and southeast coastline for the purpose of detecting and tracking aircraft. This became the Chain Home radar system that was as important in the British victory as was any aircraft.

The rapid growth of the RAF was creating organizational and command problems. The demands of managing both the bomber forces and the fighter forces had become too much for a single individual. It became apparent that new organizational arrangements were needed to provide direction and control. The solution reached was to divide the force into four operational commands: Fighter, Bomber, Training, and Coastal Commands. In 1936, Air Marshal Hugh Dowding, a longtime veteran of the research and development branch of the Air Ministry, was selected as the first commander in chief of Fighter Command. By supporting the need for technologically advanced fighters and the development of radar-based early warning, he was to become one of the people most responsible for the British victory in the Battle of Britain.

The German march to war continued. In 1935, the Luftwaffe ordered a new, advanced aircraft designed by Willy Messerschmitt, the Bf 109, following a competitive fly-off against the Heinkel He 112. Since Germany lacked engines, the early prototype was powered by a British Rolls-Royce engine. Messerschmitt evolved the 109 from an earlier model, the Bf 108, a civil touring aircraft. It was a low-wing monoplane that employed advanced aeronautical engineering to achieve its outstanding performance. By 1945, the Bf 109 was the most widely produced fighter of World War II. This evolutionary approach to design and development, where one design is the basis for the next, was a favored means of development in Germany. The evolutionary approach was used to produce many of the bomber aircraft used in 1940 by adapting designs based on earlier transport aircraft. Among these were the Junkers Ju 88, the Dornier Do 17, and the Heinkel He 111. However, of the three, only the Ju 88 performed well against British defenses in the Battle of Britain.

In June 1936, Luftwaffe Chief of Staff General Walther Wever was killed while piloting an aircraft near Dresden. This was a particularly unfortunate accident from the German perspective. General Wever, a student of strategic bombing, had been a strong advocate of the four-engine strategic bomber as the mainstay of German bomber forces. Wever sought a heavily armed, long-range bomber capable of bombing all of Great Britain from German bases. At the time of his death, Germany had the lead in bomber production; Britain had only recently issued the specification for a four-engine bomber. However, upon his death, the German High Command quickly abandoned the idea of the four-engine long-range bomber in favor of producing more two-engine bombers. Hitler had made it clear that he was more interested in the absolute numbers of bombers than in the range and bomb load capacity they possessed. Had General Wever lived, long-range bombers carrying huge bomb loads and operating from the German homeland would likely have been a feature of the Battle of Britain and, quite possibly, would have had a significant impact on the outcome.

The Spanish civil war began in July 1936. Hitler, on Goering's advice, immediately announced his intent to support Franco. It was a perfect opportunity to test the strength of the Luftwaffe, built under the cloak of secrecy for over 10 years, and to provide German armed forces with combat experience. Composed of both air and ground forces, the Condor Legion departed Germany disguised as civilians and proceeded to Spain. In April 1937, German Heinkel bombers attacked Guernica with a level of violence that even today symbolizes the horror of war. The devastation wrought by strategic bombing raised, yet again, the specter of total warfare and the targeting of civilian populations. For some, it had the effect of reconfirming the primacy of the bomber earlier suggested by Douhet. The Luftwaffe used the opportunity to develop and refine the tactics that they would use in World War II. As an example, the renowned fighter pilot, Werner Moelders, developed the loose "finger four" combat formation that is still used by tactical fighters today. Moelders himself was a flying ace in the Battle of Britain, with 54 kills (Mason 1969). Rotating assignments regularly in Spain, Germany entered the Battle of Britain with an air force whose training and experience were second to none in the world.

War really began with Germany's invasion of Austria in March 1938. Hitler's primary ambition had always been expansion eastward, so it became important to him to ensure that neither France nor England moved against him as he fulfilled his expansionist objectives. The British government, seeking to avoid war at all costs, accommodated Hitler by authorizing Prime Minister Neville Chamberlain to negotiate a nonaggression pact. Returning from the Munich Conference in September 1938, Chamberlain declared that the agreement assured "... peace in our time." In March 1939 Hitler took control of Czechoslovakia. Next would be Poland, but Hitler feared the Russians, who considered Poland to be within their sphere of influence. To prevent war with Russia, Hitler negotiated a nonaggression pact in which he promised to give Russia a part of Poland after it was defeated. The treaty was agreed to in August 1939; on September 1, Hitler attacked. On September 3, a devastated Chamberlain informed the British people that they were now in a state of war with Germany. Whatever else might be said of Chamberlain's handling of Germany, the period between the Munich Conference and September 1939 was, in some ways, a godsend. It enabled British pilots to train for the coming war; it provided time to build the airfields and aircraft that would be needed; it made possible the completion of the Chain Home radar system; and it united the British people behind the war effort to an extent that would have been impossible a year earlier.

As the dark curtain of the blackout fell across Britain, the country waited for the war that would inevitably follow. But Britain was hardly ready for war in September 1939, nor was Germany yet in a position to attack her. So, while war had been declared, the activity in England consisted mostly of a headlong rush to train and equip the forces that would defend the country when the shooting started. New aircraft were needed to replace the obsolete airplanes that comprised so much of the inventory; pilots needed training in the newer aircraft; controllers needed practice in identifying targets and guiding interceptors; and the last links in the Chain Home system had to be installed. The American press, expecting a shooting war, labeled this period the "Phony War." It could not last forever. In April 1940 Hitler invaded Denmark and Norway; in May he invaded the Low Countries and France. On May 10, 1940, Churchill became the prime minister of England.

During this period, Dowding's challenge became not only to prepare his forces for war, but also to avoid events and situations that worked to reduce his forces unnecessarily. In June, following a disastrous sequence of events that resulted in the loss of Norway to the Germans, the Allies set out to destroy the Norwegian port of Narvik, a source of iron ore for the German war machine. Dowding's task was to send aircraft to provide air cover for the attacking Allies. The Allies succeeded in taking the port, which they destroyed and then left. The RAF aircraft and men were then boarded on the carrier *Glorious* for the return trip to England. The next day they were attacked at sea and sunk by two German ships. The campaign cost Dowding a total of three squadrons of aircraft, including combat losses, plus two squadrons of trained pilots.

The impact of these losses on RAF strength, however, paled in significance compared to continuing demand for air support for the French and Allied ground forces in Europe. Four squadrons of Hurricanes were sent to support the six already there. They flew into a hornet's nest of activity. As Germany raced across Belgium, the Allies begged for still more aircraft to be sent. Churchill was inclined to support the call for more help. Finally, Dowding wrote the Air Ministry what is generally recognized as the most famous letter in the history of the RAF. In it he made the case that, if Britain continued to send fighter support to Europe in what was apparently a lost cause, the result would inevitably be that she would be unable to defend herself against German aggression. That letter carried the day, and on May 19, Churchill announced that no further fighters would be sent to France. Of 250 Hurricanes employed in France, only 66 returned; Britain had lost one-half of her Hurricanes before the battle had even begun. The need for fighter production became a crisis.

In France, the situation was deteriorating. German armored forces plunged into the heart of France, severing the links between the French forces in the south and the Allies to the north. Unable to rejoin with the French army, and with the Belgian positions crumbling on their left flank, the British Expeditionary Force (BEF) was left with but one option—retreat. In late May 1940 the extraction of the BEF from the shores of France at Dunkirk began.

As this disaster unfolded on the shores of France, the British claimed a significant victory in the communications realm. On May 22, the code breakers at Bletchley Park broke the code used by Germany on its Enigma machine. Although not as decisive as some have claimed, the access to German coded messages was significant in providing Britain with a continual flow of information about the German order of battle and intentions. It was a contributing factor that served well throughout the war effort.

The full-scale evacuation of the BEF began on May 26. Goering insisted that the Luftwaffe have the lead role in destroying the British on the beaches. The RAF met them head-on, and over the several days of the evacuation every Hurricane and Spitfire squadron in Fighter Command, with the exception of three, saw action. The lessons learned in command and control, tactics, and employment of forces served the RAF very well in the battle to come. By the time the evacuation was complete, over 338,000 troops had been evacuated and over 2700 fighter sorties had been flown (Mason 1969). As the evacuation ended, Churchill, on June 4, made one of a number of speeches regarded as among the most compelling and inspirational in the English language. He declared, "We shall fight on the beaches . . . on the landing grounds . . . in the fields and in the streets. . . . We shall never surrender." On June 18, Churchill, in an effort to prepare the British people and steel them for what was to come, said, "The Battle of France is over . . . the Battle of Britain is about to begin. . . . Let us therefore brace ourselves . . . and so bear ourselves that, if the British Empire and the Commonwealth last for a thousand years, men will still say 'This was their finest hour.' " That speech was comparable in timing and tone to that given by Shakespeare's Henry V on the eve of the battle of Agincourt. France surrendered on June 22, 1940.

With the fall of France, Britain stood alone. This raised an issue with regard to the defeated French—would Germany confiscate the ships of the French naval fleet to attack and invade Britain? Churchill, recently elected as prime minister, wasted no time in taking action to neutralize the French fleet. The largest concentration of French ships was located at the port of Mers-el-Kabir in French Algeria. On July 3, in an operation labeled "Catapult," all French ships in British territories were boarded and impounded. Later that day, Vice Admiral Somerville, commanding "Force H" in Algeria, delivered an ultimatum to the French admiral Gensoul: either bring the ships out to join the British, decommission the fleet there or at another British port, or scuttle the fleet in the harbor. When Admiral Gensoul delayed beyond the deadline established by the British, they opened fire, killing over 1200 French sailors and disabling the fleet for the remainder of the war. This action was a source of animosity between these former allies that endures to the present, but it accomplished the objective; the French fleet would not be used to carry Germans across the English Channel. It also had the effect of rallying the British and simultaneously informing the Americans that Britain intended to fight with or without their help.

One of the certainties of air combat is that aircraft will be lost. The cost in aircraft to the RAF from the continuing actions in France was a source of concern to Dowding and a serious threat to the security of the nation, given the obvious intent of Germany to engage Britain. In May, Churchill elevated aircraft production to cabinet level and named Lord Beaverbrook, heretofore a newspaper owner, as the minister of aircraft production. Among his first actions, Beaverbrook directed that five aircraft would have top national priority for production, among them the Hurricane and the Spitfire. This made the production of these aircraft a national priority, giving them first claim to materials and facilities. Fighter production nearly quadrupled during the summer of 1940, leaving Dowding with more aircraft at the end of the Battle of Britain than he had at the beginning. The limiting factor in the RAF by far was the lack of trained crews, not machines. Much credit goes to Beaverbrook in energizing British production to make the victory possible.

Like two teams readying themselves for the initial clash in an athletic contest, the opposing air forces now positioned themselves for the war that was only days away. They had trained, they were equipped, and it was now time to test themselves against each other. Germany hurried to occupy the airfields of northern France; Britain continued preparations for the attacks that would surely come from her eastern and southeastern coasts. The German air force was organized into five *Luftflotten*, organizationally equivalent to the British Groups. Three of these were expanded and arrayed against England as Germany took control of the territory along the northern coast. *Luftflotte 2*, commanded by Generalfeldmarschall Albert Kesselring, was headquartered in Brussels and deployed in the Low

Countries and northeastern France. *Luftflotte 3,* under the command of Generalfeldmarschall Hugo Sperrle, was headquartered in Paris and was located across the northern and western bases in France to Cherbourg. *Luftflotte 5,* smaller than the other two, was located in Oslo, Norway. It was under the command of Generaloberst Hans-Jürgen Stumpff and was intended to threaten the eastern and more northerly parts of England. By July 20, 80 percent of the operational strength of the Luftwaffe was deployed in the three *Luftflotten* arrayed against Britain.

The Royal Air Force was likewise organized into groups, which were assigned specific geographical responsibilities for defense. Number 10 Group, led by Air Vice Marshal Quintin Brand, was responsible for the defense of the sectors in southwestern England. Number 11 Group was under the leadership of Air Vice Marshal Keith Park, who was to become one of the primary tacticians responsible for the successful defense of the homeland. Number 11 Group had the sectors that surrounded London and extended to the south and east coasts. This would be the arena in which much of the Battle of Britain would be waged. Number 12 Group, led by Air Vice Marshal Trafford Leigh-Mallory, had the sectors north of 11 Group and the eastern coast. Finally, Number 13 Group, under the command of Air Vice Marshal Richard Saul, had responsibility for the northernmost sectors in the country.

July 10, 1940, is the day fixed as the beginning of the Battle of Britain. Although there was fierce fighting leading up to July 10, the engagements on that day were of such scope and scale that this was the date chosen for historical purposes. The day was marked by an attack on a British shipping convoy as it entered the Channel near Dover. The attacking force, a large force of Do 17 bombers escorted by several fighter squadrons, was opposed by British fighters. As each side was reinforced, a huge dogfight of over 100 airplanes—the largest engagement between the two nations up until that time—was joined.

The Battle of Britain itself consisted of a series of fairly distinct phases, based upon the changing objectives and targets of the Germans. The first phase, labeled the *Kanalkampf* by the Germans, is referred to in England as the Channel phase. The Channel phase is generally agreed to have started on July 10 and lasted through August 12. This was essentially a prelude to, and preparation for, the main assault planned on the British homeland. The primary German target during this phase of the battle was shipping in the English Channel, and much of the action took place over water. The objective was to restrict shipping, and, perhaps more important, to draw the RAF into battle and wear them down through attrition. The Germans attacked convoys of British ships in the Channel, using large formations that included bombers as the primary attacking force, supported by flights of fighters to fend off the defending British fighters.

Considerable debate has taken place over the extent to which Hitler was actually prepared, or was even inclined, to invade England. Certainly Admiral Raeder, commander in chief of the German navy, had no enthusiasm for the operation. Hitler's own expectation, strengthened by the experiences of both the Spanish civil war and the invasions that preceded the Battle of Britain, was that the English would sue for peace long before an actual invasion became necessary. Furthermore, there could be little doubt that Hitler's expansionist plans pointed eastward; the wars with France and Britain were simply means to that eventual end. That said, invasion remained a continuing topic of discussion between Hitler and his generals and admirals. On July 16 Hitler issued Directive No. 16, wherein he called for German forces to "prepare for, and if necessary carry out, an invasion of England." The operation was code-named "Sea Lion" (*Seelöwe*). On July 19 Hitler made a "final appeal to reason," calling for the British to surrender and threatening dire consequences if they chose otherwise. On July 22, Lord Halifax declined, labeling the offer a "summons to capitulate."

Air operations during this Channel phase confirmed and strengthened some expectations and weakened others. Clearly, the Chain Home radar system and its integrated fighter control system became an ever more important adjunct to the defense effort; and the Hurricanes and Spitfire aircraft were proving themselves to be among the finest fighter designs in the world. On the other hand, the German Ju 87 Stuka dive-bomber was proving to be slow and an easy target for British fighters, and the British Defiant proved itself to be an unacceptable design for air combat. In one day in mid-July, an entire squadron of Defiants was destroyed, with only three pilots surviving an encounter with a squadron of Messerschmitt Bf 109s.

On July 28, two of the greatest aces of World War II met for a moment in the air. Werner Moelders was more than an excellent fighter pilot; he was an innovator. It was he who devised the "finger four" formation while flying with the Condor Legion in Spain. This formation was so effective that it not only proved far superior to the formations that the British employed in the Battle of Britain, the same formation is used today by fighters worldwide. Moelders was the first German aviator to be awarded the Knight's Cross in World War II for shooting down more than 20 enemy airplanes. "Sailor" Malan, on the other hand, was a South African who had become one of the most deadly and effective of all RAF fighter pilots. On this day, Moelders was leading JG 51 and was the youngest Kommodore in the Luftwaffe; Malan was leading 12 Spitfires against the German group. Malan attacked one of the aircraft in the lead squadron and handily shot him down. Moelders had meanwhile turned and shot down one of the attacking Spitfires, his twenty-seventh victory. Even as Malan was finishing his kill, Moelders rolled in on his tail, preparing to finish him off. Malan immediately started the classic hard turn into the attack, continuing the turn until he brought his guns to bear on Moelders. Malan fired and his aim was true. He severely damaged Moelders's aircraft, but the machine gun shells used in British fighters lacked the necessary power to completely destroy the opposing aircraft, and he was able to nurse it back to his home base. Moelders's legs were wounded, and it would be another month before he could fly again.

The German efforts to restrict shipping in the Channel were effective. As the month wore on, the British continued to run convoys primarily as a show of defiance. They did, however, begin to run more of the convoys at night to lessen the damage that daylight shipping invited. As the Channel phase drew to a close, the British RAF situation was less than sanguine. Too often the outnumbered British fighters lost in exchanges with large formations of Bf 109s. In spite of the losses and damaged machines, Lord Beaverbrook's production and repair resources were able to keep the inventory of machines at acceptable levels. The real problem was manpower. By the end of July, the RAF had lost 220 airmen, among them 80 squadron and flight commanders. The surviving crews were exhausted from the long days of multiple scrambles and the tension that accompanies deadly engagements. And the war was really just getting under way.

One of the more shameful events that developed during this period involved Joseph Kennedy, father of the future president. At the time, Kennedy was the ambassador to Great Britain; furthermore, he was outspoken in his opinion that England could not withstand the assault of the German juggernaut. At a press conference he predicted that Hitler would be in London by August 15. Roosevelt, running for a third term as president, was in a bind. He needed the support Kennedy could mobilize for him in his campaign, yet he did not want to abandon American support for England. Roosevelt decided to send William J. "Wild Bill" Donovan, one of the original heads of the OSS (precursor to the CIA) to Britain in order to get another opinion. Donovan reported back that he was diametrically opposed to the Kennedy view. In September, Kennedy's car was damaged during an air raid. He announced that he would endure one month of bombing, seeming to blame the English for being bombed by the Germans. Kennedy closed the embassy in October and went home to resign. Although official Britain congratulated him for his service, he was roundly castigated by many at high levels within the government.

The debate on the advisability and timing of a German invasion of Britain continued within the German High Command. Finally on 1 August, Hitler issued Directive 17, ordering that German operations be focused on destroying the British air forces, their supporting airfields, and the British aircraft industry. He further directed that attacks against coastal ports be reduced "in view of our own forthcoming operations." Goering conceived a campaign—planned to require two to three weeks—during which German bombers would attack successive targets beginning 150 kilometers from the heart of London and progressing inward toward a circle 50 kilometers around London. He called the campaign *Adlerangriff* (Eagle Attack) and designated the first day as *Adlertag* (Eagle Day), the date yet to be determined and dependent on the state of preparations and the weather. This new stage of the battle would actually last nearly one month (August 13 through September 6), and it would test the resolve of the RAF as it had not yet been tested. It would bring the war to the British homeland and make it a part of British daily life in a way that the Channel phase never had. It nearly brought the RAF to its knees. Goering declared that *Adlertag* would begin on August 13.

The campaign over the English Channel had given the Germans some insight into the importance of the radar-based control system to the British defensive effort. On August 12 the Germans, in preparation for *Adlertag,* attempted to blind the British fighter control system by attacking the radar towers of the Chain Home system. Most of the stations were attacked with deadly precision, and the Germans were initially convinced that they had succeeded. However, they failed to consider the fact that lattice towers are not much impacted by blast. The stations were back in operation almost immediately—battered, bruised, but fully capable.

Eagle Day broke cloudy and gloomy—not a good day for airborne attacks that required locating and destroying relatively small targets with precision bombing. Goering issued a cancellation of the morning missions, but some had already taken off, and the combination of mass formations, coordinated fighter and bomber rendezvous, and radio problems resulted in several groups proceeding to Britain. The weather cleared later in the day and normal operations were restored. The strikes were aimed at the RAF and the southern airfields; the Castle Bromwich aircraft factory, a primary source of Spitfire production, was also attacked. During the 24 hours of *Adlertag,* Germany flew 1485 sorties against English targets. In the final analysis, 46 German aircraft were lost; 13 RAF fighters were destroyed in the air, and 47 on the ground (Mason 1969). Much of the damage done on this day achieved little of strategic or tactical importance. Given the scale of the German attack, one would have to say that the defenders won the first round of this, the second phase of the battle.

Intelligence—or the lack of it—plagued the German effort throughout the Battle of Britain. Having convinced themselves that the British had massed all of their resources in the southern part of the country to counter the raids launched from the European mainland, they planned a surprise assault on the area north of London. On August 15, the Germans launched a massive bomber attack from *Luftflotte 5,* located in Norway and Denmark. Foreseeing no significant opposition, the Germans planned no fighter escorts, reduced the bomb loads, and pulled the tail guns out of the bombers to reduce weight and extend range. The plan was to feign an attack with a small force on the Scottish coast north of the intended main target area to draw off any potential interceptors, then strike with the main force in the area to the south.

Fortunately for England, Dowding was not a man given to sudden whims and changes of direction. Early on he had insisted that aircrews be periodically rotated to the northern 13 Group areas to give them some respite from the exhausting tempo of battle in the south. As the British fighters climbed out to meet the Germans, they were told that the attacking forces numbered about 30 aircraft. When visual contact was established, the actual count was over 100. The German main attacking force had become lost en route and was 75 miles north of the intended track, placing them in the vicinity of the diversionary attack. The result was that the planned feint succeeded only in alerting the British fighters and providing them time to intercept the primary force with maximum deadly effect. With insufficient fighter support and reduced defensive gunnery, the German formations were easy targets for the experienced British fighter crews. *Luftflotte 5* lost nearly 20 percent of the aircraft engaged in the raid. This experience was a sobering lesson for the Germans. Churchill congratulated Dowding on his steadfast resolve in holding a reserve in the north, saying "... the generalship shown here [is] an example of genius in the art of war" (Hough and Richards 1990). Clearly, fighter reserves in the north meant there were no free lunches when it came to targeting Great Britain, and Germany was confronted with the reality that bomber attacks without single-seat fighter support were impractical. In fact, on this day (15 August) Germany flew 1786 sorties, of which only 520 were bomber flights (Deighton 1979). The implication—that each bomber sortie required two fighter support sorties—was disheartening. The Luftwaffe lost a total of between 55 and 75 aircraft (accounts differ); later on, August 15 was referred to by the Germans as "black Thursday." The day was described by Churchill as "one of greatest days in history."

As worrisome as the loss of fighter aircraft was to Dowding, by far the greater concern was the loss of airmen, especially experienced leaders. By late August 80 percent of RAF squadron commanders had been lost. The RAF used various innovative devices to maintain combat capability. They recruited intensely from other nations, especially from the old British Empire countries—Australia, New Zealand, South Africa, Canada, and Rhodesia. There were 129 New Zealanders who flew as Fighter Command aircrew in the Battle of Britain. Among the most capable recruits were 141 Poles and 87 Czechs who brought personal animosities to bear on their German adversaries that sharpened their already excellent skills as airmen. One of these, Sergeant Josef František, a Czech pilot who had fought with the Polish and French air forces before joining the Royal Air Force, was the RAF's highest-scoring pilot, with 28 kills at the time of his death on October 8, 1940. Among other standouts, A. G. "Sailor" Malan, a South African, became one of the leading aces in the Battle of Britain.

The situation was desperate; extreme measures were taken to put pilots into cockpits. On August 10, Dowding agreed to cut the time for transition from basic training into fighters to two weeks. The transition course had previously been six months in duration. Flight leaders at times had no combat experience when they led their first intercept, and pilots were routinely given only 10 hours of flying time in the high-performance airplanes before being cleared for combat. Needless to say, the lack of flying proficiency and combat experience often left these hapless young men easy prey to the more experienced Germans.

During mid-August Goering took the unusual step of ordering his Bf 109s to escort the bombers more closely, apparently in response to criticisms by the bomber crews that they seldom saw the escorts. By requiring the 109s to fly both lower and slower, this order effectively required them to forgo the natural advantages of speed and altitude that give the fighter its edge in aerial combat. Until this time, the Spitfires had typically been dispatched to combat the high-flying Bf 109s, and the job of intercepting the bombers had been tasked to the Hurricanes. Now Park ordered his controllers to dispatch both Spitfires and Hurricanes against the bombers. The Germans then began to send more 109s as escorts, some with the bombers and others to fly high combat air patrol, or *cap*. The British then reverted to the earlier policy—Spitfires targeted against the high flyers, Hurricanes against the bombers, but, given that 109s were in both formations, Hurricanes often encountered 109s. Such are the moves and countermoves typical of tactics in aerial warfare.

The fighting on August 16 was as intense and ferocious as it had been the preceding day. Wave after wave of German aircraft appeared. Over 1700 German sorties were flown. The sky was full of attacking fighters and defenders alike. One pilot recollected that the likelihood of a midair crash was as high as was the probability of being shot down. Another recalled seeing nine pilots parachuting at one time (Hough and Richards 1990). Flight Lieutenant James Nicolson was attacking a flight of Ju 88s when his own aircraft was hit by an Me 110 fighter. The bullets hit Nicolson and also ripped through his center fuel tank, which burst into flame. With his aircraft now on fire, Nicolson prepared to bail out, when he saw his attacker diving below him. Nicolson, willing himself to ignore the pain of the burns he was experiencing, dove after the Me 110 and shot it down. During the engagement, Nicolson's wingman had also been shot down. Seeing two parachutes descending, an artillery officer near the village of Millbrook in Hampshire concluded that a parachute assault was in progress and directed artillery fire at the two. In one of the sad, but not uncommon, mistakes of war, his wingman was killed by friendly fire. Nicolson himself, already wounded and severely burned from his aircraft, was again wounded by his own countrymen. For his actions, he became the only fighter pilot awarded the Victoria Cross during World War II.

Intense combat continued into August 18. On this day the Germans mounted the strongest effort to date to destroy the RAF airfields. They concentrated their attacks on two primary Fighter Command bases, Biggin Hill and Kenley, both in the 11 Group area. The RAF responded with equal fervor and the fight was on. One of the notable outcomes of the combat of this day was that the Ju 87 Stuka dive-bomber proved so vulnerable—30 of them were lost—that Goering soon decided to withdraw them as primary combat platforms. Time after time, small units of RAF fighters flung themselves into the massive formations of attacking bombers. Twilight comes late in England in midsummer, and the combat continued on into the evening until the weather mercifully began to deteriorate. This day became known as "the hardest day" of the Battle of Britain, although one would likely be hard-pressed to differentiate it from most days of this week of daily tests.

As the intense aerial combat continued, Goering gave orders that no more than one commissioned officer was to be included in any bomber crew. The manpower situation was becoming a problem on the German side, but the situation was far graver across the Channel. The specter of a force that was gradually being wasted away confronted the British commanders. At 11 Group, Park ordered his controllers to engage closer in to the British homeland and to limit nonbomber engagements as much as possible. He intuitively understood that the objective was to kill bombers, and that objective had to be kept sacrosanct.

By this time, the fighter pilots on both sides had become the superstars of their time. They were featured in movie newsreels, papers, and radio—much as the athletes and entertainers of today are. Of course, theirs was a much more serious pursuit, and the cost of failure was death for the individual . . . and possibly the nation. On August 20, Churchill recognized the contribution of these young people and the debt that England—and the world—owed them when he observed that, "Never in the field of human conflict was so much owed by so many to so few."

In late August, the weather closed in. The British used the opportunity to repair and make modifications to the fighter fleet. Goering, foreshadowing the nature of the next phase of the battle, spoke of the need to attack the aircraft industry. He further stipulated that such attacks should be conducted at night, since Britain's night defenses were weak. The British antiaircraft guns were without radar, and the Blenheim, a light bomber serving in a night fighter role, proved totally inadequate to the task assigned. Goering noted that no restrictions on targets should be observed, although he did reserve for himself the order to attack London and Liverpool. In spite of what was said, Liverpool's ordeal started at this time—well before the phase of the battle when cities were routinely targeted.

As the end of August approached, the ferocity of the attacks on the airfields of 11 Group increased to almost unbearable levels. On August 24, huge formations massed over the French coast and proceeded toward England. The daylong contest continued into the night, and as fighters from both 11 and 12 Groups engaged the German bombers, a number of bomb loads were jettisoned by the German bombers in the area of the Thames estuary, near London. Over 200 bombs, each weighing 250 kg, fell on Portsmouth and the East End of London, killing over 100 people and wounding more than 300 (Hough and Richards 1990). This was the first time since the Gothas had bombed London in 1918 that the city had been under attack. This attack, although accidental and primarily caused by the disorientation of the aircrews engaged in night combat, presaged the Blitz that would characterize the third and final stage of the Battle of Britain. Total aerial warfare had again returned to England.

Although largely accidental, the bombing of the British capital angered Churchill. He had already queried the Air Staff regarding the possibility of bombing Berlin and was assured that Bomber Command could, if required, do so. Twenty-four hours after the bombs fell on London, 80 British bombers attacked the outskirts of Berlin. Then, for several nights, they returned to rain bombs on the German capital. While not particularly effective in terms of damage done or targets destroyed, these raids stunned the German people, who had been assured by their Nazi leaders that this would never happen. They also had the effect of providing the impetus that Hitler needed to turn to the bombing of London sooner, rather than later—a move which saved Fighter Command from annihilation.

The fighting during this period also closed the book on the Defiant as a day fighter. With one squadron already virtually destroyed, another was thrown into the airfield defense effort. Within a matter of three days, 12 aircraft and 14 crew were lost. The Defiant was transitioned to a night fighter role, but it performed inadequately in that role as well.

During these last weeks of the second major phase of the battle, a disturbing trend was developing. The Germans were sending many more Bf 109s than had been the case earlier, and the British found that the exchange ratio (German fighters killed versus RAF fighters lost) from August 26 through September 6 was distressingly low. On August 29 the Germans sent over 700 fighters into the fray with the specific mission of coaxing the RAF fighters into a deadly exchange. The British controllers, adhering to Dowding's dictum to concentrate on the bombers, ignored the bait. This led the chief of Kesselring's fighter forces to claim that air superiority had been gained. This had been one of the early preconditions for consideration of a German invasion. The intense effort to destroy the RAF continued. On August 30, Fighter Command flew 1054 sorties, with some squadrons flying up to four sorties on one day. The pilots and supporting crews were exhausted. On August 31, the RAF lost 39 aircraft while destroying 41 German aircraft. In terms of aircraft lost, this was the single worst day of the battle for the RAF. The whittling away and eventual annihilation of the RAF appeared to be a distinct possibility if the situation continued.

Kesselring, commander of *Luftflotte 2,* bragged that he would break the back of the British defenses by September 1. In the attempt, he made the last days of August particularly painful for the RAF. Massive bomber formations, escorted by huge formations of Bf 109s, attacked the British 11 Group. August 29, 30, and 31 were marked by combined bomber-fighter formations that numbered in the hundreds of aircraft. Wave after wave approached, timed to catch the defending forces at the most inopportune moments as they returned to bases to refuel and rearm. At one point, 48 different observer posts were reporting attacking forces overhead (Hough and Richards 1990). The major fighter bases—Biggin Hill, Croydon, Tangmere, and others—all received major damage.

Likewise, the attacks on aircraft production facilities were equally effective. Notably, an attack on the Vickers plant produced over 700 casualties in a single attack. The massive attacks against the RAF, the airfields, and the aircraft industry continued into the first days of September. The losses on both sides were nearly equal. The agony continued on into September 5 and 6. The RAF was very nearly defeated.

In these critical days, however, several factors were working together to save Fighter Command. Hitler was enraged that Britain had bombed Germany; he desperately wanted revenge. Bombing the British cities seemed to be one avenue to repay them for their attacks on German cities. In addition, Kesselring was strongly advocating switching the focus of attacks to concentrate on London as a means of forcing the fighters into more of the deadly exchanges that would finally bleed them dry. And, of course, there was the final objective of the entire campaign in the west—to either invade and conquer England or have her capitulate. On September 3, the chief of staff of the Wehrmacht issued a schedule that would govern the invasion. The earliest date for Operation Sealion was delayed one week and set for September 21. At a rally at the Berlin Sportspalast on September 4, Hitler worked the crowd into a frenzy as he publicly described the coming destruction of British cities and the invasion that would follow.

The Germans had been assembling an armada of barges, tugboats, trawlers, and motorboats since mid-July. The makeup of this fleet alone speaks to the lack of real preparedness on the part of the Germans for a cross-Channel invasion; they really did not possess the infrastructure necessary to do the job. However, the British had accomplished the miracle of Dunkirk with a similar force, so invasion was at least a credible threat in spite of the lack of assets. By the end of August, these boats and barges began moving along the inland waterways toward the invasion departure ports. This sudden movement and buildup on the European side, plus intelligence reports that reinforced the belief that invasion preparations were under way, made the threat real—something that would have to be dealt with immediately. On September 5, Bomber Command initiated the first of a series of attacks on the Belgian and Dutch invasion ports. The next day the entire RAF was placed on Alert No. 2 status—"attack probable within the next three days."

Measures and countermeasures in the air war continued. The Germans placed armor plating in their bombers and fighters to protect aircrews against the British fighter attacks from the rear. As this made the Germans somewhat less vulnerable to rearward attacks, some RAF pilots began to employ head-on attacks against the German bombers to counter the defensive armor plating. By the end of August, the British had placed armor plate in their own fighters, and, in addition, they also finally added rearview mirrors to all fighters.

With Fighter Command now down to its last gasps, all that remained was for the Luftwaffe to administer the coup de grâce . . . and then the German High Command made the decision that allowed the fighters to survive. On September 7, Goering, speaking to the German people on the radio, announced that he had taken personal control of the air war and was redirecting the campaign away from the RAF and the airfields. The Germans would now focus on the city of London. This was a strategic mistake of huge proportions—a turning point in the Battle of Britain. With Fighter Command essentially defeated, the German decision to abandon the attacks on the fighters and turn toward

London gave the exhausted RAF the respite so desperately needed to rest and regroup. Within a matter of weeks they were reenergized and back in the battle with the full vigor and effectiveness they had demonstrated from the beginning. The airfield phase of the battle had cost the Germans 670 aircraft lost; the British, 400. More troubling for both was the loss of experienced aircrews and leaders. But now Fighter Command would survive, even as London burned.

On this same day, September 7, the British chiefs of staff issued to the London region and other areas in the south and east the code word "Cromwell," the highest state of alert, requiring all troops to proceed to battle stations. From the Fighter Command perspective, the day opened on a rather calm note, with little action in the early part of the day. Then in the afternoon, the Chain Home system reported massive enemy buildups over the Pas de Calais. Finally, over 1000 German aircraft, among them more than 400 bombers, proceeded toward the British coast. The RAF prepared to meet them. But as the formations continued inbound, the fighter pilots realized that this was not to be another day of attacks on the airfields and aircraft of Fighter Command. It soon became clear that the target was London. The third phase of the Battle of Britain—the London phase—had begun. Before the bombing ended on that day, some 333 tons of high explosives and 13,000 incendiary weapons had been dropped in the London metropolitan area. Goering was ecstatic; he telegraphed his wife that the surrender of Great Britain was imminent.

Little did they realize it, but London's citizens were in the first days of what would be fifty-nine nights of uninterrupted bombing. During the first week, the average number of bombers attacking nightly was about 200. Four hundred people were killed and 750 wounded on the second night, and worse was yet to come (Hough and Richards 1990). But as bad as the situation was for the civilian population, the circumstances of the fighter pilots improved with each day that passed. By mid-September it was as though Fighter Command had passed through a dark tunnel and then had emerged again into the sunlight. The health of the men and machines was once again on the increase; confidence had returned. They had survived.

September 15 began with a Chain Home detection of a large raid forming over the Pas de Calais. By the time the attacking force—over 100 Dornier 17s—had massed and proceeded to England, the fighters of 11 Group, reinforced by 12 Group had climbed to altitude and were ready to engage. They were able to successfully break up the bomber formations, causing many of the German bombers to release their bombs before they reached London. Suburbs and towns south of the city were bombed randomly as the bombers maneuvered to escape the fighters, but the raid was largely ineffective. However, given the time of day, all were certain that this was not the end of that day's action. Shortly after 2:00 P.M. another huge assault force crossed the Channel en route to England. Once again, the fighters of the RAF met them in force, combining the efforts of both 11 and 12

Groups. The Luftwaffe had been fighting for years. Spain, Poland, and France—all had fallen before the force of their assaults. Now they had been fighting the RAF since July. The Germans had been told again and again that the British had nearly exhausted their inventory of fighters. They expected token resistance.

Churchill had chosen this day to visit the headquarters of 11 Group to observe the day's action. Park gradually fed one squadron, then another, and yet another into the fray, until a total of 24 RAF squadrons were committed. As Churchill watched, he was moved to ask of Park, "What other reserves have we?" "There are none," Park replied. But the pilots in the air sensed something different on this day. The RAF fighters registered kill after kill, until finally, according to some reports, it was as though the German pilots who had fought a brave and disciplined campaign finally decided that it was all to no avail. They confronted the combined efforts of some 300 RAF fighters on this afternoon. Discipline broke and the engagement became a rout. In the final analysis, the RAF lost 26 aircraft; the Luftwaffe lost 60. It not only left the German forces shaken; it was followed two days later, on September 17, by Hitler's decision to postpone the invasion indefinitely. It is generally agreed that the Battle of Britain was finally won on September 15, 1940. This day was later commemorated as Battle of Britain Day.

By mid-September the Germans had amassed almost a thousand boats and barges in the European ports, with 1500 more en route. The bombing of the invasion fleet by the British had continued since September. These raids were increasingly successful, and on the evening of September 13, the British sank 80 barges in the port of Ostend. Finally, on September 19, the German High Command ordered the dispersal of the invasion fleet. During September, Bomber Command flew more than 1600 sorties against the invasion ports, dropping over 1000 tons of bombs. As Germany withdrew the fleet from the ports, British reconnaissance observed the lessening invasion threat. By September 21, in view of the shorter days, the deteriorating weather, and the reduction of the invasion fleet, the British canceled the "Cromwell" alert status and reverted to Alert No. 2 status.

Meanwhile the Luftwaffe had apparently turned once again to the attempt to destroy British aircraft production facilities. From September 24 through 26, numerous destructive attacks were mounted against the factories that produced Spitfires and other aircraft. These attacks may have had serious consequences had they been carried out earlier, but the net effects of these raids were marginal at this point. Production was quickly restored, since production capacity was by that time substantial and well dispersed among many plants throughout Britain.

On September 30, two large groups of German bombers were detected inbound to the English coast. By virtue of some miscommunication, fighter escort had never rendezvoused with the bombers, so they were initially without escort. The Spitfires and Hurricanes intercepted them and were destroying them handily until some Bf 109 escorts finally arrived. Many bombers were downed, and the others either jettisoned their loads or turned back toward Europe. Germany lost a total of 50 aircraft on this day and completely failed to accomplish any of its objectives. This was the last of the great day-bombing raids mounted by Germany during the Battle of Britain. After this the bombing would continue, but the English would no longer see the massive formations during daylight. From here until the end, any bombing by the Germans would take place under the cover of darkness.

With the beginning of October, the centerpiece of the war became the nightly bombing attacks on the city of London—the fourth phase of the Battle of Britain. An average of 150 bombers hit the city every night during the month except one. The weather was becoming less and less favorable, impeding fighter operations on both sides. The Germans began to send smaller raids, many consisting of single fighters armed with bombs, to conduct nuisance raids. Interception was difficult, but so was the problem of finding targets. The RAF was mounting substantial numbers of sorties in the attempt to intercept them, but neither side was accomplishing much in the fighter battle. In all air-to-air engagements during this period, the British continued to enjoy a numerical advantage in the kill ratios. As this went on, Hitler met with Mussolini on October 4 and remarked that the only thing that had frustrated his invasion plans had been the lack of five consecutive days of good weather. On October 5, a day of drizzle and fog, several engagements took place resulting in the downing of four Me 110s and six Bf 109s that were attacking British airfields. On this day, Fighter Command mounted 1175 sorties—proof that the command had recovered and was growing in strength daily and proof that England had, indeed, won the Battle of Britain.

On October 12, Hitler ordered the planned invasion to be postponed until spring or early summer of 1941. His order made it clear that further orders would be issued to invade only if the military conditions could be improved to an extent that would make invasion advisable. To anyone reading the order, it was clear—Britain had won the battle. Hitler did, however, emphasize that appearances of preparations for invasion were to be maintained in order to bring political and military pressure on Great Britain. Surely this idea was supported in no small part by the fact that the American ambassador was, at this very time, closing the embassy in London and proceeding homeward.

Even as the fighter battle waned, the intensity of the bombing attacks on London grew. On the evening of October 15, over 400 bombers attacked the city. Before October ended, more than 7000 tons of bombs had fallen on the city. In addition, Liverpool, Manchester, and other areas were subjected to heavy bombardment. The fires from the bombing raids became the means by which the bombers located the target areas, making the night targeting an easier task in the poor weather. The British often set decoy fires to attract the bombers to areas away from the cities. Over 13,000 British civilians were killed in October, over 20,000 wounded. The raids would continue until May of 1941, when Hitler finally turned his attention to the east and Russia. On October 18, Goering, on meeting with a group of his pilots, congratulated his fighter pilots for the losses they were inflicting on Fighter Command and praised his bomber pilots for reducing the British to a state of "fear and terror."

Rain, fog, and drizzle were by this time the normal condition, limiting fighter operations to the rare periods when the weather improved. Under these conditions, losses from accidents began to exceed combat casualties on both sides of the conflict. On October 22 five German crashes resulted in the deaths of several senior German officers. In a bit of comic relief just as the month ended, Mussolini persuaded Goering to allow a force of Italian bombers and escorting fighters to attack the British in retribution for the earlier Bomber Command raids on northern Italy. On October 29, a force of 15 Italian bombers crossed the channel, flying in wingtip formation. They were escorted by a force of approximately 70 CR 42 biplane fighters of World War I vintage. The armada dropped a few random bombs, then departed, having satisfied the propaganda objective. A couple of weeks later a similar force undertook a similar raid, this time against Harwich. However, by this time the humor of the British had worn thin; 13 Italian aircraft were shot down. This ended the Italian air offensive.

The official date chosen to mark the end of the Battle of Britain is October 31, 1940. On this date the weather was so poor that neither side flew. The battle had clearly been won earlier; this date simply acknowledges the obvious—that the fighter war had been won by Fighter Command. While the bombing of London would continue on into the following spring, the threat of invasion was past and Germany's opportunity to conquer England had been squandered.

There are those who belittle the importance of the Battle of Britain, arguing that Hitler never really engaged personally in the battle as he had in earlier battles and that the significance of the effort was small when gauged in terms of the worldwide strategic thrusts of the Allied and Axis war efforts. Those who make that case fail to understand or acknowledge that the Battle of Britain literally turned around the entire war dynamic. At a time when Hitler appeared invincible, a tiny group of men flying their fighters and supported by the entire British population served notice to the world that Britain would resist to the end of her ability to fight. They did so and they prevailed, giving the rest of the world the push needed to join together in an allied effort that finally broke and destroyed Germany in 1945. Without England as the base from which the invasion of the continent was launched, D-Day would certainly never have happened. Had the Battle of Britain not been won by the Royal Air Force in 1940, the history of civilization in the twentieth century would be a vastly different story.

CHAPTER 2

THE TECHNICAL TIMELINE

TIMELINE OVERVIEW

Like the flow of political and historical events that began in World War I and eventually culminated in the Battle of Britain, many of the technologies used in the battle had roots in those earlier times. In the pages that follow, we will examine a number of technological events and products that influenced the outcome of the Battle of Britain. Some of them can be characterized as interim steps in a chain of discovery, while others represent the culmination of years of research, resulting in revolutionary new capabilities. Still others are the result of innovations developed in the heat of combat to meet an immediate need. Each was significant in shaping the events of the summer of 1940 when two of the finest air forces in the world met in the skies above Britain.

One of the most important elements in the British victory in 1940 was the highly effective early warning and control system Britain used to intercept attacking bombers. This technically advanced system had, in fact, been developed and refined from defenses used in the early days of World War I. Germany, an early proponent of air power projected through the employment of long-range aircraft, first used Zeppelin airships to bomb the British homeland in that war. Improved defenses and the vagaries of wind and weather soon convinced the Germans to turn to airplanes—Gotha bombers—to carry the attack to the British. By 1917 they were bombing British ports and cities with impunity. As damage mounted, the British civilian population demanded that the government take action, and the need for improved defenses became both a military and a political imperative. To defend against the raids, Britain employed both technological and organizational innovations. Improvements were made in antiaircraft artillery, new methods of operating were explored, and improved organizational structures evolved. New and important technologies emerged, such as the HF radiotelephone, which, by 1918, became a standard feature in the fighter aircraft used for daylight interception. These and other

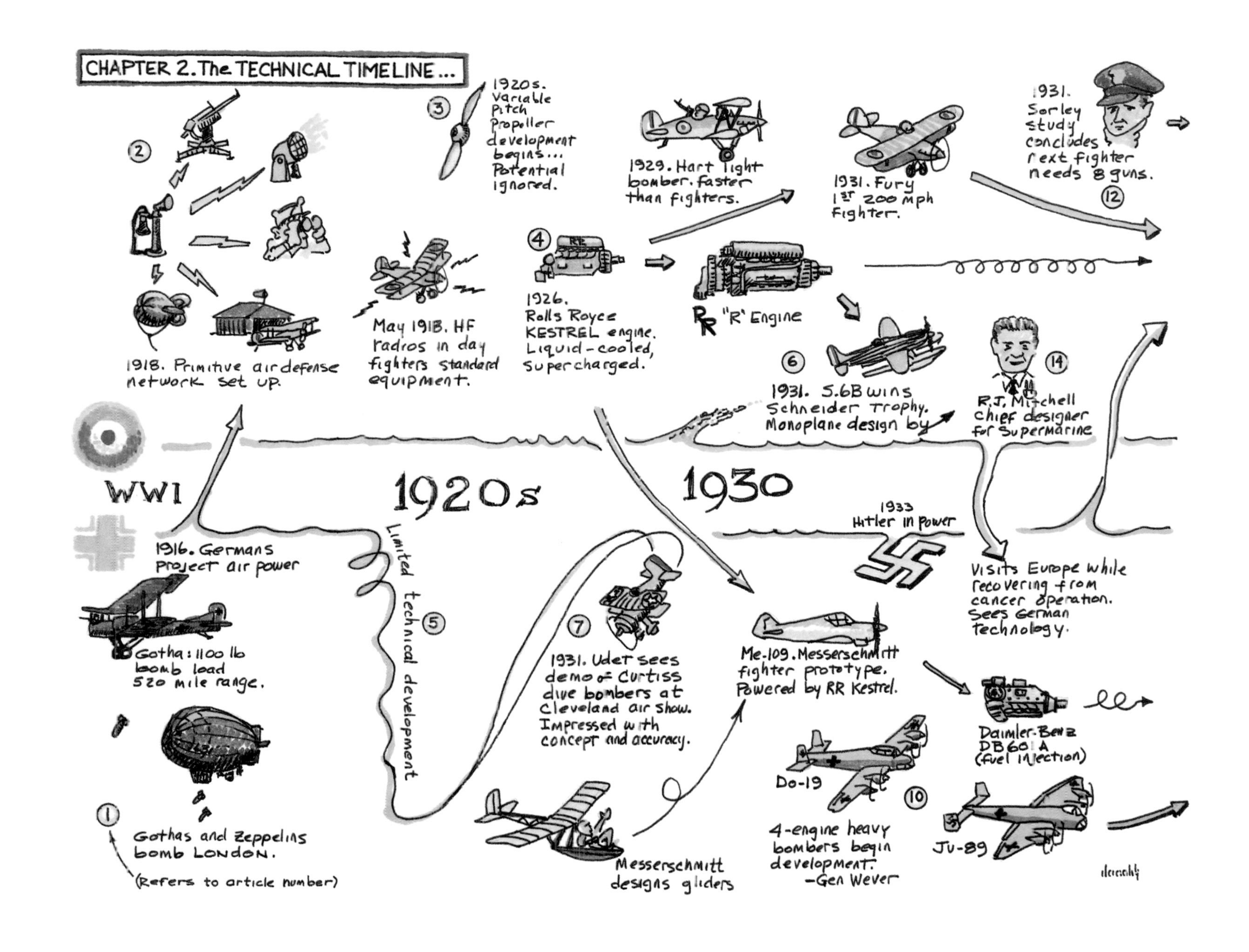

CHAPTER 2. The TECHNICAL TIMELINE...
1918. Primitive air defense network set up.
1920s. Variable Pitch Propeller development begins... Potential ignored.
May 1918. HF radios in day fighters standard equipment.
1926. Rolls Royce KESTREL engine. Liquid-cooled, supercharged.
1929. Hart light bomber. faster than fighters.
"R" Engine
1931. Fury 1st 200 mph fighter.
1931. Sorley study concludes next fighter needs 8 guns.
1931. S.6B wins Schneider Trophy. Monoplane design by
R.J. Mitchell chief designer for Supermarine
WWI
1920s
1930
1916. Germans project air power
Gotha: 1100 lb bomb load 520 mile range.
Gothas and Zeppelins bomb LONDON.
(Refers to article number)
Limited technical development
1931. Udet sees demo of Curtiss dive bombers at Cleveland air show. Impressed with concept and accuracy.
Messerschmitt designs gliders
1933 Hitler in power
Me-109. Messerschmitt fighter prototype. Powered by RR Kestrel.
Visits Europe while recovering from cancer operation. Sees German technology.
Daimler-Benz DB601A (fuel injection)
Do-19
4-engine heavy bombers begin development. -Gen Wever
JU-89

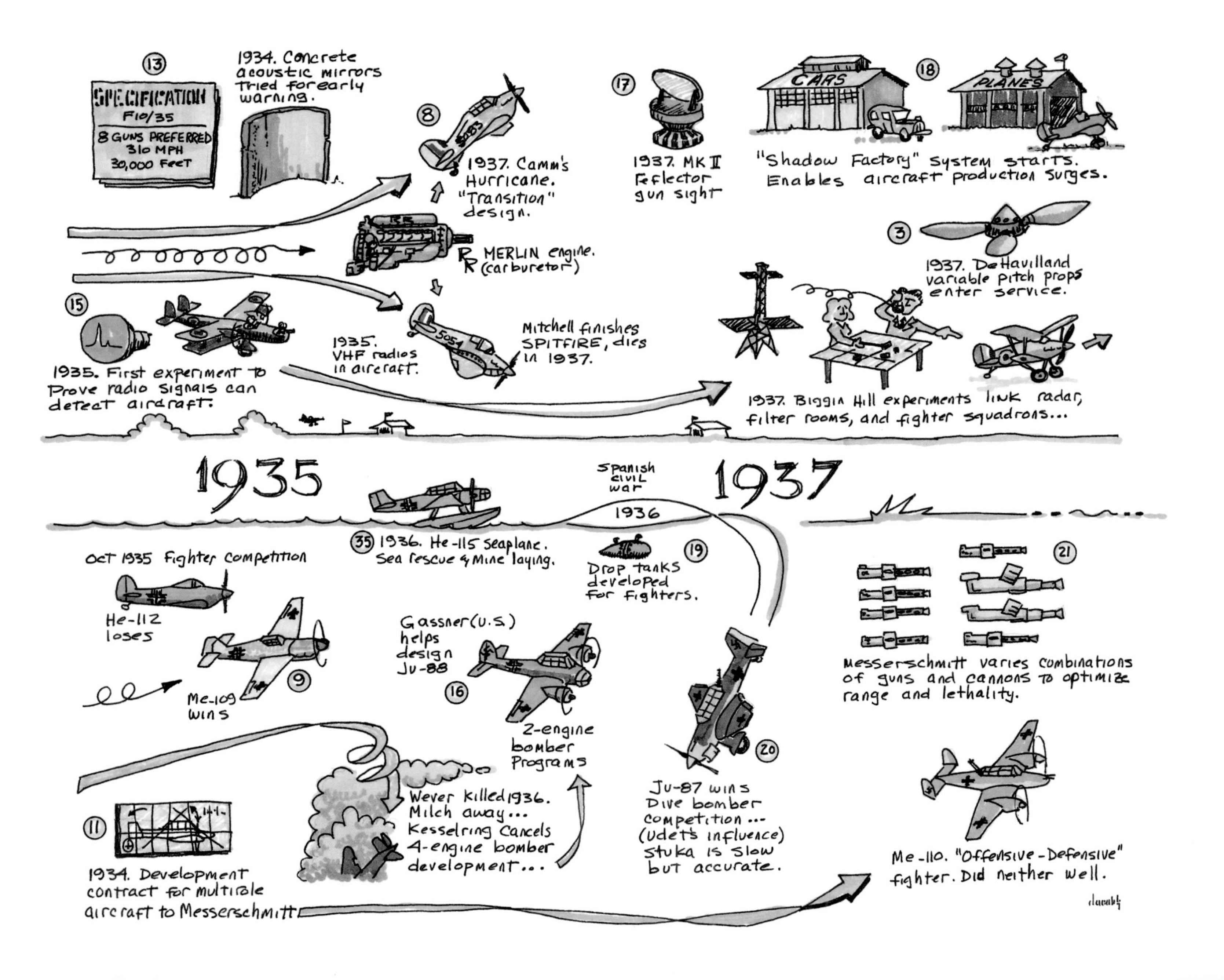

13
SPECIFICATION F10/35
8 GUNS PREFERRED
310 MPH
30,000 FEET
1934. Concrete acoustic mirrors tried for early warning.
8
1937. Camm's Hurricane. "Transition" design.
RR MERLIN engine. (carburetor)
17
1937. MK II reflector gun sight
18
CARS
PLANES
"Shadow Factory" system starts. Enables aircraft production surges.
3
1937. DeHavilland variable pitch props enter service.
15
1935. First experiment to prove radio signals can detect aircraft.
1935. VHF radios in aircraft.
Mitchell finishes SPITFIRE, dies in 1937.
1937. Biggin Hill experiments link radar, filter rooms, and fighter squadrons...
1935
Spanish civil war
1936
1937
35
1936. He-115 seaplane. Sea rescue & mine laying.
19
Drop tanks developed for fighters.
21
Oct 1935 fighter competition
He-112 loses
9
Me-109 wins
Gassner (U.S.) helps design Ju-88
16
2-engine bomber programs
Ju-87 wins dive bomber competition... (Udet's influence) Stuka is slow but accurate.
20
Messerschmitt varies combinations of guns and cannons to optimize range and lethality.
11
1934. Development contract for multirole aircraft to Messerschmitt
Wever killed 1936. Milch away... Kesselring cancels 4-engine bomber development...
Me-110. "Offensive-Defensive" fighter. Did neither well.

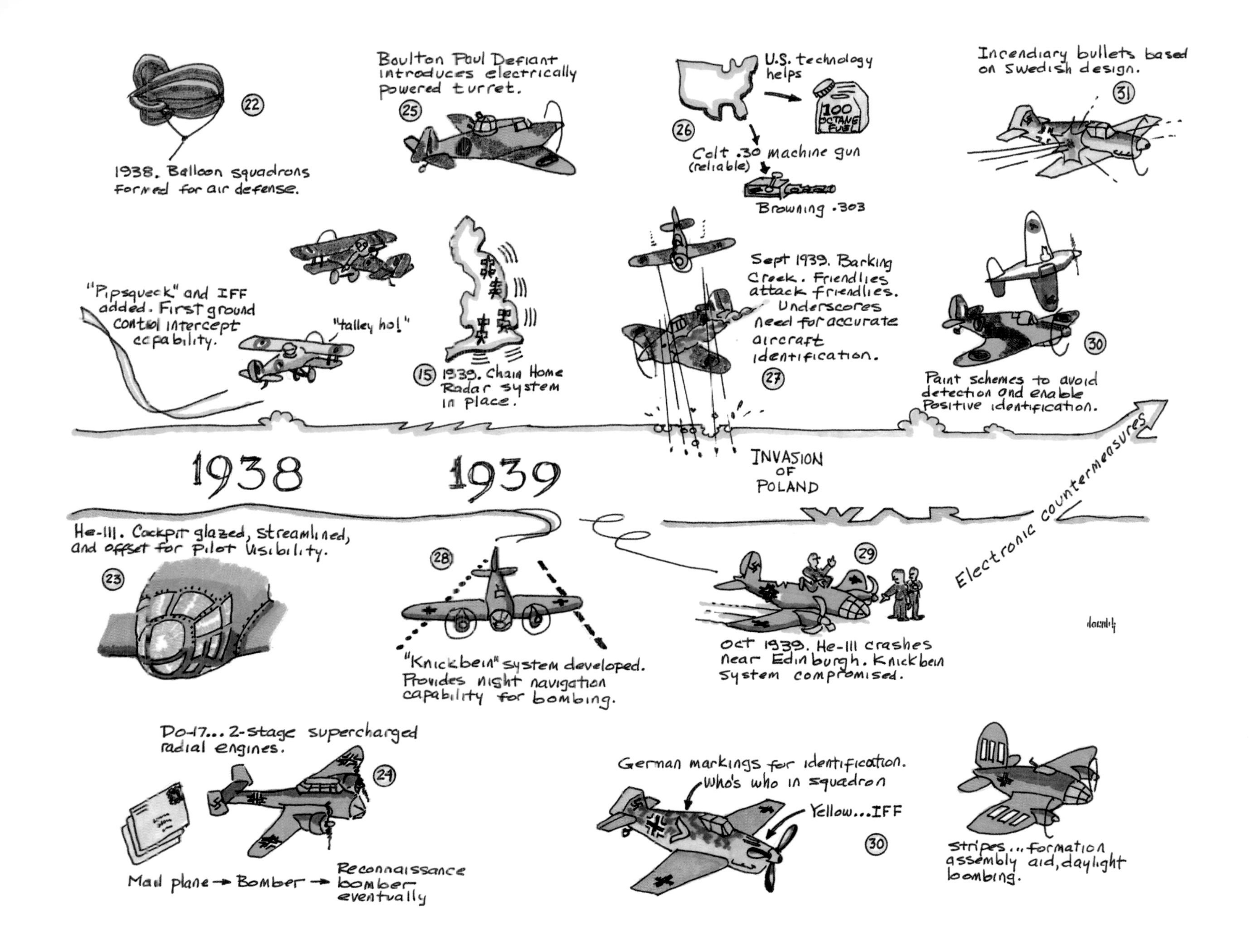
1938. Balloon squadrons formed for air defense.
22
Boulton Paul Defiant introduces electrically powered turret.
25
U.S. technology helps
100 OCTANE FUEL
26
Colt .30 machine gun (reliable)
Browning .303
Incendiary bullets based on Swedish design.
31
"Pipsqueck" and IFF added. First ground control intercept capability.
"talley ho!"
15
1939. Chain Home Radar system in place.
Sept 1939. Barking Creek. Friendlies attack friendlies. Underscores need for accurate aircraft identification.
27
30
Paint schemes to avoid detection and enable positive identification.
1938
1939
INVASION OF POLAND
WAR
Electronic countermeasures
He-111. Cockpit glazed, streamlined, and offset for pilot visibility.
23
28
"Knickbein" system developed. Provides night navigation capability for bombing.
29
Oct 1939. He-111 crashes near Edinburgh. Knickbein system compromised.
Do-17... 2-stage supercharged radial engines.
24
Mail plane → Bomber → Reconnaissance bomber eventually
German markings for identification. Who's who in squadron
Yellow...IFF
30
Stripes... formation assembly aid, daylight bombing.

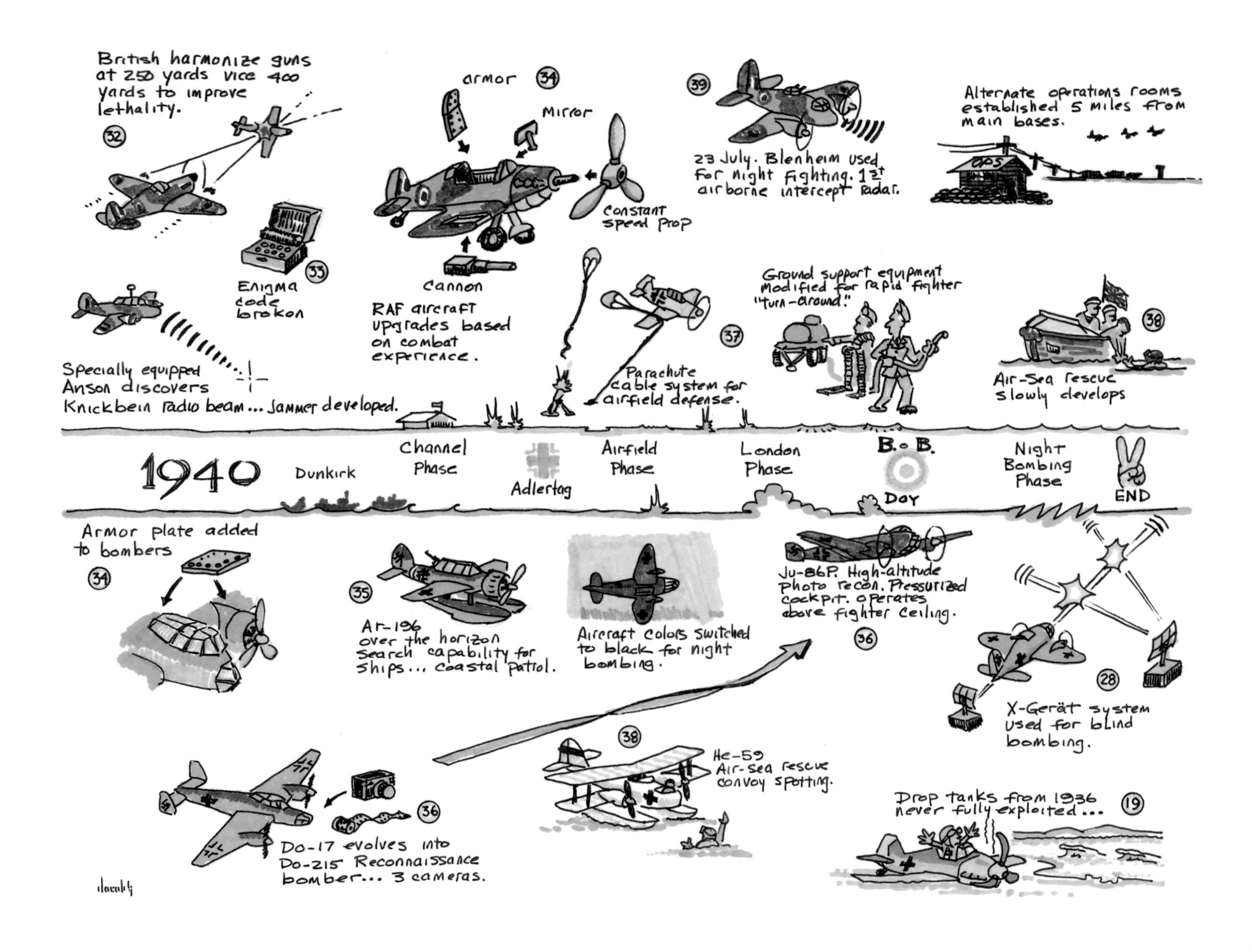

British harmonize guns at 250 yards vice 400 yards to improve lethality.
32
33
Enigma code broken
armor
34
Mirror
cannon
Constant speed prop
RAF aircraft upgrades based on combat experience.
39
23 July. Blenheim used for night fighting. 1st airborne intercept radar.
Alternate operations rooms established 5 miles from main bases.
OPS
Specially equipped Anson discovers Knickbein radio beam... Jammer developed.
37
Parachute cable system for airfield defense.
Ground support equipment modified for rapid fighter "turn-around."
38
Air-Sea rescue slowly develops
1940
Dunkirk
Channel Phase
Adlertag
Airfield Phase
London Phase
B. o B.
Day
Night Bombing Phase
END
Armor plate added to bombers
34
35
Ar-196 over the horizon search capability for ships... coastal patrol.
Aircraft colors switched to black for night bombing.
Ju-86P. High-altitude photo recon. Pressurized cockpit. operates above fighter ceiling.
36
28
X-Gerät system used for blind bombing.
36
Do-17 evolves into Do-215 Reconnaissance bomber... 3 cameras.
38
He-59 Air-sea rescue convoy spotting.
Drop tanks from 1936 never fully exploited...
19

Thompson and Rubensdoerffer *On August 15, the Luftwaffe attacked several British airfields. RAF Croydon was hit by the Me 110s of Erp. Gr. 210, led by the brilliant pilot, Walter Rubensdoerffer. John Thompson led his 111 Squadron Spitfires against the attacking 110s as the Germans were making their postattack withdrawal. Thompson and Rubensdoerffer got caught up in an extended low-level chase over the Sussex countryside. Thompson was impressed with the great skill shown by Rubensdoerffer in taking advantage of whatever cover he could find. Thompson remarked that he knew he was getting low when he saw his tracers removing shingles from a farmhouse roof. Eventually Thompson struck home, and Erp. Gr. 210 lost its most gifted leader and pilot*

The Air Was Alive with Tracer Bullets *On July 1, 1940, Pilot Officer William Walker of 616 Squadron was flying a training sortie with two squadron mates. He had very little time in Spitfires and was flying an older Mark I version. While airborne, his section was directed to intercept a "bandit." Walker lagged behind and arrived at a Dornier bomber seemingly alone. He took up a good firing position and pressed the trigger. Having never fired a Spitfire's guns, he was surprised at how little vibration there was. He did note that the "air was alive with tracer bullets." The bomber caught fire before his eyes and went down. Back on the ground, while bragging to his mates, a flight sergeant came in and asked if he realized that his guns were not loaded. It then dawned on Walker that the tracers filling the air were going the wrong way! Apparently, Walker's wingmen had previously inflicted mortal damage to the Dornier, and he had showed up just as the plane caught fire (Hough and Richards 1990).*

Simpson Plays Through *Golf course fairways were very attractive to pilots with damaged or dead engines. On August 17, Peter Simpson of 111 Squadron was shot down by Do 17s he had been attacking. He managed to crash-land on a golf course, much to the aggravation of the local golfers. The offended players threatened him with their clubs, thinking that at worst he was a German, and at least he had torn up the fairway. Simpson resorted to producing a pack of Players cigarettes to establish his true identity. In this depiction, no doubt, with typical British aplomb, the pilot is assuming that he should be commended for stopping his Hurricane before tearing up the green, while the golfers will no doubt insist that he repair his substantial divot on the fairway (Hough and Richards 1990).*

Mackenzie Knocks One Down *On October 7, Pilot Officer Ken Mackenzie was flying on one of his first operational sorties. He exhausted all his ammunition attacking an Me 109. He then attacked another 109 that had been damaged and was trying to make it back across the Channel. He flew alongside and signaled the pilot to ditch. The German refused. Mackenzie, running out of ideas, thought he could lower his gear and use it to batter the 109's fragile tail. Lowering the gear slowed the Hurricane too much, so he resorted to using his wingtip to batter the Messerschmitt's tail. The 109 went down, and Mackenzie crash-landed most of his Hurricane (minus a wingtip) on the Folkestone coast.*

innovations were woven together into a somewhat primitive (by later standards) early warning and control system that served the British very well until the end of the war. The highly advanced and effective system that the Germans confronted in 1940 descended directly from the system developed during World War I.

Following World War I, development slowed on both sides. Technical development in Germany was extremely limited during the 1920s as a result of harsh restrictions in the Treaty of Versailles that denied access to militarily applicable technologies. Aircraft and engines were among the key technologies denied the Germans. The British, for their part, turned quickly to reorder their national priorities and move to a peacetime economy. They immediately decreased the size of their fighting forces, reducing the RAF to less than 20 percent of its operational strength at the end of World War I. The RAF entered the 1930s with aircraft that represented only slight improvements over the equipment that had flown in World War I. On both sides the struggle for the air services during the 1920s became a fight for survival, rather than a period of progress. The Germans turned to gliders and civil aviation to keep the flame of aviation alive. They struggled to rebuild a military in secrecy, while in Britain the battle was to keep the Royal Air Force from being reabsorbed into the army or navy and thus disappearing amidst the budget wars then in progress. The economic depression that descended on Europe and the rest of the world in the late 1920s served only to exacerbate the tenuous positions that the air forces held in both Germany and Great Britain.

Even though few new aircraft systems emerged, the 1920s was a period of technological base building. Much of the success of the new aircraft designs that emerged in the 1930s is specifically attributable to the developments in propulsion technologies that took place in the 1920s. Development of variable-pitch propellers began during the 1920s, although the applications and advantages associated with them were largely ignored until the late 1930s. The use of tetraethyl lead to suppress engine detonation, blended fuels to achieve high octanes, and improvements in superchargers were all products of 1920s technological development.

Like the underlying component technologies, engine designs were also evolving. The pervasive aircraft engine design during and after World War I was the *radial engine,* which had a large frontal area to accommodate the pistons that were arrayed outward from the center of the engine block. The American Curtiss D-12 engine, introduced in 1921, was a 12-cylinder, V-shaped engine that enabled the large frontal area to be reduced, thus substantially reducing drag. In 1925 a British light bomber, the Fairey Fox, had used the D-12 engine in the absence of British engines of equal quality and had demonstrated superior performance. This seems to have spurred and focused development at Rolls-Royce, because in 1926 Rolls-Royce introduced a new liquid-cooled, supercharged, in-line engine called the F.XI "Kestrel." It can be argued that this was the single most significant technical event of the 1920s in Britain. This design became the basis for the development of the R engine, which, in turn, was the forerunner of the world-renowned Rolls-Royce PV1200 Merlin engine, the most famous aircraft engine in history.

During the 1920s, a building process in the field of airframe design was also under way in Britain. Two brilliant young designers, Sydney Camm and Reginald J. Mitchell, were developing the knowledge base from which they would, in just a few short years, launch some of the finest designs of the twentieth century. In the late 1920s, Camm, the chief designer for the Hawker Engineering Company, produced the Hawker Hart, a light biplane bomber. Powered by the Rolls-Royce Kestrel, this aircraft achieved a maximum speed of 184 mph, faster than any fighter at the time. Then, in 1931, Camm designed and produced the Hawker Fury, a fighter version of the Hart bomber. The Fury had a maximum speed of over 200 mph, the first fighter to achieve that level of performance.

If development slowed in Britain during the 1920s, it was essentially halted in Germany. Denied open access to aeronautical technologies and prohibited from militarily applicable development under the terms of the Treaty of Versailles, the Germans turned to those avenues left open to them—gliders and civil aviation. The Weimar government encouraged Germans to participate in sports aviation—glider clubs and other activities. By the late 1920s, there were 50,000 active members of gliding clubs throughout Germany; this activity created a pool of experienced fliers that would later become instrumental in building the new Luftwaffe. When the Allies lifted the restrictions on the size and number of German civil aircraft, the government immediately moved to create a state-owned airline, Lufthansa. By 1932 Germany was a world leader in civil aviation, with routes that spanned the world. Furthermore, the Germans had developed the organizational structures, training methods, and flying procedures to support their aviation industry—all of which would serve them well in the air war to come.

As Britain entered the 1930s, the Schneider Trophy, established in 1912 to reward the development of high-performance floatplanes, had become a focus of great international competition and prestige. In 1931, a monoplane design produced by R. J. Mitchell, chief of design at Supermarine, won the trophy for the third time (Britain had previously won in 1927 and 1929) and, with that victory, won for Britain the right to hold the trophy permanently. This was the Supermarine S.6B airframe in combination with the Rolls-Royce R engine. It achieved a speed of 340 mph and led to the development of the Spitfire aircraft a few years later. But at that moment in the early 1930s, even though significant advances were being made in civil aeronautics, the RAF still possessed only the wood-and-fabric biplanes that represented World War I technologies. Clearly, with aeronautical technology once again on the move, drastic improvements were needed if British military aircraft were to be capable of defending the country.

It is interesting to note the extent to which imported technologies benefited developments in both Britain and Germany. The impact of the V-shaped, in-line design of the Curtiss D-12 on Rolls-Royce engine design has been mentioned. Germany also took full advantage of developments that were in progress in other countries. Ernst Udet was one of Germany's highest-scoring aces in World War I, second only to von Richthofen. He was invited to fly at the U.S. National Air Races in Cleveland in 1931, where he saw a Curtiss Gulfhawk in a dive-bombing demonstration. Udet later purchased two Curtiss Hawk aircraft, brought them to Germany, and demonstrated the highly accurate dive-bombing and strafing capability of the aircraft to the German Air Ministry. The demonstration convinced them that a modernized dive-bomber could be an effective weapon in support of army forces. Soon after, the Air Ministry established a new requirement for a dive-bomber for the German air force, still being rebuilt in secret. This requirement would lead to the Ju 87 Stuka dive-bomber, which was one of the icons of the blitzkrieg style of warfare introduced by Germany in the late 1930s.

In the fall of 1931 the British Air Ministry, awakening to the need to update their antiquated air forces, issued Specification F.7/30, outlining the requirements for a new fighter to replace the outdated Bristol Bulldog. These were very tight times for aircraft manufacturers; consequently, there was great interest in the competition for the contracts expected to follow the issuance of the new specification. Competing firms submitted eight designs, hoping to win the development contract to build the new fighter. Demonstrating that the age of the biplane had not quite yet passed, five of the competing designs were biplanes, and the winner selected by the RAF for development was the Gloster SS37, a radial engine biplane. One of the unsuccessful entries was a design labeled the Type 224, submitted by R. J. Mitchell of Supermarine, the designer of the S.6B aircraft that had won the Schneider Trophy. In spite of the fact that the Type 224 was not chosen as the winner in the RAF competition, Vickers, the parent company of Supermarine, agreed to allow Mitchell to continue work to improve the design.

The Type 224 became the first in a series of prototype aircraft that would lead to a new British fighter. In 1933, while working on another prototype of his new fighter, labeled the Type 300, Mitchell fell ill with cancer and entered the hospital for an operation. Afterward, he vacationed in Europe while recovering. In Europe, he had the opportunity both to see Germany's new fighter, the Messerschmitt Bf 109, in flight and to speak to some young German pilots. That experience convinced Mitchell that war with Germany was inevitable, and it further convinced him that a high-performance airplane of the type he envisioned would be required if Britain was going to prevail. Mitchell returned to Britain, refused all further treatment of his condition, and devoted his remaining days to the design, development, and production of the aircraft that today is most identified with the Battle of Britain—the Spitfire.

In Germany, efforts to circumvent the restrictions of the Treaty of Versailles continued. After 1922, the development of civil transport aircraft was permitted. This, together with the continuing efforts to build and train an air force, encouraged many gifted German designers to explore the design and development of civil aircraft as a path toward the development of militarily useful aeronautical technologies. A premier example of this phenomenon was the development by Willy Messerschmitt of the Bf 108 "Taifun" aircraft in the early 1930s. A low-wing, single-engine monoplane, the Bf 108 was a record-breaking sports aircraft and was also employed as a liaison and utility aircraft in World War II. However, its true significance is that it was the forerunner of the Bf 109, Messerschmitt's famous design that became Germany's premier frontline fighter in the war.

Due to the restrictions that prevented overtly building military aircraft, German engine development in the 1920s had focused on propulsion for the heavy transports used by Lufthansa and other airlines of the time. The result was that German engines had not developed technically on a par with British engines. Even into the later 1930s, German engine development lagged well behind that of the Rolls-Royce Merlin engines, then being demonstrated in Britain. However, by 1937 Daimler-Benz had finally produced and refined the DB 601A engine, which then began to replace the Jumo as the preferred German engine. The DB 601A was an excellent design—even more advanced technically in some respects than the Merlin. It delivered 1100 horsepower, far surpassing other German engines in terms of horsepower to weight, a common metric in comparing fighter engines.

In England, the issue of guns—how many and what type—was one of the primary differences between the various aircraft specifications that were evolving. In 1934 Squadron Leader Ralph Sorley, the officer in charge of the operational requirements section of the Air Ministry, performed analyses showing that the average fighter pilot in modern air warfare would be able to bring the plane's guns to bear on an aerial target for only two seconds or less. His analysis concluded that a fighter would need at least six, and preferably eight, guns to bring the opponent down, given the short period of the firing window.

The RAF had issued a new specification (F.37/34) to govern the development of Mitchell's new fighter at Supermarine. Although the specification required only four machine guns, the RAF had, as a result of Sorley's analysis, concluded that fighters should have eight guns to effectively bring down enemy aircraft in combat. As the urgency of the need to develop high-performance fighters grew, the Air Ministry was fast moving to contract for a production fighter based on Mitchell's experimental design. In 1935, the ministry issued another specification (F.10/35) describing the requirements for the production version of the Spitfire:

- not less than six guns, with eight desirable;
- an airspeed of 310 knots at an altitude of 15,000 feet;
- a service ceiling [the altitude at which the rate of climb at maximum power is 100 feet per minute] of not less than 30,000 feet. (Price 1982)

The production contract for the Supermarine aircraft specified eight guns, but allowed a reduction in the amount of fuel carried as a trade-off. Contractor testing began in March 1936, and the Air Ministry placed its first order for 310 Spitfire aircraft in June 1936. Mitchell had given Britain one of the finest high-performance aircraft designs that the world has known.

In 1935, the German government formally acknowledged the existence of the Luftwaffe. The question of force structure—how many aircraft and of what types—was not the kind of analytically arcane issue that Hermann Goering cared to deal with. Furthermore, Hitler had made it clear that he was only interested in producing the greatest number of aircraft possible within the shortest time. This and other factors combined to make the issue of design and production of a four-engine strategic bomber one that had little resonance among the top brass of the German High Command. The notable exception was General Walther Wever, the first chief of staff of the Luftwaffe. Wever strongly advocated that Germany develop a four-engine bomber with sufficient range to attack cities as far away as Scotland from bases inside Germany. Under Wever's sponsorship, prototypes of two different four-engine bomber designs were developed. But in 1936, Wever was killed in a flying accident. Neither Kesselring nor Goering had any enthusiasm for pursuing the strategic bomber project—it would have lessened total aircraft production output. While Milch, who might have urged the continuation of the project, was traveling, Kesselring cancelled four-engine bomber development. Germany thus restricted bomber production strictly to two-engine designs, a decision that, in the long run, considerably reduced the total damage that German bomber forces were able to inflict on Britain once the London phase of the Battle of Britain got under way.

During the mid-1930s, as aircraft and engine designs were advancing, significant progress was also being made in early warning and control. Beginning as early as the 1920s, Britain had experimented with acoustic reflectors as a means of detecting aerial targets. Several "acoustic mirrors" were built on Britain's south coast. These curved concrete structures were designed to focus the reflected sound waves toward a center focal point. Performance was marginal at best; they detected aircraft only at short ranges, and then gave no information as to azimuth, altitude, or range. Furthermore, the devices did not detect targets at all if they were outside the radius of the structure's curvature. Clearly something else was needed.

In late 1934, the Air Ministry set up the Committee for the Scientific Survey of Air Defence under the direction of Henry Tizard, a World War I pilot turned scientist. This small group was to examine various theories and recommend solutions to the air defense problem. Tizard immediately summoned Robert Watson-Watt, a scientist who headed the radio research branch of the National Physical Laboratory. Watson-Watt was first asked about the feasibility of a "death ray" weapon that might destroy aircraft in flight—an idea he quickly dismissed as a schoolboy fantasy. The committee then asked for recommendations. Watson-Watt recommended three areas for investigation—reflected radio waves to detect aircraft, radiotelephone communications to link aircraft to ground controllers, and coded signals to enable friendly aircraft to be differentiated from enemy aircraft. In retrospect, these recommendations essentially laid out the architecture on which the British integrated aerial defense system was based. It was brilliant.

Watson-Watt suggested that reflected radio waves could be used to detect attacking aircraft. After a demonstration in which the technique was used to detect a bomber in flight, a series of tests was approved. Within weeks, Watson-Watt demonstrated that he could track aircraft at a range of 40 miles. Sir Hugh Dowding, at that time the air staff member in charge of research and development, threw his wholehearted support behind the effort. Within nine months after the first meeting of the committee under Tizard's leadership, the British Air Council recommended construction of a chain of stations to develop and employ this new technology. Furthermore, in 1935 very high frequency (VHF) radios were installed in British fighters, giving them a substantially increased communications range that would complement the increasing detection range offered by Watson-Watt's new system. This illustrated the balanced building-out process that attended the early warning and defense system. Britain continually improved not only the detection capability of the system, but also the communications network and associated operating procedures that permitted command and control of interceptor aircraft. It was this total systems approach that gave the Chain Home system the decisive capability that it demonstrated during the Battle of Britain.

When one thinks of the superior British aircraft designers who most influenced the outcome of the Battle of Britain, one automatically thinks of R. J. Mitchell, the designer of the Spitfire, and Sydney Camm, designer of the Hurricane. Germany also had her own superstar designer—Willy Emil Messerschmitt, designer of the Bf 109, Me 110, and Me 262, among many others. Messerschmitt had been working on aircraft design and development since his childhood, and he had been a pioneer in the field of glider design in the 1920s. By 1935 rearmament was fully under way in Germany, and it was clear that a new, modern fighter was needed. In October 1935 a competitive fly-off was conducted to select the aircraft design that would be produced in quantity. Four designs were initially considered. Two were quickly rejected, leaving the choice between the Heinkel He 112 and Messerschmitt's entry, the Bf 109. The Bf 109 won the competition and became Germany's frontline fighter for the Battle of Britain and the duration of World War II.

In his design for the Bf 109, Messerschmitt created the lightest, smallest design possible that could accommodate the most powerful engine then under development in Germany, the Daimler-Benz 601A. But during the time of the competition to select Germany's new production fighter, the DB 601 was still in development; therefore, the competitors needed an engine of similar size and performance. In another case that illustrates the flow of technology across national borders, Germany took advantage of the advanced state of engine development in Britain. During the initial competition and testing, both the Heinkel and the Messerschmitt entries were powered by British Rolls-Royce Kestrel engines.

This phenomenon of the flow of technology across national boundaries was—and is—a continuing theme in modern arms development. Commercial transactions result in technology developed in one country flowing to another; air shows and visits between militaries transfer technology; and individuals with expertise contribute to developments in other countries. In addition to the many examples already noted, a German, Edgar Schmued, who had worked at both the Fokker and Messerschmitt companies, played a major part in designing the American P-51 Mustang aircraft. Likewise, an American, Alfred Gassner, was a codesigner of the Ju 88, Germany's best bomber at the time of the Battle of Britain.

The Messerschmitt Bf 109 design, with its thin wings and high wing loading, demanded advanced aerodynamic design innovations to ensure adequate performance across the entire flight envelope. The aircraft employed leading-edge slats for additional lift at low airspeeds and slotted ailerons interconnected to the flaps. These were very advanced developments in 1935, especially for fighters. In spite of the fact that the cockpit was cramped and visibility was poor, the aircraft was a fine performer in flight, with altitude and airspeed capabilities similar to those of the Spitfire. The most significant weakness in the Bf 109 design was the lack of range, which prevented it from probing as deeply into British airspace as would have been preferred and limited its loiter time in the combat arena. By 1939 the 109E, called the "Emil" by German fighter pilots and powered by the DB 601A engine, was being delivered to German squadrons at the same time that the Spitfire was being deployed to British squadrons. By the end of the war, the Messerschmitt Bf 109 was the most produced airplane on either side, with over 35,000 delivered to the Luftwaffe.

While design and development of the Bf 109 was in progress, Messerschmitt received the go-ahead to design and develop a long-range escort aircraft—a "strategic fighter"—that was to have the capability of escorting bombers on deep penetration raids into enemy territory and also of engaging enemy fighters when encountered. Messerschmitt designed the Me 110 twin-engine fighter, a low-wing cantilevered monoplane, with a crew of two. The first Me 110 prototype flew in 1936. In spite of the weight and maneuverability penalties paid for the additional engine and crewmember, the production version of the 110, powered by the same DB 601 engine as was used in the Bf 109, was faster in straight and level flight than either the Bf 109 or the Hurricane. Unfortunately for the crews of the aircraft, fighter escort missions demand high maneuverability and rapid climbing rates to successfully engage opposing fighter interceptors. The Me 110, a compromise between the characteristics of both a bomber and a fighter, did not fare well against either the Hurricanes or the Spitfires in the Battle of Britain. Multirole designs have seldom been highly successful. In the effort do all things adequately, they are too often unable to compete successfully with designs that are optimized to perform a single mission well. This can be deadly in combat.

The parallels between the development of the British Hawker Hurricane and the Spitfire are remarkable. In the same sense that Specification F.7/30, the specification that opened the competition to replace the Bristol Bulldog biplane, spurred development of the Spitfire, it was also the spark that gave birth to the Hawker Hurricane. Hawker Aircraft Ltd., led by Chief Designer Sydney Camm, submitted two designs in response to the specification, one a monoplane and the other a biplane. As in Mitchell's case, the Air Ministry did not choose either airplane for further development. Camm, like Mitchell at Supermarine, then began an effort to design a monoplane fighter supported by private funds from his company, Hawker. He ignored Specification F.7/30 and proceeded on a path of his own choosing. He based his new design on his previous success with the Fury biplane fighter. In essence, the Hurricane was a transitional design that represented a step between the old wood-and-fabric biplanes and the all-metal monocoque frame design that Mitchell employed in the Spitfire. While not as technically advanced as the Spitfire, the fabric-covered wood-and-metal frame of the Hurricane was able to withstand considerable enemy fire with minimum damage. The Germans found it very difficult to destroy the steel girder structure of the Chain Home radar towers; it was also difficult to do mortal damage to an aircraft with a similar structural design.

By late 1935, the Rolls-Royce PV 1200 Kestrel engine had been redesignated the Merlin and was in the final stages of a development program to improve its already impressive performance. As it became increasingly clear that the other available engines of British manufacture would not measure up to the Merlin, both Mitchell and Camm opted to use the Rolls-Royce engine in their two fighters, both of which were by then nearing production. Powering both the Spitfire and the Hurricane aircraft in the Battle of Britain, this liquid-cooled, supercharged, carburetor-fed engine became the most famous aero-engine of World War II.

As previously observed, Mitchell proceeded to complete the design for the Spitfire, and it flew for the first time in 1936. The Air Ministry placed an order for a production run of 310 aircraft in June 1936, with delivery of the first article anticipated in October 1937. However, in 1937 Mitchell finally succumbed to the cancer that had plagued him for years. He died at age 42 and never saw the production version of the Spitfire.

As for the Hurricane, when the RAF published Specification F.10/35, Camm had essentially completed the design work for the Hurricane. The only significant change required was the mounting of the eight-gun array. Camm had new wings built and installed to accommodate the guns, and in November 1935 the prototype of the Hawker Hurricane with the Merlin C engine flew for the first time. Hawker Aviation knew they had a winner; they authorized the construction of tooling to produce 1000 aircraft before they received the contract for the first 600 units. In the Battle of Britain, even though the Hurricane was not equal to the Messerschmitt Bf 109E in aerodynamic performance, it was credited with 80 percent of the kills that Fighter Command claimed. In a real sense, Camm's vision of a highly producible aircraft whose design and performance were "good enough" provided Fighter Command with the weapon that did much of the heavy lifting in the Battle of Britain. Camm himself remained active in aircraft design until well after World War II. He died in 1966.

As the premier fighters were emerging in both Germany and Great Britain, other types of aircraft were also being developed for the many and varied tasks and missions anticipated during a war. Many of these were bombers of various sorts, but Germany also developed several floatplanes, or seaplanes. One of the best examples was the Heinkel He 115, a versatile seaplane capable of conducting bombing missions, minelaying missions, and aerial torpedo deliveries. Most of the records that mention the He 115 indicate that its primary function was minelaying in coastal estuaries, although it was also used in German air-sea rescue missions. The He 115 came to be regarded as one of the best floatplane designs from any of the combatant nations in World War II.

From the early days of World War I aerial combat, the problem of aiming the guns of fighter aircraft had been handled in a manner similar to aiming a rifle. The aiming apparatus typically consisted of an iron ring near the pilot position, with a bead further out on the nose of the aircraft. The pilot would line up the bead in the center of the ring, and the guns would be situated such that, at a specified range, the bullets would converge on the centerline of the ring-bead axis. In the fast-maneuvering, high-speed environment normal to the new generation of fighters, it was very difficult to bring all of the parameters together at the right place and time. In 1937, a reflector gunsight was produced that represented a significant improvement over the ring-and-bead sight, and installation was begun on RAF fighters. In spite of the fact that this improvement was technically ready for use, supply problems interfered with its prompt delivery to the RAF, and many Spitfires and Hurricanes were forced to continue using the ring and bead through the Battle of Britain.

Production of fighters in the quantities needed was a continuing problem as Britain rushed to prepare for the coming war. Spitfire production was lagging, a function of both the complex design of the aircraft and the production methods used by Supermarine. Hurricane production rates were adequate since the Hurricane was a much more producible aircraft. In addition, Hawker Aviation, using standardized tooling and production-line techniques, was better prepared for volume production than was Supermarine, whose experience was primarily in low-volume seaplane manufacture. In the late 1930s, Britain, concerned about the ability of the defense industry to produce aircraft at the rates necessary for the war, developed what were described as "shadow factories." The fundamental concept was to employ the large production facilities of the automobile industry to supplement and boost the production rates of the aircraft industry. These shadow factories became an important source of manufacturing production as the war effort progressed.

There are times when a specific technology is proven, tested, and ready, yet is not adopted for reasons that are obscure and difficult to comprehend. The variable-pitch propeller is one of those. The military airplanes of World War I and those developed through the 1920s all employed fixed-pitch propellers. Even before 1920, the limitations of such propellers were recognized, and substantial work had been done to design propellers with adjustable blades. That work continued into the 1920s, and working designs for variable-pitch propellers, developed most notably by de Havilland of Britain and Hamilton-Standard of the United States, were patented, produced, and available for use in aircraft designs by the late 1920s. That fact notwithstanding, none of the primary fighters that would fight in the Battle of Britain—the Spitfire, the Hurricane, and the Bf 109E—opted for a variable-pitch propeller in their initial designs. However, in each case, as the designs first emerged and were then reengineered and improved to obtain maximum possible performance, they all turned to variable-pitch propellers. In 1940, the RAF requested that a Rotol constant-speed propeller be fitted to the Spitfire. This change provided a substantial increase in performance for the Spitfire, thus improving its performance relative to the Bf 109E at high altitudes.

Of course, the British were hardly alone in their somewhat slow acceptance of a technological innovation that might have been of considerable benefit. The combat range of the Bf 109 was highly restricted due to lack of fuel. An obvious solution would appear to have been incorporation of external fuel tanks that could be dropped when the aircraft engaged in combat. Germany had developed drop tanks in 1937, during the Spanish civil war. The Heinkel He 51 was the first aircraft to be fitted with fuel lines and carrying connections for tanks. However, for reasons not fully understood, the Bf 109E models that fought in the Battle of Britain never were configured with drop tanks, and the lack of range continued to be among the most serious of the relatively few shortcomings associated with that airplane.

In 1933, Udet's demonstration of the potential offered by the dive-bomber had resulted in the development by the German Air Ministry of a new requirement for a next-generation dive-bomber. Two designs quickly emerged as the primary competitors—the Heinkel He 112 and the Junkers Ju 87 Stuka. Udet himself flew and crashed the He 112, after which the Ju 87 was declared the winner. The Stuka entered service in 1937. It was a low-wing monoplane with nonretractable landing gear. Sirens mounted on the landing gear pylons gave the aircraft the high-pitched wail that became a hallmark of blitzkrieg warfare in the late 1930s. The aircraft performed well in the Spanish civil war, but, with airspeeds generally in the 230-mph range, it was far too slow to survive for long in the Battle of Britain. By August 1940, the decision had been made to withdraw it from the battle.

Messerschmitt experimented with a variety of gun-cannon configurations on the Bf 109; however, the combination most used by the Germans at the time of the Battle of Britain employed two cannons, one in each wing, and two 7.92-mm machine guns mounted in the nose of the aircraft. With the exception of a combination that featured only machine guns, every German gun configuration was more effective than the British eight-gun array. Machine guns had neither the range nor the penetration power of the cannons.

Meanwhile, in Britain, development of the networked air defense system was continuing at a rapid pace. In 1937, experiments at Biggin Hill linked the Chain Home radar stations, filter rooms, and fighter squadrons together for the first time. Much of the success of these experiments came from the deliberate involvement of representatives from both the operational and the scientific communities. It was this unique interplay of diverse functional skills that enabled the development of radar from a scientific concept into an operationally viable system within the span of just a few short years.

As early as World War I, the British, defending themselves against the attacks of the Gotha bombers, had discovered the benefits gained by using "balloon barrages"—blimps with attached cables hanging from them. The balloons forced enemy bombers to attack from higher altitudes and also helped make antiaircraft artillery more effective. These earlier lessons were not lost on the British as they organized for the defense of their country once again in the late 1930s. In November 1938, Air Commodore Boyd was appointed to lead the RAF Balloon Command. Balloons were quickly deployed in defensive positions around London and in other key locations along the anticipated routes of attacking aircraft. Balloons were flown at altitudes up to 5000 feet and were particularly dangerous for low-level bombers and dive-bombers, especially at night.

As preparations continued for the air war, Britain worked to refine and complete her defensive command and control system. One of the last problems to be solved was the ability to identify and vector British fighters to the right areas so that they could visually acquire attacking airplanes and take over the intercept from there. In 1939, Britain had added the "Pipsqueak" capability to the system. Using Pipsqueak, the ground stations could identify British formations, and they could direct flights toward the positions of attacking bombers detected on radar. This was one of the first attempts to positively identify friendly aircraft in flight, an issue that would grow in importance very shortly. This concept of fighters under positive ground control was to become extremely important to the efficiency and success of the RAF during the battle.

While most of the attention is focused on Messerschmitt in the treatment of German aircraft design during this period, no discussion would be complete without including the accomplishments of Junkers, Heinkel, and Dornier. These were the designers and producers of the bombers that rained such destruction on the British homeland during and immediately after the Battle of Britain. The Junkers Ju 87 Stuka was a qualified success; it proved vulnerable to the high-speed fighters encountered in the Battle of Britain. The Junkers Ju 88, on the other hand, was the best bomber that Germany produced. It went from drawing board to first flight in a single year in 1936. It was a fine airplane, and it filled a number of roles, from dive-bomber to minelayer. The airplane went into operational service in 1939 and performed well throughout the war. To repeat an interesting side note, an American, Alfred Gassner, was a codesigner of the Ju 88.

If there was a single airplane that Britons would come to associate with German bombers, it was the Heinkel He 111. The unique throbbing sound of its unsynchronized twin engines made it highly recognizable as formations approached target areas; furthermore, there were more He 111s in the German fleet than any other bomber employed. Designed in the early 1930s as a high-speed transport aircraft, it had the characteristics necessary for military missions from the start, including a glazed, streamlined cockpit with the pilot position offset for improved visibility. The He 111s were used quite successfully in the Spanish civil war; they were able to outrun the fighters employed to defend against them, which led to erroneous and overly optimistic conclusions about how they would likely perform when employed against the defenses in England. By September 1940, the He 111 had been relegated to night bombing due to losses incurred.

The third of the German twin-engine bombers employed in the Battle of Britain was the Dornier Do 17, known as the "Flying Pencil." Originally designed in 1934 to carry airmail, the airplane had been pulled from active flying and was in storage when the Luftwaffe reactivated it after going public in 1935. Like the He 111, the Do 17 was forced to use a less than optimum engine so that the DB 601s could be used in the Bf 109s. It had two Bramo 232P nine-cylinder engines—a two-stage, supercharged radial design. Again like the He 111, it proved too slow for the aerial environment of the Battle of Britain and was easy prey to the Hurricanes and Spitfires.

Britain also had less well-known active aircraft during the Battle of Britain. The most interesting of these was the Boulton-Paul Defiant. The airplane had nice lines, with an electrically powered turret mounted behind the cockpit and good handling qualities; at the time of its deployment in 1939, it was hailed as the "fighter with the sting in its tail." The Defiant had a few initial successes, primarily because it caught a few attacking Bf 109s unaware as they approached from the rear. Once the Germans had analyzed the configuration and found its weaknesses, the Defiant proved an easy target. It was insufficiently maneuverable to be effective and too slow to defend itself; furthermore, it was also very difficult to escape from. The aircraft suffered disastrous losses and was soon consigned to a night-fighter role, a mission that it performed with marginal success, as well.

The build-out of the fully networked integrated air defense system based on Watson-Watt's radar was moving to completion in 1939. Having originally built three stations to prove the concept, the Air Ministry now moved to construct the remaining 17 Chain Home stations initially planned. Located over a stretch of the British coast from Land's End to Newcastle, the stations were networked and linked to plotting rooms, filter rooms, and fighter control operations. With each station identifiable by its two steel girder towers, one a transmitting antenna and the other a receiver, the Chain Home system was quite possibly the single most important technical advance that influenced the outcome of the Battle of Britain.

The flow of technology from country to country, friend to friend, and friend to foe alike is an interesting phenomenon that was amply demonstrated during the 1930s as both Germany and Britain developed the capabilities that they would take to war. Several cases have already been mentioned—the fact that an American helped design the Ju 88 bomber; the sale by the United States of dive-bombers to Germany; the use of the British Kestrel engine in the prototypes of the German Bf 109; and the licensing of the manufacture of the Merlin engine to an American manufacturer, which led to the use of the Merlin in the American P-51. But the flow of technology from America to Britain deserves specific mention. In several instances, America provided key technologies to Britain at critical times, which enabled and facilitated developments that were important to the outcome of the battle. For example, the Curtiss D-12 engine used in the Fairey Fox was influential in the configuration of the Merlin engine, as well as the timing of its emergence. Then, in 1938 to 1940, as the British sought to improve the performance of the Merlin by increasing the boost pressures in the engine, the American oil company, Eastern States Standard Oil (Esso), made the improvement possible by selling Britain its 100-octane gasoline. Another key transfer was the agreement by the Colt Arms Company to license manufacture of its highly reliable .30-caliber machine gun to the British. The gun was bored out to .303 caliber and was produced in Britain as the Browning machine gun. It was used on all British fighters in the Battle of Britain.

The last piece of the mosaic that defined the complete Chain Home air defense system involved the problem of positively identifying each aircraft in flight. Pipsqueak had provided a start toward this problem, but in the fall of 1939, no effective identification-friend-or-foe (IFF) capability had been installed in British fighters, and the confusion over whether radar returns represented friends or foes sometimes led to deadly results. In an incident wryly referred to within the RAF as the Battle of Barking Creek, Spitfires shot down two Hurricanes, killing a British pilot. By June 1940, all operational British aircraft had been equipped with IFF.

As Germany prepared for the coming battle for Britain, she developed an ingenious system that would enable her bombers to drop bombs with acceptable accuracy even when the ground was obscured. The system was known as "*Knickebein.*" An audible signal was electronically transmitted. As long as the plane was on course, the pilot heard a continuous signal. If the plane drifted left or right, the pilot would hear Morse code. From a different, geographically separated, site, a second signal was broadcast that intersected the first near the target. When the plane, flying the inbound course, crossed the second signal, the pilot would release the bombs. The British were aware that the Germans were using some sort of electronic means to navigate, but they knew little of the detailed nature of the *Knickebein* system. In June 1940, a flight of Anson trainers was established with the specific mission to intercept the mysterious navigational signals, and, in doing so, to enable the development of countermeasures. By July, the RAF had developed transmitters capable of jamming and otherwise distorting the *Knickebein* signals. With the signals rendered sufficiently unreliable that they could no longer be employed for precise bombing, the system was made essentially useless. This was one of the earliest examples of electronic warfare and illustrates the thrust-and-countermeasure nature of technological warfare.

Identification of aircraft in flight by ground-based controllers was a complex technical problem, as is demonstrated by the account of the development of the integrated air defense system. No less important was the ability of pilots in a flight or squadron to identify each other in close combat situations. However, rather than electronic interrogation-and-response methods, in-flight identification during the heat of air-to-air combat was generally accomplished by the colors and markings of the airplanes. In the Battle of Britain, German aircraft used a more complicated system of colors and markings than the British, but the objectives of both systems were the same—concealment from foes and identification to friends. German aircraft were generally painted a dark green and black camouflage pattern on the tops of the fuselage and wings, with light gray-blue underneath. If viewed from above, the aircraft would appear to fade into the patterns of the ground below, while, when viewed from below, it would be difficult to distinguish the aircraft from the sky around it. By the end of September 1940, as the decision was made to concentrate on night bombing of the cities, the German aircraft were painted black and other dark colors to reduce their visibility in night combat conditions. The Royal Air Force employed coloring schemes similar to those used by the Germans. The tops of the aircraft were painted a green-over-brown camouflage pattern, with light-colored undersides. The British national symbol was a bulls-eye-type pattern with a red circle at the center. Like the German *Balkankreuz* symbol, these were carried on both the tops and bottoms of the wings and on the sides of the fuselage. On the other hand, the marking system used by the British was considerably less complicated than the German system.

Britain had, by 1938, begun the search for ways to improve the lethality of the eight-gun arrays that the fighters carried. One technology that appeared to have promise was the incendiary round. This offered several benefits: it was more lethal in terms of the damage done by the round on impact, and it further enabled pilots to easily gauge where their bullets were striking so they could adjust their aim as they fired. While the incendiary round used in the Battle of Britain was usually referred to as the "de Wilde" round, so named for a Swiss inventor who had proposed a design for an incendiary round to the British Military Purchasing Commission, the round actually used was in fact developed by C. Aubrey Dixon, a small-arms expert who worked in the design department of the Royal Arsenal.

Another issue associated with air-to-air gunnery and the lethality of the armament systems carried has to do with the aiming of the guns on board the fighter. In order to concentrate the fire from multiple guns, all guns are typically "harmonized" to a single focal point some distance in front of the aircraft. The regulation standard for British fighter aircraft at the time of the Battle of Britain was 650 yards. In too many cases, direct hits did not bring German aircraft down. To deliver their bullets with maximum effect, top British fighter pilots quickly found that they needed to be much closer to the target. Soon the de facto standard had been shortened to 250 yards, as pilots requested that their armament officers reharmonize their guns for close-in combat. The practice was eventually recognized as the standard for the RAF. This practice would continue until Britain began to arm fighters with cannons, the heavier rounds and extended ranges of which made kills at increased distances possible. However, the addition of cannons to British fighters came a little late to significantly impact the Battle of Britain.

In May 1940, the tragedy of the defensive effort in France was grinding toward Dunkirk and the end of the Allied effort. As this defeat was unfolding, a significant victory was quietly taking place in the technological war. The Government Code and Cipher School at Bletchley Park had intercepted the codes used by the Germans during the Norwegian campaign. Using a device referred to as the *Enigma machine,* they cracked the newest codes enciphered by the Germans to transmit top-secret messages. It took some time before Britain was able to exploit this capability to its fullest extent, but from that time forward the Allies were able to begin building a database of information regarding the patterns of German actions, organizational structures, and future intentions. While the breakthrough did not materially alter the outcome of the Battle of Britain, it was a very significant development from the standpoint of the overall conduct of the continuing war effort.

As the Battle of Britain was joined, it did not take long for experience in combat to begin to influence both the tactics and the technology associated with air warfare. For example, by August 1940, all Hurricanes and Spitfires had armor plate installed behind the pilot's head and shoulders and all squadrons had installed rearview mirrors so that pilots could see anything behind them at a glance. These were two simple additions that saved many lives. Among the more complex changes were the additions of cannons and constant-speed propellers to the fighters. By 1939, it had become apparent that the eight-machine-gun configuration carried by the fighters was not sufficiently lethal, especially when German bombers began to carry armor. Testing had already shown that cannons were more effective than machine guns. By the spring of 1940, the technical problems of the wing-mounted cannon had been largely solved, and the first operational wing of Spitfires was equipped with a two-cannon variant, labeled the Mark IB Spitfire. Constant-speed propellers represented the final step in the evolution that brought propellers from fixed-pitch designs to limited variable-pitch and finally to constant-speed designs. By August 1940, Britain was well on the way to installing constant-speed propellers in the Hurricanes and Spitfires, although many of them fought the Battle of Britain with variable-speed propellers.

As the British rushed to make the changes that would create an advantage for their interceptors, the Germans likewise were making alterations to improve the survivability of their bomber fleet. Since most fighter attacks came from the rear, German bombers began to carry substantial armor plating for the protection of the airplanes and their crews. As is normal in combat, action demands reaction. In this case, many British fighter pilots changed tactics, approaching the bombers head-on, rather than the tail approach that had been rendered less effective due to the armor plating.

Reconnaissance is a necessity for any attacking force; it enables planners to assess the impact of previous efforts and to identify potential future targets. The Germans invested heavily in improving their aerial reconnaissance capability. In 1939 the Arado Ar 196 floatplane became operational. This single-engine seaplane was assigned to Germany's principal warships for use as an onboard reconnaissance vehicle. With range in excess of 600 miles, the Ar 196 provided an over-the-horizon search capability, not unlike the aircraft carrier air fleets of today. Aerial reconnaissance of British targets was a daily activity for the Germans. The Luftwaffe generally used bomber aircraft for the purpose, usually Ju 88s or Do 17s. They also employed a derivative of the Dornier Do 17, labeled the Do 215. Three cameras were mounted on board, thus permitting the aircraft to photograph target areas and to assess bomb damage while simultaneously performing bombing missions.

The Battle of Britain is generally agreed to have begun on July 10, 1940. As the Channel phase (*Kanalkampf*) ended, the battle for the airfields began. Just as combat experience led to innovation in aircraft technologies, combat experience on the ground led to innovations in airfield defense. The British found themselves under carefully coordinated air attacks that often featured bombers approaching at altitudes of 1000 feet or less. One of the more unusual techniques developed to defend against these attacks was the parachute-and-cable (PAC) rocket. A simple device, the PAC was a rocket with 1000 feet of trailing cable that could be fired into the air as bombers approached. A small parachute was deployed, and, as the parachute descended, the bombers were confronted with multiple steel cables hanging in their path. While hardly a decisive weapon, the PAC rocket was one example of those developments that are born of necessity in the heat of battle. More than one German bomber crashed after becoming tangled in the drifting cables around the target airfields during this phase of the battle.

As might be suggested by the development of German seaplane technology, Germany had developed a substantial air-sea rescue capability. This was an area that had been largely ignored by the British, with the consequence that their ability to return pilots downed at sea to active duty lagged well behind that of the Germans. The Germans, to the contrary, developed a soundly designed system that served them well during the Battle of Britain. The Germans employed the Heinkel He 59 floatplane as a sea rescue aircraft that doubled as an observation plane. The aircraft was painted white and carried the Red Cross symbol. After the He 59s were regularly observed in the vicinity of convoys, the British approved attacks on the aircraft engaged in the rescue missions—a decision that caused considerable consternation over the issue of whether the aircraft deserved the protections normal to the Red Cross. The Germans finally repainted the He 59s with a camouflage color scheme and provided them with armament for self-defense as they continued their duties in the Channel area.

Germany's need for accurate target intelligence was growing as the scope and intensity of the bombing raids increased. Furthermore, the loss rate among the single bombers sent out to reconnoiter and film targets was growing. To meet the requirement for aerial reconnaissance, Germany turned to the Ju 86P, a bomber that had been used in the Spanish civil war and had also been used as a passenger aircraft. It was far too slow for combat use in the theater where the Hurricanes and Spitfires operated, so the Germans equipped it with double supercharged diesel engines and a pressurized cabin, which permitted it to operate above 30,000 feet. At this altitude it was well above antiaircraft artillery fire and the maximum operating altitude of most fighters. It could, therefore, operate with a degree of impunity. However, the state of the art in optics at the time was such that aerial photography from these altitudes could not provide the level of detail necessary for useful aerial intelligence. The inability to provide good target resolution, and hence good intelligence, often led German planners to focus on decoy targets and to ignore primary facilities.

By August the need for organized air-sea rescue had finally become apparent to the British, and they began to develop a rudimentary capability. Until this time they had only a fleet of 18 motorboats assigned to the task of searching for downed airmen along the entire British south coast. Without the well-developed system of floatplanes and ships that the Germans had, many rescues of British airmen were accomplished almost by chance when a civilian boat happened to be conveniently located nearby and was able to assist the downed pilot.

It was only natural that the success enjoyed by the British with radar, as implemented in the Chain Home system, would lead to further development aimed at other applications. Airborne radar was a natural extension of the technology. The Bristol Blenheim was fitted with the first radar small enough to be carried on board an aircraft and was assigned the mission of intercepting the German bombers that were by then a regular nighttime feature of the battle. Unfortunately, the Blenheim was too slow to catch most German bombers. In spite of the addition of the new radar, it was essentially ineffective in every mission assigned, and it played no significant role in the battle. It demonstrated the futility of simply trying to field quantities of aircraft in the hope that sheer numbers, or special equipment, would make up for the lack of performance in those key characteristics that determine mission outcomes. Quality matters.

As the battle progressed into the intense stages of late August and September, which were marked by attacks on the RAF and its airfields, further innovations emerged of the sort exemplified by the parachute-and-cable rockets—innovations born of necessity. The British made a number of advances in servicing airplanes to return them to battle as quickly as possible and repairing damaged aircraft. Along related lines, to ensure the continued operation of the critical command and control system, which was coming under increasingly intense attacks, the British established fully equipped alternate operations rooms within 5 miles of the sector stations. These alternate operations rooms could immediately reestablish sector control, should a station be destroyed in an attack. These rapid adjustments in methods and procedures are representative of an environment in which people are committed to a mission and are given the authority and resources to make immediate changes as they are needed.

The failure of Germany to employ external drop tanks to carry more fuel—a technical innovation that might have made a significant difference—has continued to be a subject of discussion for historians and participants alike. It has been estimated that drop tanks might have permitted the Bf 109 to reach any target in Britain, rather than having been restricted to targets in the southern and eastern quadrants of the country. The use of tanks certainly would have permitted the aircraft to loiter in the combat area for longer periods and would have reduced the need to shuttle flights of fighters in and out of the theater continuously. Erhard Milch later identified the failure to use external tanks on aircraft as one of the key mistakes that led to the German failure in the Battle of Britain (Air University 2001).

As the focus of the German attack turned from the RAF and its airfields to the bombing of the cities, the means for bombing targets with accuracy even when the target was not seen became more important. Night bombing and missions flown in poor weather presented substantial challenges to crews that not only had to find targets under difficult visual conditions, but who were under attack themselves. Even while using the *Knickebein* system, the Germans had already begun to develop an improved system. This improved system, known as "*X-Gerät*," was similar to the *Knickebein* system, but it featured improvements that made it more accurate. A special operations group, *Kampfgruppe 100,* was trained and equipped for the *X-Gerät* missions. This group acted as a pathfinder for the large bomber formations that descended on London. The tactics employed had *Kampfgruppe 100* carry incendiary bombs, which were dropped in the target area. The bomber formations that followed would then use the fires ignited as the aim points for their bombing. It would be well into 1941 before the British were able to jam and distort the *X-Gerät* signals sufficiently to render them ineffective—about the time that the long blitz ended.

The period that began in the last years of World War I and which continued through the Battle of Britain was, indeed, one of the most interesting periods in history from a technological perspective. It started with primitive systems that bore little resemblance to the systems of today, and within a matter of just a few short years progressed to systems that are remarkably similar to modern systems, both in terms of the technologies used and the concepts of employment. From balloons and biplanes to high-speed monocoque aircraft designs and complex radar-based command and control systems in less than two decades—breathtaking development by any standard! The period was, furthermore, characterized by very distinct phases that marked development. It began with the building of foundation technologies that, while not as visible and romantic as the aircraft which would follow, was critical to their development. This was followed in the 1930s by the explosion of revolutionary new aircraft designs and radar. The five-year period from 1934 to 1939 was indeed one of those periods when technological development suddenly blooms into significant new products at a rate that marks the period as unique. The British Spitfire, the Hawker Hurricane, and the German Bf 109—all three airplanes were produced and initial deliveries were in the hands of the pilots who would fly them by 1939. And, of course, the British development and refinement of radar into a workable adjunct to the newly deployed fighters made that country's air defense system into a modern marvel that is still a model for such systems today. The final phase, during the Battle of Britain, was marked by innovative developments that are perhaps less complex technically than the earlier developments, but which are important because they solved real problems rapidly under combat conditions. During this phase, innovations such as armor to protect aircraft and crews and devices such as the parachute-and-cable rockets were developed. This was also a phase in which there was rapid acceptance of technologies that, for various reasons, had been ignored until the heat of combat became too intense to ignore them—the constant-speed propeller, for example. Indeed, there is much still to be learned from the period 1916 to 1941, and much of it deals with the way technology is developed to support military capability and how technology transitions from concept to capability and from country to country to provide what is needed, when it is needed.

1. Zeppelin and Gotha Raids*

Intimidating by its sheer size—they were often over 500 feet long—a giant Zeppelin airship plies the night skies over London in 1915, ushering in a new concept in warfare. For the first time in history, a country could project military power well beyond its own borders without the use of land forces. The impact was profound. A nation at war could include strategic objectives like civilian populations and production facilities among its targets, and resources would henceforth have to be devoted to a new and evolving aspect of warfare—air defense.

For years, militarily aggressive nations in Europe used mounted cavalry, land armies, and later tanks to carry the fight to other nations. Britain, as an island nation, relied primarily on her navy to project power and had a long history of dominating other countries through the strategic employment of naval armadas. Until World War I, Britain had been largely able to choose to either fight or not, attacking when conditions favored the employment of naval power. World War I saw the first instances of the organized use of the airplane in war between nations. Initially employed primarily in reconnaissance missions, the airplane soon graduated to use as an offensive weapon employed by both sides to support and advance the causes of the ground forces engaged below. The exploits of early German pilots are legendary; names like von Richthofen, Immelmann, and Boelcke are still remembered among the great tactical flyers of history.

Germany was one of the first nations to see the strategic potential of the airplane and to project military power by means of air forces. In 1915, the Kaiser approved attacks on the city of London. As Germany looked toward Britain, she chose to bypass the power of the British armada and instead attacked from the air, exploiting Britain's lack of defenses against air attack—what today is referred to as *asymmetric warfare.* Initial attacks on Britain were conducted using the huge Zeppelin airships. Carrying bombs that weighed up to 600 pounds, these airships were typically several hundred feet long and they flew at speeds in the 50- to 80-mph range. Not only did these aerial attacks bypass the navy; they further differed from the more traditional forms of warfare at that time in that they were directed at the civil-industrial center of London, rather than the military forces of the opponent. This approach to war, which includes all the resources of an enemy nation as viable targets—military, industrial, plus civilian populations—is often referred to as *total warfare.* The zeppelin attacks on London in 1915 were unique both because they represented one of the first examples of the projection of military power by means of air forces and also because these attacks were among the earliest examples of air power employed to escalate national conflict to the level of total war.

By 1916, the Germans began to shift from the airships to airplanes as the primary means of aerial attack. The zeppelins were difficult to handle; they were susceptible to the

*Section numbers refer to timeline illustration at beginning of chapter.

Zeppelins over London, 1915

weather and winds; and the British had developed defenses that were beginning to take a toll on them. The Germans chose the Gotha IV bomber as the primary aircraft to carry the war to the British. The airplane carried a crew of three: the aircraft commander, a pilot, and a gunner. Powered by two six-cylinder Mercedes engines, the Gotha was designed specifically for the long-range, heavy bombing role. With a maximum speed of 87 mph at 12,000 feet, it carried bomb loads in the range of 600 to 1100 pounds. Machine guns were also mounted in the nose and at midship. It had a combat radius (the distance it could travel to a target and return) of 520 miles. The normal bomb load on a mission to England was six 110-pound bombs, and typical raids were conducted by formations of about 20 bombers. The Gotha IVs were effective weapons. In a memorable first raid, they attacked the village of Folkestone in May 1917. This particular raid so incensed the British civilian population that a public outcry arose demanding that the government develop an effective air and civil defense capability. Although British aircraft were launched to defend against the Gothas, there was little organization to the effort. Interceptor airplanes were launched in a random fashion, but, once airborne, they had no method of finding the bombers; furthermore, the pilots were untrained in any tactical procedures that might have enabled them to down a bomber had they found one. The air defensive effort was completely ineffectual. Reports in the popular press highlighted the many shortcomings of the existing system, and by mid-1917 the public demand for improvements had grown to crisis proportions. This outcry was the impetus for the development of a defense system that would provide the foundation for the formidable early warning and defense system that would serve the British so well in the Battle of Britain.

After World War I, various theorists, among them Giulio Douhet, advanced the idea that the day of the army and navy as the primary means of conducting war had passed and that a new age—the age of the airplane—would render those nations that depended on ground and naval forces vulnerable to nations that could mount aerial campaigns. While it is clear that the advent of strategic air forces and power projection by means of aerial vehicles did not render the armies and navies of the world irrelevant, it is equally clear that it ushered in a new age—one in which air power would have an important role in both the offensive and the defensive roles that are routinely required of military forces.

2. Initial British Early Warning System

The spotter searching the skies for enemy aircraft underscores the primitive nature of the British air defense system that confronted the German air forces during the early days of World War I. The system was fragmented, the equipment was either poor or nonexistent, and those charged with detecting and intercepting aggressor aircraft were incapable of doing the jobs they were charged to do. It was not until public outcry energized the politicians of the day that the defenses began to improve significantly.

On May 25, 1917, a formation of 23 Gotha bombers departed Belgium to bomb London. Twenty-one of them actually crossed the English coast. Finding the city obscured by weather, the planes wandered over the countryside, occasionally dropping bombs when they found a likely target. They finally dropped their bombs on the unfortunate city of Folkestone as they headed toward Dover and the Channel en route to their home bases. They left 95 dead and 260 injured in their wake (Mason 1969). Antiaircraft artillery batteries at Dover were warned, and they set up a barrage to meet the raiders. Interceptor aircraft were launched. Unfortunately, the artillery shells were set to explode at altitudes well below that of the Gothas, and none of the interceptors ever came into contact with the now fast disappearing bombers. There was no effective warning that the attack was coming, there was no coordination of defensive efforts, there was no effective control of the fighters—there was no defense. The public criticism of the lack of effective air defenses, fanned by the press, immediately made the attack a political issue. Similar raids followed with similar results until civil demonstrations ignited the issue and made it a political imperative.

Prime Minister Lloyd George named General Jan Smuts to head a committee to study and reorganize home defense. Smuts immediately recognized that a primary weakness in the system was a lack of centralized command of air defenses, and he recommended the formation of a single defense command, the London Air Defense Area (LADA), to oversee and control all defenses south and east of London—the quadrant from which attacks from Europe would normally come. The LADA was given control over all guns, observers, defense fighter aircraft, and warning systems. Major General Ashmore, an artillery officer who was also qualified as a pilot, was made the commander of the new organization.

The defense was organized around London. A gun belt was established at a distance of 25 miles from the city extending to the north, south, and east. Fighters of the Royal Flying Corps (the RAF had not yet been established as a separate service) patrolled the space between the gun belt and the city. In addition, gun defenses were organized inside the city itself. The fighters that were dedicated to the defenses were upgraded. Newer and faster Sopwith Camels and Bristol fighters replaced the older fighter aircraft.

As the capabilities of the fighters increased, it soon became evident that the greatest weakness in the system was the lack of control of the fighters once they were airborne. This led to the development of early means of ground controlled interception (GCI). One such attempt was the Ingram system, which consisted of large placards with Morse code symbols placed on the ground; pilots could read these from the air to determine the direction of the bombers they were trying to intercept. Another was a system of white pointer arrows placed on the ground so that defense stations could indicate to pilots the direction in which attacking aircraft were flying. But these systems were hardly adequate to the task of giving precise directions so that fighters could intercept bombers in the air. The need for ground-to-air communications had been identified as early as 1914, but little attention was directed to it due to budget constraints at the time. As enthusiasm in Parliament for improved air defenses grew, development of the high frequency (HF) radiotelephone was reenergized. By May 1918, an operating HF communications system was standard equipment on all day fighters and was also installed in the aircraft of patrol leaders of night fighters.

Radar Precursor..
1917 Observer Corps

General Ashmore established a separate telephone system independent from the public system. Sound locators were positioned along the routes normally flown by bombers inbound from bases in Belgium to assist in directing pilots toward the bomber formations. Flying squadrons had a telephone operator standing by. When a warning message was received that raiding aircraft were inbound, the telephone operator would use a Morse code key that sounded klaxon horns located in the quarters and mess facilities of the fighter squadrons. On receipt of the warning, pilots were to run to their airplanes and warm them up. If a second signal followed confirming the attack, pilots were to launch. Concentrated training was provided to pilots to improve their night-flying skills and to teach them fundamental tactics—what to do when they did, indeed, find an enemy aircraft and engaged it in combat.

Organization of the defenses received special attention. Each station in the defense—whether it was an aerodrome, a gun station, a searchlight, or an observation post—was linked by phone to a subcontrol center. There were 25 such subcontrols; they were linked, in turn, to a central control. At central control there was a large map table with a grid overlay. Ten plotters sat around the table, each connected by direct open lines to two or three subcontrol stations. As an attacking aircraft proceeded inbound from the English coast, stations would report to subcontrols, which reported to these plotters around the table at central control. Positions were typically updated every 30 seconds or so, and positions were plotted on the grid map. Colored markers were used, and positions were updated and removed when appropriate to keep the board from becoming congested and confusing. Ashmore, as commander, sat in a position above the map so that he could observe and direct the campaign. He had communications switches that enabled him to cut into the lines between the plotters and subcontrols so that he could speak to his subordinate commanders. By his side was the commander of the air forces, who had a direct line to the squadrons under his command. In addition, there was a direct line to a wireless transmitter at Biggin Hill, by which orders could be relayed to pilots in the air over their HF radiotelephones.

As the threat from the zeppelins decreased and raids were increasingly conducted by the Gothas, Ashmore realized that the aircraft were operating at the limits of their ranges. He understood that this fact had the effect of forcing the aircraft to fly predictable paths when inbound to London. Seizing on this, he began to redeploy his guns to the south and east along these paths and he reoriented his fighter patrols to focus on the avenues of approach. In addition, he introduced barrage balloons as an added complication for the German attackers. Ground defenses were also improved, as telephone communications enabled the coordination of searchlights and guns; plus, the grid map system allowed central control to direct barrages toward specified grids, even though those on the ground might not actually see the targets at which they were firing.

In May 1918, a large formation consisting of 28 Gothas and 3 German Giant aircraft (a huge 4- to 6-engine aircraft with a wing span in the 125-foot range) crossed the

English coast headed for London. They were met by an artillery barrage of over 30,000 shells fired during a span of just over two hours. Three bombers fell to the guns. British fighters, Sopwith Camels and Bristol Fighters, intercepted and attacked the formation, downing three more Gothas. With 20 percent losses and the raid scattered and too disorganized to inflict substantial damage, the raid was a significant defeat for the Germans. The air defense system was clearly a success. As events transpired, this would be the last time that enemy bombers crossed into English airspace for 22 years.

The architecture of the air defense system developed in Britain during the waning days of World War I was sound. It featured the fundamental detection, tracking, and interception functions required of such a system, and it integrated them all together through a communications system that enabled the entire system to be centrally coordinated and controlled. There is no doubt that the implementation of some of the functions was relatively crude—ground observers with binoculars are a poor substitute for the radar systems that would follow a few years later—but the structure itself was excellent. The structure and methods employed in the last year of World War I provided the foundation upon which the defense system used in the Battle of Britain was built. While the technologies advanced and the later system would feature some of the most interesting technical advances known to history, the underpinnings were remarkably similar and clearly owe their heritage to the organization, methods, and procedures that Generals Smuts and Ashmore gave the British as World War I ended.

3. Propeller Development

Propeller development was characterized by three distinct phases. The illustration of the Rotol constant-speed propeller represents the third stage in the evolution of propeller design. Propellers were initially fixed; the angle at which they operated relative to the airstream could not be changed. Later, as the aerodynamics of the propeller became better understood, variable-pitch propellers were developed that allowed the angle to be controlled and varied in flight. The third and final stage, the constant-speed propeller, integrated a governor into the control mechanism so that the propeller could be automatically adjusted as the airspeed of the aircraft varied.

A propeller, viewed in cross-section, is an airfoil. It looks just like a small section of an aircraft wing. As the engine turns the propeller, this airfoil develops lift as a result of its movement through the air. This lift vector acts along the longitudinal axis of the airplane, pulling the aircraft forward. The more efficiently the propeller develops lift, the better the aircraft performs in terms of speed and acceleration. In the airplanes of World War I and continuing into the aircraft of the 1930s, aircraft propellers were made of wood and were of the *fixed-pitch* variety—the angle at which they bit into the windstream could not be varied.

In considering the propeller, it is useful to think about the concept of *relative wind.* Relative wind is the apparent wind that results from motion in more than one direction; it is a common concept in flight. When an aircraft is holding ready for takeoff, the propeller is turning but there is no forward motion of the aircraft, so all the wind that the prop "sees" comes from the turning motion of the prop itself. The situation changes once the aircraft has accelerated to flying speed. The wind from the turning motion of the prop is still present, but now there is another component to the wind—the forward motion of the airplane creates a wind that appears to come from a position in front of the plane. The wind that the prop actually experiences is the result of these two components—the wind from the spinning of the prop through the airstream and the wind from the forward motion of the aircraft. This relative wind appears to come at the propeller airfoil from a direction that is somewhat different than that experienced when the aircraft was not moving forward.

Since a fixed-pitch propeller does not have the capability to change the angle at which the airfoil approaches the airstream, designers typically choose the flight conditions the propellers will most commonly encounter and set the pitch angles of the airfoil to operate in those conditions. This represents a compromise among the many choices available. However, the angle at which the airstream approaches the propeller changes considerably as the aircraft accelerates; consequently, the fixed-pitch propeller operates with maximum efficiency at only a single point of the aircraft's performance envelope. This is particularly objectionable in a fighter aircraft, where high rates of acceleration and deceleration with rapid changes in airspeed are common. The ideal propeller would have the ability to vary the pitch angle under different flight conditions.

This need to be able to control propeller settings for varying flight conditions led to the *variable-pitch* propeller. Initially, two types of variable-pitch propellers were developed—one that had to be set by a flight mechanic prior to flight, and another that could be adjusted by the pilot, using either electrical or hydraulic controls in the cockpit. The earliest efforts were aimed at developing propellers that could be varied between two predetermined settings, typically referred to

the ROTOL variable-Pitch Propeller.

as "fine" and "coarse." This was an improvement over fixed-pitch designs, because, at a minimum, the propeller could be varied to a setting that somewhat accounted for the differing speeds at which the aircraft would operate. The two settings for propellers of this type were normally optimized for takeoff and cruise conditions.

Work on variable-pitch propellers was under way in various countries as early as 1920. For example, NACA (National Advisory Committee for Aeronautics, the organization that is today NASA) Memorandum No. 2, dated September 1920, is titled "Variable Pitch Propellers." The two companies most identified with early research and development of propeller technologies are Hamilton-Standard, a U.S. company, and de Havilland, a British company. The fully controllable variable-pitch propeller was followed by the development of a propeller assembly that was capable of automatically varying the pitch, based on the flying conditions at any time. This permitted the engine-propeller combination to operate at a constant speed in revolutions per minute (rpm) whether the airplane was maneuvering, accelerating, or in cruise. Frank Caldwell of Hamilton-Standard is credited with having invented both the ground-adjustable-pitch propeller and the constant-speed propeller. He patented a version of the constant-speed propeller in 1928 and was awarded the Collier Trophy for outstanding achievement in aeronautical technology for his work on it in 1933. This constant-speed propeller both substantially improved the efficiency of the engines and extended their service lives. The constant-speed prop operates automatically, changing the pitch of the propeller with changes in the power requirements of the aircraft. For example, if the aircraft is climbing, the pitch is decreased, which allows the propeller to slip through the airstream with less resistance, which, in turn, allows the engine crankshaft to turn faster. Conversely, in a dive the pitch is increased, thus slowing the crankshaft's rotational speed. In this way the speed of the engine propeller combination can be maintained at a selected rpm, and the engine is prevented from overspeeding. The constant-speed propeller also enables the inclusion of an *autofeather* feature that automatically aligns the blade of the prop with the airstream on engine failure, thus substantially reducing the drag produced by the prop in emergency situations. The constant-speed prop was in routine use in commercial aviation by the mid-1930s.

In spite of these advances, the adoption and use of this emerging technology was quite slow in coming. The variable-pitch propeller was developed and available in the early 1930s, and constant-speed props were an accepted feature of commercial air carriers by the mid to late 1930s. Despite the fact that the technology was available, both the Spitfire and the Hurricane were initially produced with wooden fixed-pitch propellers. The transition to variable-pitch props began in 1937. Different configurations were tried—two-bladed, three-bladed, and four-bladed variations. The first 77 production Spitfires had a two-bladed wooden propeller, after which a three-bladed variable-pitch de Havilland prop became the standard (Price 1982). By the time of the Battle of Britain, both the Hurricanes and the Spitfires were using propellers capable of varying the pitch angles of the blades. Some of these were variable-pitch props (pilot controlled) and some were automatic constant-speed props.

Do these differences—fixed pitch, variable pitch, constant speed—really matter? Does the type of propeller used really make much of a difference in performance? When the Battle of Britain began, the early Spitfires had a fixed-pitch propeller, while the Bf 109 was fitted with a variable-speed propeller. The performance advantages associated with variable-pitch propellers were immediately evident, and, beginning with production aircraft 78, all Spitfires were equipped with the de Havilland three-bladed variable-pitch metal prop. This addition brought the Spitfire up to a level of performance essentially equal to that of the Bf 109, although the Messerschmitt could still outperform the Spitfire at higher altitudes. The RAF requested that de Havilland install constant-speed props on the Spitfire in June 1940. The constant-speed prop so improved the performance of the aircraft that the Air Ministry subsequently authorized the installation of the constant-speed props on all Hurricanes and Spitfires; a total of 1050 were installed by August 1940. Addition of the constant-speed props, which were manufactured both by de Havilland and by Rotol, provided as much as 7000 additional feet to the service ceiling of the aircraft (Deighton 1979). Since the currency of air-to-air combat is kinetic energy—the airplane with the most energy (altitude and speed) typically has the advantage—any change that results in additional altitude and airspeed is highly valued by pilots. With regard to propeller technology, the wonder is not that the fighter pilots welcomed the advances in the technology, but, rather, that it took as long as it did for the designers and producers to adapt the technology to routine use in fighter aircraft.

4. Engine Development

When the Rolls-Royce Kestrel engine appeared in the 1920s, it quickly became the primary power plant for the RAF. It powered both the Hawker Hart light bomber and the Hawker Fury fighter. In the Hart it exposed the serious technical and tactical problem illustrated here—the Bristol Bulldog, the frontline pursuit aircraft, was incapable of intercepting the class of bombers it was charged with defeating. The Kestrel helped to address this mismatch when it was used to power the Hawker Fury, Britain's first 200-mph fighter. In no uncertain terms, engine development was the technical enabler that put Britain on the path that led to the new generation of British fighters represented by the Spitfire and the Hurricane.

Most aircraft flown in World War I used radial engines, which are easily recognizable by the large, round appearance of the engine when viewed from the front. The cylinders and pistons of these engines radiate outward from the center crankshaft. The radial

1930's dilemma. Front line fighters like this Bristol Bulldog were slower than the bombers they were designed to intercept.

engine has both the advantage and the disadvantage of placing the cylinders directly in the airstream through which the airplane flies. The advantage is that the air rushing directly onto the cylinders provides ample cooling, so the engine requires no liquid coolant. On the other hand, the radially mounted cylinders create a substantial obstacle to the wind rushing into the engine; thus, the drag associated with radial engines is quite high, making airplanes that employ them slower than they might otherwise be.

In 1917, J. G. Vincent of Packard Motors and E. J. Hall of Hall-Scott Motors developed the Liberty engine in the United States. Building on an experimental design developed earlier by Packard, this engine took a different approach to engine design. Rather than having cylinders that radiated outward from the crankshaft to form the circular shape characteristic of radial engines, the Liberty engine was designed as a 12-cylinder, V-shaped engine with two banks of cylinders. The in-line design of the cylinder arrays substantially reduced the frontal area of the engine, which reduced the drag and produced much higher airspeeds than were common for radial engines of the day. Production of the Liberty engines was handled by several different automobile manufacturers, among them Packard, Ford, Lincoln, and General Motors. During the war, 20,478 of the new engines were produced.

As the Liberty engine was achieving its successes, another designer, Charles L. Kirkham of the Curtiss Aeroplane and Motor Company, was developing another V-shaped in-line engine to compete with it. His engine, labeled the K-12, was designed as a high-performance engine, and it required a small reduction gear between the crankshaft and the propeller to allow the engine to develop maximum power without exceeding the optimum speed for the propeller. This gear caused immense problems for the K-12; Curtiss finally withdrew the engine from the market and initiated redesign efforts. The problem was finally solved when Dr. S. A. Reed developed a new propeller design that could operate efficiently when driven at speeds consistent with the optimum rpm range of the Curtiss engine. This engine was dubbed the Curtiss D-12 (*D* for *direct drive*), and it was one of the finer engines ever developed.

To demonstrate the new product, Curtiss entered the Pulitzer Trophy race with a biplane powered by the D-12. The Curtiss entry won the races easily, and in 1922 airplanes powered by the D-12 won all four top places in the same races. These successes did not go unnoticed by the army and the navy. They were sponsoring entries in the Schneider Trophy races, the famous seaplane races conducted annually in Europe. In 1923 a pair of Curtiss CR-3s powered by the D-12 with a Reed propeller won both first and second places in the Schneider Trophy race. In 1925, Army Lieutenant Jimmy Doolittle, flying a Curtiss R3C-2 seaplane racer with D-12 engines and Reed propeller, won the Schneider Trophy with an average speed of over 230 mph. He flew the same airplane over a straight course the next day, achieving an airspeed of 245 mph. From that moment forward, in-line engines were evaluated based on comparisons with the Curtiss D-12.

Richard Fairey, an Englishman who headed the Fairey Airplane Company, concluded that the combination of the D-12 engine and the Reed propeller were the secret to the success of the Curtiss design. He subsequently visited the Curtiss factory and bought a D-12 and a Reed propeller. He carried them back to England in his stateroom to preclude any damage to them. Fairey modified one of his biplanes to accept the Curtiss engine-prop combination and called the airplane the Fairey Fox. The airplane flew in 1925 and reached an airspeed of 156 mph—50 mph faster than any aircraft flying in Britain at the time.

Fairey attempted to interest the British Air Ministry in his aircraft, but neither the British aircraft industry nor the Air Ministry were interested in sponsoring development of an aircraft that was dependent on an American engine. Only 28 Fairey Fox production articles were purchased, and it is said that this was due only to General Trenchard's support. One aero squadron was outfitted with the Foxes. The Air Ministry solicited English engine producers Napier and Rolls-Royce to develop an engine superior in performance to the D-12. Napier declined, but Rolls-Royce accepted the challenge.

The extent to which the Rolls-Royce engines were direct descendants of the Curtiss D-12 is a subject for debate. Certainly the performance of aircraft powered by the D-12 had caught the attention of air enthusiasts in both America and Europe and had demonstrated that, to be competitive, high-performance aircraft would henceforth require the V-shaped in-line engine design that the Liberty and D-12 engines represented. Whatever the genealogy, the fact is that Rolls-Royce introduced its new F.XI engine one year later. Similar in concept and appearance to the D-12, the major innovation of the new engine was the casting of the cylinder banks from a single piece of aluminum, which, in addition to the reduced frontal area common to the American engines, achieved substantial weight savings. The result was an engine with a weight-to-horsepower ratio of 2:1 (weight-to-horsepower is a common metric for comparing the performance of high-performance rotary engines). The new engine was called the *Kestrel* by Rolls-Royce. Its development was arguably the most significant technological development that occurred in Britain during the 1920s.

Sydney Camm, chief designer at H. G. Hawker Engineering Company, had complained about the inadequacies of traditional radial engine designs. The high drag profiles of the engines tended to defeat the designer's best efforts at aerodynamically streamlining designs to achieve higher performance. Recognizing the potential represented by the Kestrel engine, he proposed two related—but different—designs to the Air Ministry. One was for a light bomber, the Hawker Hart, and the other was for an interceptor; both were to be powered by the Kestrel. The bomber flew first and became operational in 1929. It logged a top speed of 184 mph, 10 mph faster than the top RAF fighter of the day, the Bristol Bulldog. The fact that a bomber could so handily outperform the best of the RAF's fighters caused the requirements directory of the Air Ministry to begin to reconsider the requirements for future fighters. This eventually led to the decision to issue Specification F.7/30, which set in motion the developments that eventually led to both the Spitfire and the Hurricane aircraft. Camm's other design, the fighter version of the Kestrel-powered aircraft, was dubbed the *Hawker Fury.* The Fury flew at a top speed of 207 mph. The Air Ministry enthusiastically received the Fury. Production began in the early 1930s and before the production line was finally closed in 1937, the RAF had purchased 214 of them. The fighter squadrons that flew the Fury were the elite corps of the RAF, the first to fly fighters capable of speeds in excess of 200 mph.

Like Camm at Hawker, Reginald Mitchell, chief designer at the Supermarine Company, was also vitally interested in the new Rolls-Royce engine. Supermarine was a producer of seaplanes. The company was very much involved in designing aircraft to compete in the Schneider Trophy races, the highly prestigious seaplane races held annually in Europe to encourage and advance seaplane designs. Mitchell had designed the winning entries in the Schneider Trophy races in both 1927 and 1929. The 1927 entry had been powered by a 900-horsepower Napier engine, but in 1929 Mitchell introduced his updated design, the S.6, which he powered using a Rolls-Royce R engine. The R engine represented an evolution of the Kestrel. On September 13, 1931, RAF Lieutenant John Boothman flew the R-engine-powered S.6 to a third win in the Schneider Trophy, averaging 340 mph over the designated course. This third win allowed Britain to retire the cup and retain it permanently. Later that year, the S.6 set a world air speed record of 407 mph. Every Mitchell-designed aircraft from that point forward featured a Rolls-Royce engine.

Meanwhile, further development was under way at Rolls-Royce. Under the guidance of Sir Henry Royce, and through the engineering efforts of A. J. Rowledge, A. G. Elliott, and E. W. Hives, the company was developing a new engine, which they called the *PV 1200*. Developed using the private funding of the Rolls-Royce Company (the *PV* indicated *private venture*), the new project had as its objectives the development of an aeroengine that would deliver 1000 horsepower with a weight-to-horsepower ratio lower than that achieved by any other engine of the day. This was the power plant that became the famous *Merlin* engine. With a supercharger, it would eventually reach well over the objective 1000 horsepower. It was rated at 1.3 pounds per horsepower, an extremely low weight for the power delivered. The engine is no doubt the most well-known aircraft engine ever developed. Its history is intimately linked with the Battle of Britain; it was the engine chosen to power both the Spitfire and the Hurricane. In September 1940, the Packard Company in the United States agreed to build Merlins for both U.S. and British aircraft. Full production began in the United States in 1942. Before World War II ended, Packard had produced more than 16,000 Merlins. The Merlin was used to power the P-51 fighter, one of the best American fighters in the war. The British used the engine in their Mosquito and Lancaster bombers, in addition to the Spitfires and Hurricanes. Today the Merlin engine is recognized worldwide as one of the truly important developments that enabled Allied aircraft to compete with and eventually dominate the German fighters in the European theater. In many museums of the world, not the least of which is the National Air and Space Museum in Washington, D.C., the Merlin sits proudly by the Spitfires and Hurricanes on display—an equal partner in the British victory in the Battle of Britain.

5. Germany in the 1920s

A young glider pilot waves to friends while soaring above the slopes in Germany during the 1920s. Gliders represented one of the few outlets for those interested in aviation in Germany due to the restrictions imposed by the Treaty of Versailles. Many Luftwaffe pilots cut their aviation teeth learning to fly in glider clubs, and designers, including Willy Messerschmitt, learned aeronautical engineering by designing gliders. While flying and airframe design skills could be honed through gliders, German engine development was another story, lagging well behind engine development in Allied countries.

Glider clubs keep the aviation interest alive in post-WWI Germany.

World War I had been labeled "the war to end all wars." The Treaty of Versailles set forth restrictions that were intended to define the limits of acceptable German activities during the years following the armistice. It had two primary objectives: first, to ensure that Germany was denied the means for making war in the future; and second, to punish her for starting what had been among the most horrible wars in the history of the western powers. The terms of the treaty were aimed at curbing the ability of the German nation to exercise both the economic and the military elements of national power. The punishments were applied with extreme prejudice and the restrictions administered with such damaging consequences that the treaty all but guaranteed the Germans would rise up again at some future date. The seeds of World War II were surely sown at the Versailles Convention where the treaty governing the terms of peace during the 1920s and early 1930s was drafted.

To a large extent the military clauses of the Treaty of Versailles were aimed directly at denying the Germans access to the technologies that are key to building military forces.

The size of the German armed forces was to be limited to a maximum of 100,000. The German navy was to be destroyed; it was scuttled in Scapa Flow. The air force was to be surrendered to the Allied Control Commission, whereupon 15,000 aircraft and, perhaps even more important, 27,000 aeroengines were turned over to the Allies. No air force was to be supported under the terms of the treaty, nor was Germany to import or manufacture military aircraft or their components.

Interestingly, the limitations on civil aviation were not nearly as restrictive as were those on military air forces. Civil aviation was really in its infancy, and in all likelihood the crafters of the treaty did not take the time to understand the parallels between civil and military aviation. This failure to restrict civil aviation in any meaningful way became one of the important loopholes through which Germany maneuvered to build a significant new Luftwaffe while giving the appearance of observing the Treaty of Versailles. The restrictions imposed on civil aviation limited production for a period of only six months. It was just enough time for Germany to do the planning necessary to plot the path by which the Luftwaffe would be rebuilt and modernized.

The terms of the treaty caused a substantial degree of dislocation among the designers and manufacturers who dominated the German aircraft industry at the end of the war. Claudius Dornier initially left Germany to set up factories in Italy and Switzerland, but he returned later to build seaplanes at a factory in Friedrichshafen. Hugo Junkers built a company at Dessau in 1920 to produce civil aircraft, and in addition established factories in Sweden and Turkey. Ernst Heinkel located his factory at Warnemunde on the Baltic Sea and also had facilities in Sweden. Heinrich Focke and Georg Wulf formed their company (Focke-Wulf) in 1924.

In 1922 the Allies, through the Paris Air Agreement, a postscript to the Treaty of Versailles, had set forth certain limits on commercial aircraft. Among these restrictions were limits on the size and type of commercial aircraft that Germany would be allowed to manufacture and use, but by 1926 these restrictions had all expired. Germany proceeded then to establish Deutsche Lufthansa, the state airline under the guidance and management of Erhard Milch. Junkers was forced to join and subsidize the new organization. Lufthansa quickly grew into a major force in commercial air, with routes all over Europe. The airline developed methods and procedures for training pilots and crewmembers, for handling large-scale operations of large aircraft, and for operating under the poor weather conditions normal to northern Europe. It was fertile training ground for the development of both aircraft technologies and crews that would serve Germany well when the time came to turn overtly to the production and preparation of bomber forces.

Paralleling the activity in commercial air, Germany maximized the opportunities available through another loophole, recreational air sports. The official German air sporting organization grew to 50,000 members between 1920 and 1928. This organization was devoted to encouraging the German population to support and participate in activities such as flying gliders and light aircraft, parachuting, and model building.

In 1926, Willy Messerschmitt bought the Bavarian Aircraft Works (Bayerische Flugzeugwerke) in Augsburg, Bavaria, to build sport aircraft. If the Treaty of Versailles was very explicit regarding the restrictions on military air technologies, it said little about restrictions on civil air and even less about limitations on recreational and sporting aircraft. Dornier, Heinkel, and Junkers would escape through the loopholes in the treaty regarding civil aviation; Messerschmitt would accomplish the same thing by concentrating on sports aviation—gliders and light aircraft. Messerschmitt had learned the art of glider design at the feet of Friedrich Harth, one of the pioneer designers in Germany. During the 1920s he designed and built some of the most advanced gliders of the day—record-breaking designs that were far ahead of their time in terms of design and performance. The salient features of the Messerschmitt designs—light structures with highly advanced aeronautical designs—would show up later in his other designs, notably in the Bf 109, the principal German fighter that flew not only in the Battle of Britain, but through the entirety of World War II.

Although hardly at the explosive pace that would characterize the 1930s, England was developing new aircraft with a degree of regularity in the 1920s in spite of the limited government resources allocated to the pursuit. The Fairey Fox, the Hawker Fury, and, of course, the Supermarine S.5 and S.6—all were products of new development and innovation during that period. Germany, on the other hand, found herself stymied by the terms of the Treaty of Versailles. The new developments in Germany were restricted from being overtly intended for military applications. Few new aircraft developments were undertaken, and those were carefully designed to have what is today referred to as *dual use applications*—aircraft that were ostensibly intended for commercial uses such as passenger or cargo transportation, but which, with careful design, might have military utility when the need arose. The advances in Germany were primarily directed toward development of methods, procedures, and technologies to improve the far-flung airline operations of Lufthansa.

Possibly the greatest impact of the treaty in Germany was the lack of engine development. While Britain was making rapid advances in the development of high-horsepower-to-weight engines suitable for use in fighter aircraft, German engine development, restricted at the outset by the forfeiture of all aeroengines to the Allies, was further retarded during the 1920s by being restricted to large civil aircraft applications. Large aircraft applications did not demand the focus on high-horsepower-to-weight and reduced-drag profiles that were so necessary in fighter applications. German engine development would lag far behind British development into the 1930s—to the point that, even as the Bf 109 was competing for selection as Germany's frontline fighter, the aircraft would initially fly with a Rolls-Royce Kestrel engine due to the nonavailability of acceptably advanced German engines. The Daimler-Benz DB 601 engine, which would become the premier German engine during the Battle of Britain, would not be ready for operational use until the late 1930s.

6. The Schneider Trophy Races

Trailing a cloud of water vapor, the Supermarine S6.B begins to accelerate on its takeoff run at the 1931 Schneider Trophy race. Winning this race was a crowning technical achievement for the British in the postwar era. During the 1920s and 1930s, air competition was often the impetus that spawned technical focus and accelerated development. The Schneider Trophy race series advanced British developments in both engine and airframe technologies and led to development of the Merlin engine as well as the stressed-skin, monocoque technologies and monoplane designs that were used in the Spitfire.

During the 1920s, no single event or institution was more important to the development of high-performance airplanes than the Schneider Trophy. Schneider Trophy races were held almost every year. The trophy was awarded to the seaplane that won a 150-nautical-mile race over a prescribed course. The races and the trophy awarded to the winner were the brainchild of Jacques Schneider, a Frenchman who was convinced that the future of aviation lay with airplanes that would operate from the water. Schneider Trophy races were held at various sites in Europe. Although the races had been suspended during World War I, they recommenced in 1919 following the signing of the armistice that ended the war.

In 1919, at the age of 24, R. J. Mitchell was elevated to the post of chief designer at Supermarine, a British manufacturer of seaplanes. Supermarine's marketing strategy included a company-financed commitment to the Schneider Trophy races as a means of demonstrating and publicizing the superiority of their designs, much as today's car manufacturers often enter races like Daytona and Indianapolis to demonstrate the capabilities of their cars. In 1922, a Supermarine biplane flying boat, the Sea Lion II, won the race held in Naples, covering the course at an average speed of 145 mph. American Curtiss designs won the next race, held in 1923; in fact, Curtiss designs took both first and second place. In 1925, a Curtiss R3C-2 flown by America's Jimmy Doolittle won the trophy again. This had the effect of raising the prestige of the Schneider Trophy to immense proportions, and the aircraft manufacturers involved were committing hundreds of thousands of dollars to the event. The prestige and money spent produced innovation on a substantial scale, as well as a forum for the exchange of ideas among designers. After 1925, the U.S. Navy withdrew its support from the races and the contest became largely a competition between the English and the Italians.

As anticipation of the 1926 race loomed, Mitchell, at Supermarine, had produced a radically new design in the S.4. The S.4 was a monoplane floatplane with very clean

the S.6B on takeoff run at the 1931 Schneider Trophy Race.

lines—a very low drag profile. The airplane set a world speed record for seaplanes of 226 mph; however, as the aircraft was being prepared for the 1926 Schneider Trophy race, it crashed. An Italian design, the Macchi M-39 went on to win the 1926 race with a design based on the Curtiss design from the previous year. These events left no doubt that the low-drag designs made possible by the in-line V-12 engines that had descended from the Liberty engine were inherently superior to radial designs then common in high-performance aircraft. More horsepower was not necessarily the key to winning, unless it was also combined with low-drag, streamlined airframe and engine designs.

A Mitchell-Supermarine design, the S.5, won the 1927 Schneider Trophy with a speed of 281 mph. The S.5, like the S.4, was a low-wing monoplane, highly streamlined and with the airframe set well above the pontoon floats. To compensate for the torque of the engine, the stanchions that supported the airframe above the pontoons were lengthened on the right side to account for the tremendous torque that the engine produced and that tended to rotate the airplane to the right during takeoff. The race was next run in 1929. Mitchell unveiled his new design, the Supermarine S.6. Mitchell had been using a Napier engine to power his designs, but was convinced to switch to a Rolls-Royce R engine for use in the S.6. The S.6 lines clearly showed the genesis of the design that would some time later emerge as the Spitfire. Although there were substantial problems involved in the integration of the new engine into the airframe, Mitchell persisted. Finally successful, the S.6 won the 1929 Schneider Trophy race with a speed of 328 mph.

As the 1931 event approached, the British prime minister announced that there would be no government support for the Schneider Trophy race. The races were a popular event and the public outcry was immediate. The government was unyielding, however, so in January 1931, Lady Houston, widow of a shipping magnate, put up £100,000 of her own funds to ensure that the race would be held. Mitchell came forward with a modification of his 1929 design, designated the S.6B. The aircraft covered the course at an average speed of 340 mph, securing the Schneider Trophy in perpetuity for Britain. A couple of weeks later, the S.6B flew to a new world airspeed record of 407 mph. These high-performance seaplanes would later give birth to the Spitfire that so captured the imagination of aviation enthusiasts worldwide and which played such an important role in the Battle of Britain less than a decade later.

As a footnote, in spite of the superior performance of their seaplane designs, the S.4, S.5, and S.6 designs produced little in the way of commercial success for Supermarine. There was no market, commercial or military, for high-performance seaplanes. Total production of all designs was eight aircraft. The company made most of its money doing overhaul work on floatplanes. In 1928, economic pressures finally forced the company to agree to be acquired by the Vickers Aviation Company, one of the largest producers of aircraft in the world at the time. After that, the company would retain the Supermarine name, but as the Supermarine division of Vickers, not as an independent producer.

7. Ernst Udet: Dive-Bombing

Like its namesake, Al Williams's specially modified Curtiss Hawk begins a steep dive to demonstrate a new concept in aerial warfare. After leaving the navy, where he had been an advocate of dive-bombing, Williams spent much of the 1930s flying aerial demonstrations that featured dive-bombing exhibitions. As a spokesman for Gulf Oil, Williams flew a modified Curtiss Hawk painted the orange and black colors of the company and was a regular at the Cleveland Air Races. In 1931, one member of the audience was more impressed than most. Ernst Udet, a famous German fighter pilot and internationally recognized aviator and stunt pilot, realized that diving steeply toward a target and releasing a

Al Williams impresses Udet with dive bombing at the '31 Cleveland Airshow

bomb at the last minute could provide accuracy that was not achievable using the more conventional level bombing techniques favored at the time.

Ernst Udet was a fighter pilot's fighter pilot. In World War I, he had initially joined the army as a motorcycle messenger, but switched to the air force in 1915. His first air victory came in March 1916, and by the end of 1917 he had 15 official kills, although it is said that he had many more unconfirmed kills to his credit. Udet was quickly recognized and promoted. By 1918 he was commander of *Jagdstaffel* (Jasta) 37, a fighter squadron. In March 1918 Manfred von Richthofen, the Red Baron, invited Udet to join his Flying Circus, *Jagdgeschwader 1.* One of Udet's fellow members of the Flying Circus was one Hermann Goering. When Udet scored his twentieth victory, von Richthofen made him commander of Jasta 11. Widely recognized as one of the best pilots in the German air force, he was credited with 62 air combat kills before the war ended, second only to von Richthofen himself.

There are several great stories about Udet's exploits as a fighter pilot in World War I. In the spring of 1918, he entered an engagement between seven Fokker D-7s and eight Nieuports, a plane used primarily by the French at the time. He managed to shoot down one of the Nieuports, and the airplane crash-landed behind German lines. Seeing that the airplane appeared to be under control at impact, Udet descended and landed beside the wrecked Nieuport to inspect for himself. He found the pilot alive and—to his surprise—an American. The pilot was Walter Wannamaker from Akron, Ohio. Udet spoke to him and the Germans then took him prisoner. Udet took the serial number from Wannamaker's aircraft as a souvenir. In a day when there was no gun camera film to record combat engagements, what better way to prove that one has a verifiable kill?

After the war ended, Udet could not bring himself to quit flying. In 1922, he formed a manufacturing company, Udet-Flugzeugbau, with several partners. The company planned to build light sports aircraft and trainers, the only pursuits open to aircraft manufacturers under the conditions of the Versailles treaty. Udet being no manager, the company soon failed. He then set out as a vagabond flyer, going wherever there was work or play that provided him the opportunity to fly. Working as a stunt pilot, Udet made movies in both Africa and the Arctic. He became known as the "Flying Fool," performing as a barnstormer and doing stunt flying and aerial demonstrations all over the world.

In 1931 Udet was invited to the U.S. National Air Races in Cleveland, Ohio. He delivered an exceptional aerial flying demonstration, and afterward he was introduced onstage by no less than American ace Eddie Rickenbacker. Also introduced onstage was the same Walter Wannamaker whom Udet had shot down in the war. The two greeted each other like long-lost friends. To bring the story full circle, Udet then presented Wannamaker with the serial numbers cut from his airplane 13 years earlier.

While at the National Air Races, Udet saw a demonstration of a Curtiss Hawk named the "Gulfhawk" and flown by Al Williams, a former navy test pilot. This very airplane resides today at the National Air and Space Museum Annex in Washington, D.C. Williams was a man much like Udet himself, a flying enthusiast who regularly flew demonstrations, wrote columns, and who was the idol of children and fellow aviation enthusiasts everywhere. His typical air show featured aerobatics and concluded with a dive-bombing demonstration as the finale. To make the show even more impressive,

Williams used small, live practice bombs and his target was a beaverboard fort in which was positioned black powder. When the practice bombs hit the fort, the powder would ignite, giving the audience the noise, smoke, and fire of simulated combat.

Udet may not have been a brilliant manager, but he knew air combat, and he was impressed with the accuracy of the bomb delivery that Williams achieved using the Hawk in a dive-bombing attack. Germany had always seen the air force as an adjunct to the army, and highly accurate bomb delivery would have been a capability that was very attractive to the German military. It would permit airplanes to be employed in direct and close support of ground units. Until that time, any bombing attempted by German aircraft had been done from straight and level flight, which at best was inaccurate and at worst dangerous to the troops one was trying to help. Airplanes were simply not built to withstand the forces involved in diving toward the ground and then pulling out of the dive at high airspeeds, nor were pilots exposed to dive-bombing as a tactical maneuver in general. The Curtiss Hawk offered the promise of changing the equation and making the airplane an integral part of a ground offensive. Udet understood this immediately. Returning to Germany, he discussed his ideas with his old squadron mate, Hermann Goering, who was by that time rising rapidly in the Nazi Party hierarchy. Goering agreed to support funding the purchase of a couple of Curtiss Hawks on the condition that Udet agree to join the newly forming Luftwaffe.

In 1933 Udet returned to the United States to visit the Curtiss factory in Buffalo, New York. There, on September 27, he flew and then bought two Curtiss Hawk IIs. The aircraft were disassembled and shipped to Germany. They were unloaded at Bremerhaven in October. Udet had them reassembled, after which he demonstrated the dive-bombing techniques he had witnessed to members of the German Air Ministry. The results were sufficiently impressive to convince the staff that they needed to generate a new requirement for an aircraft modeled on the Curtiss Hawks that would be designed to perform the dive-bombing tactics. The Junkers Ju 87 Stuka, the airplane that came to symbolize the blitzkrieg style of warfare that the Germans introduced in Europe in the late 1930s, grew out of this requirement.

8. Sydney Camm: The Hawker Hurricane

The illustration captures the main features that Sydney Camm designed into his Hawker Hurricane. Camm's approach to meeting the requirements for the new fighter was to upgrade an existing design rather than to create a revolutionarily new design. He chose to retain the best qualities of his Fury design and to improve where technology would allow. This approach resulted in a transitional design—partially old biplane technology and partially new monoplane. Camm compromised somewhat on performance in order to

HAWKER HURRICANE

Design based on Camm's Hawker Fury.

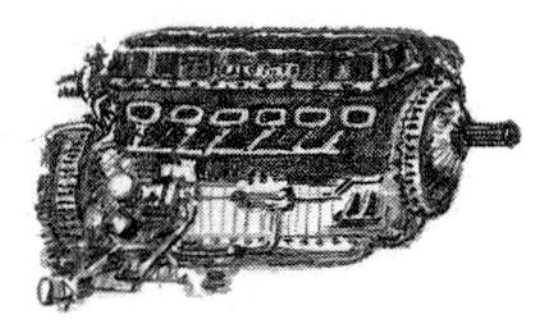

Rolls Royce Merlin engine derived from the S.6.B's "R" engine.

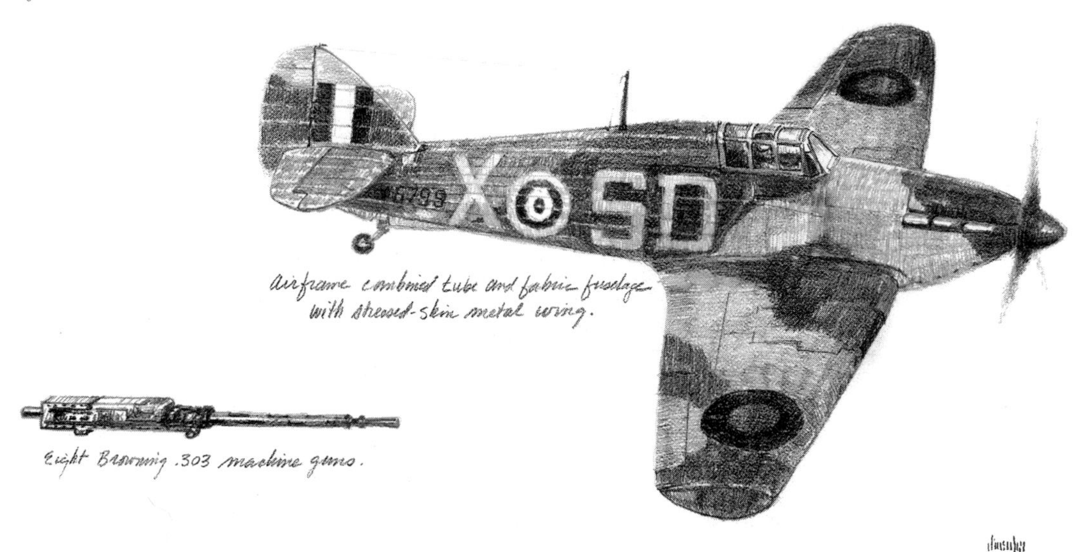

airframe combined tube and fabric fuselage with stressed-skin metal wing.

Eight Browning .303 machine guns.

build an airplane that was easily manufactured and repaired. The Hurricane was inexpensive to build and support with performance that, if not the maximum possible, was good enough to make it the backbone of the British defenses during the Battle of Britain.

Born in Windsor, England, in 1893, Sydney Camm was a meticulous and demanding designer who oversaw the development of a whole generation of fighter aircraft as the chief designer at Hawker Aviation. He joined the company in 1923. The parallels between Camm at Hawker and Mitchell of Supermarine were remarkable. Camm, like Mitchell, was recognized early in his career as a premier designer. He became the chief designer at Hawker in 1925 at the age of 32, and, like Mitchell, he submitted designs to the British Air Ministry in response to Specification F.7/30, hoping to win a production contract to produce Britain's next-generation fighter. Again like Mitchell, Camm's designs were rejected.

The Type 224 aircraft, designed and built by Mitchell of Supermarine in response to the specification, gave few hints that it would one day lead to the Spitfire. It offered neither the speed nor the climb rates desired in a new fighter. Concerned about the slow progress being made toward an acceptable new fighter aircraft, the Air Ministry approached Camm in 1933 to discuss the possibility of using Camm's earlier Fury biplane design as the starting point for development of a monoplane that could meet or exceed the specified requirements. Hawker Aviation, at Camm's urging, decided to pursue this approach. By encouraging both Mitchell and Camm to pursue their efforts, the Air Ministry had effectively set in motion two parallel developments, both underwritten by private investment, which would eventually lead to the two frontline British fighters.

Camm's design was very different from Mitchell's aircraft. Whereas the Spitfire was a revolutionarily new design, Camm's design was a logical transition from the older biplane technologies. It was an interim step—more advanced than the biplane designs but not as advanced as the Spitfire and Bf 109 designs. Initially called the Fury Monoplane, the fuselage included a tubular steel frame around which was built a wooden structure of forms and stringers. A fabric skin completed the construction. The initial design incorporated a Goshawk engine, but Camm, like Mitchell, changed his design to accommodate the new Rolls-Royce PV-12 (Merlin) engine when it became apparent that the engine would deliver high power at relatively low weights and offer substantial opportunities to streamline the profile of his airplane.

Models tested in August 1934 at the National Physics Laboratory suggested that the new design would be capable of airspeeds in excess of 300 mph. With that, the Air Ministry developed a specification (F.36/34) explicitly for the development of the new Hawker aircraft. When the design was submitted, the Air Ministry, influenced by the analytic work of Squadron Leader Ralph Sorley, insisted that the design include eight machine guns and gave the approval to build a prototype. The first Hawker Hurricane flew on November 6, 1935, just four months before the Type 300 Spitfire prototype flew at Supermarine.

Hawker Aviation was busily engaged by this time in delivering the Hawker Hart bombers and Hawker Fury fighters. The company was building over 200 of each at its plant, and it had subcontracted the production of hundreds of aircraft to other companies. The company understood volume production and the intricacies of managing subcontracted production. It only made sense that they would apply these lessons to the development and production of the Hurricane. Camm designed the airplane to be produced in large quantities employing the jigs and tooling that were in use in ongoing production at Hawker. He used easily produced features which required little in the way of advanced manufacturing techniques—like the straight-wing profile and the wooden frame over which fabric was stretched. These features made the aircraft easy to produce, relatively inexpensive, and easy to repair when damaged. A significant side benefit was that, in combat, the Hurricane was able to withstand substantial damage without being destroyed. The fabric-covered frame presented an opposing fighter a problem much like trying to destroy a metal tower. One might hit the target, but most of the blast or bullets would tend to pass right through the frame, doing little real damage.

Camm and Hawker realized early on that they had designed a winner and that the Air Ministry was anxious to build the fighter inventory in the face of the German buildup. By this time, the Luftwaffe was publicly recognized and Germany was participating in the Spanish civil war. In 1936, Hawker's board of directors authorized the purchase and buildup of tooling to produce 1000 aircraft. Three months later the Air Ministry contracted for an initial buy of 600 Hurricanes. By September 1939 the orders had grown to 3500 and Hawker had delivered 497 airplanes to equip 18 squadrons of Fighter Command with the new airplanes. Before the war ended, over 14,000 Hurricanes were produced.

Early in 1940 the Hurricane was upgraded with a constant-speed propeller, and the fabric-covered wings were covered with a stressed metal skin. Armed with eight Browning .303 machine guns and powered by the Rolls-Royce Merlin III engine, the Hurricane was capable of 328 mph at 20,000 feet, about 30 mph slower than the Spitfire. It had a range of 505 miles and a service ceiling of 34,000 feet. Although not equal to the Spitfire in terms of performance, the Hurricane was truly the workhorse of the British fighter fleet in the Battle of Britain. Between July and October 1940, 1715 of them participated in the battle, logging 80 percent of the kills claimed by British pilots. In the first year of the war, Hurricane squadrons accounted for more than 1500 confirmed victories against the Luftwaffe—about one-half of all British victories during the period. Czech Sergeant Josef František, the highest-scoring Allied pilot during the Battle of Britain with 17 victories, was a Hurricane pilot.

The Hurricane flew with various armament configurations, including a version with 12 machine guns and several with cannons. During the Battle of Britain, no other aircraft contributed more to the success of the defensive effort. Later in the war, as the action moved to the European continent, the Hurricane was assigned the role of a fighter-bomber, as its performance margin made it increasingly vulnerable to the higher-performance fighters. It was fitted with bomb racks and could carry both 250- and 500-pound bombs in addition to fuel tanks. It operated in almost every major theater during the war.

Sydney Camm continued to work actively in design through World War II and afterward. He welcomed the advent of the jet engine, and he was responsible for the designs of both the Sea Hawk naval fighter and the Hawker Hunter, one of the most successful fighters ever produced. Camm conceived and was active in the design and development of the P.1127 vertical takeoff and landing (VTOL) aircraft. This airplane became the first operational VTOL fighter as the Harrier jump jet. Camm was knighted in 1953 in recognition of his sustained contributions to British aviation. He died in March 1966, one of history's great aeronautical engineering designers.

9. Willy Messerschmitt: Bf 109

The illustration shows some of the features that made the Bf 109 Willy Messerschmitt's most complete engineering effort. During World War II there were more Bf 109 fighters produced than any other single combat aircraft. Something on the order of 34,000 of them were built in a succession of models and variants. Willy Messerschmitt, the designer and developer of the Bf 109, was the premier designer of high-performance airplanes in Germany—the genius behind revolutionary gliders, high-performance propeller-driven aircraft, and the originator of the first operational jet fighter. Known after 1939 as the Me 109, this fighter combined high performance, firepower, and excellent supportability characteristics. It was a full match for the British Spitfire and more than a match for the

Hurricane; Messerschmitt himself was probably a notch above either Mitchell or Camm in terms of pure aeronautical engineering genius.

Born in 1898, Willy Emil Messerschmitt, the son of a Frankfurt wine merchant, was captured by the romance of flight early in his childhood. While Messerschmitt was still a boy, Friedrich Harth, one of Germany's early air pioneers noticed him and took an interest in his education in the aeronautical sciences. Under Harth's mentorship, he had built a glider to Harth's specifications by the time he was 16 years old. Messerschmitt joined the Schleissheim military flying school at age 18 and worked in partnership with Harth on aircraft design until the older man was injured in a serious crash.

In 1926, with the financial aid of his wife and the state of Bavaria, Messerschmitt bought an aircraft company. Bavaria also had an interest in a second company, Bayerische Flugzeugwerke (BFw). Unable to keep both companies independently operational in the post–Versailles treaty economy in Germany, the government insisted that the two merge into a single firm. Messerschmitt would provide design and engineering expertise, and BFw would provide manufacturing capability. The company was headquartered in Augsburg. [The designation of German aircraft typically consisted of two letters that identified the manufacturer, plus a number that identified the particular aircraft; thus an Arado 80 would be designated as the Ar 80. In the case of Messerschmitt's aircraft, the designation "Bf" identified the Bayerische Flugzeugwerke; hence the Taifun built by Messerschmitt would be designated the Bf 108. In 1938 the German government insisted that the company name be changed to Messerschmitt AG to better publicize his association with the company. From that point forward, the airplanes were often referred to as Me 109s, although the official German handbooks retained the Bf 109 designation.]

The first airplane that the new company built was the M20, an all-metal, single-engine monoplane that carried eight passengers. The German state airline, Lufthansa, ordered two of the M20s, but when development troubles slowed delivery and then a crash resulted in the death of the pilot, the director of procurement for Lufthansa, Erhard Milch, canceled the purchase. Messerschmitt and BFw prevailed on Lufthansa to reinstate the purchase, but a second crash angered Milch and convinced him to cancel the order and demand repayment of all cash advances. Thus began a lifelong uneasy relationship between Messerschmitt, the premier designer, and Milch, who would eventually rise to the rank of Generalfeldmarschall in the Nazi hierarchy.

During the 1920s, Germany was mired in the restrictions on aviation imposed by the Treaty of Versailles, so Messerschmitt turned to the one area that was both acceptable under the terms of the treaty and encouraged by the German government—gliders. Using the skills taught him by Harth, Messerschmitt designed a series of record-breaking gliders. These aircraft were excellent performers, but they shared a tendency to be designed with little margin for error. They were easily damaged, sometimes dangerously so—a characteristic that would show up in later Messerschmitt designs.

By 1931, BFw was back in business, again with Messerschmitt as the chief designer. While the rebirth of German nationalism and the rise of the Nazi Party might have seemed to auger well for a German aircraft manufacturer, the animosity between Messerschmitt and Milch, who had risen to the post of secretary of state for aviation, ensured that BFw would get little or no work from the government. To survive, the company sought business outside Germany. They landed two contracts, one for a commercial transport and the other for a trainer, from the Romanian government. Milch labeled the transaction treason, but members of the board of directors of BFw interceded with Hermann Goering, and the company was allowed to produce the airplanes.

In 1934 the Air Ministry decided to showcase German aviation by entering a high-visibility airplane race to be held in France, asking several manufacturers, among them BFw, to design and build air racers. BFw had actually built racers to enter in this same race in 1932, but two of the four aircraft built had crashed and been disqualified—testimony again to designs that left little margin for error. Messerschmitt decided to design a new racer based on the trainer he was then producing for the Romanians. The result was the Bf 108A, a two-seat, low-wing monoplane with dual controls and a V-8 engine. The airplane had a top speed of 200 mph and was highly maneuverable. Just before the race, one of the five Bf 108s crashed, killing the pilot. Messerschmitt and his team worked feverishly to ensure that the remaining four aircraft qualified with no incidents. The Bf 108A won third place in the competition, not as high as Messerschmitt had hoped, but high enough to demonstrate the excellence of the concept and to encourage the belief that the company could build an aircraft that might be competitive for the new fighter that the Luftwaffe had announced interest in building.

The Bf 108A evolved into the Bf 108 Taifun, a four-seat touring aircraft in which Messerschmitt had combined the most powerful engine available with the lightest air frame possible. The concept included such advanced design features as leading-edge slats and slotted flaps to compensate for the high wing loading characteristic of the design. The Taifun was used by Germany in World War II as a utility aircraft and trainer and was eventually recognized as one of the better light civil aircraft ever designed. With a maximum speed of 196 mph and range equal to 870 miles, the Taifun would be an impressive light airplane today, almost 70 years after its introduction.

The German Air Ministry issued the specification for a new fighter in 1934, establishing a requirement for a fast monoplane that would carry two 7.9-millimeter machine guns and could be fitted with the new inverted V-12 engines then being developed by Daimler-Benz. The solicitation was sent to several manufacturers, including Arado, Focke-Wulf, Heinkel, and Junkers, but Milch blocked it from being sent to BFw. The combination of an intense lobbying campaign directed at Goering and enthusiasm on the part of certain Luftwaffe officers for the work done by Messerschmitt on the Bf 108 Taifun finally overcame Milch's refusal to allow BFw to compete, although he stated that there would be no production contract for BFw under any circumstances. The company was given a contract in 1935 to build an aircraft to fly in the competition that would define and select Germany's new frontline fighter.

Messerschmitt had actually begun to design the new fighter as he was developing the Taifun. Just as Mitchell would borrow from his earlier work on the S.6B in designing the Spitfire, Messerschmitt borrowed from the Taifun in building his new fighter, designated the Bf 109. It was a single-seat, low-wing monoplane with retractable landing gear and an enclosed canopy that swung to the side to open. Since the new Daimler-Benz engine was not yet available, the first Bf 109 prototype used the 695-horsepower Rolls-Royce Kestrel engine, as did the other competitors. The Bf 109 was first flown in mid-1935 and was delivered to the Luftwaffe Test Center at Rechlin in September of that year. The other competitors were the Arado Ar 80V1, the Focke-Wulf Fw 159 V1, and the Heinkel He 112V1. [The *V* designation indicates a prototype.]

The design of the Bf 109 was the result of a focus on producibility and maintainability. The wing, designed with a straight leading edge, was easy to produce within the tolerances required. In addition, the engine was easily accessible and could be removed and replaced in just 12 minutes by trained crewmembers. The landing gear were hinged at the fuselage and closed by folding outward into the wings—precisely opposite from the Spitfire. This arrangement had its heritage in Messerschmitt's glider designs, where the wings did not have the strength to take the stresses of landing and supporting airframes. Having the gear attached to the fuselage enabled the wings to be removed while the aircraft was parked and resting on the gear; it required no special jacks or other equipment to support the airframe. The downside of this arrangement was that the gear had a very narrow base—the distance between them was very small. This made the airplane relatively unstable on the ground, and landing accidents were a common occurrence. Like the Spitfire, the Bf 109 had a long nose that, given the tail-wheel landing gear design, obscured the forward vision of the pilot while taxiing on the ground. Approximately 1500 Bf 109s were lost in takeoff and landing accidents in the years between the beginning of the war and autumn 1941. It was, however, a superb performer once airborne. The light design with high lift devices, coupled with high-powered engines, made the airplane a fast, agile performer.

As the competition progressed, both the Arado and the Focke-Wulf entries fell by the wayside, leaving the Messerschmitt Bf 109 and the Heinkel He 112. Having flown the initial prototype with a Rolls-Royce Kestrel engine, Messerschmitt switched to the Junkers Jumo 210A engine in his second prototype, delivered in October 1935. It was fast becoming clear that the Bf 109 was the superior design. This fact was reinforced when German intelligence reported that the British were developing a similar advanced fighter concept, the Spitfire. In March 1936, the Bf 109 was declared the winner of the competition, and, in spite of all the problems encountered because of the relationship between

Milch and Messerschmitt, BFw was awarded an initial production contract for 10 of the new German fighters. The early version Bf 109B was estimated to have a top speed of 292 mph at 13,000 feet and was able to climb to 20,000 feet in 8 to 9 minutes.

When intelligence reports on the new fighters under development in Britain indicated that the Spitfire would carry eight guns, the German government insisted that Messerschmitt increase the armament of the 109. The problem was that the wings of the aircraft could not easily accommodate more weight. They were already close to the structural margins that would present a danger to the pilot flying the aircraft under the stresses of aerial combat. Messerschmitt was forced to redesign and strengthen them. Various combinations of armament were tried on the aircraft; in fact, the armament configuration of the Bf 109 was one of the enduring issues during the life of the airplane. Significantly, the Germans opted to incorporate cannons well before the British did in the Spitfire, and they were, in general, more effective than machine guns in air-to-air combat. Early Bf 109B aircraft were flown in the Spanish civil war with excellent results. In that war, German pilots learned the strengths and weaknesses of the airplane and developed tactics to take advantage of its superior performance at altitude. A variable-pitch propeller was installed on the airplane beginning in July 1937.

The Bf 109E was the first truly mass-produced model of the aircraft. Deliveries to the Luftwaffe began in early 1939. Often referred to by German pilots as the "Emil," it was the model that most often confronted the English pilots over Britain. Powered by the Daimler-Benz 601 engine, the airplane was capable of a maximum speed of 357 mph at 12,300 feet and had a range of 412 miles, with a service ceiling of 36,000 feet. It was comparable to the Spitfire between 12,000 and 17,000 feet, and it was superior at altitudes above 20,000 feet. The DB 601 offered an additional advantage in that it was a fuel-injected engine rather than carburetor fed like the British Merlin. German pilots quickly discovered that they could "bunt" the 109 (push the stick forward to induce a negative-*g* condition) and cause the Merlin to skip and misfire if it tried to follow, thereby providing the German a little more speed and acceleration when in danger of being fired upon from behind. The Bf 109 went on to become the signature German fighter airplane of the war. So great was the appetite of the Luftwaffe for the 109E that production was spread out among a wide array of German producers, apparently with considerably less difficulty than the British experienced with the Spitfire as production was expanded to meet war needs.

Messerschmitt continued to produce revolutionary and record-breaking designs even late in the war. His Me 262, the first jet fighter, pointed the way to the American F-80 and the generations of jet engine–powered fighters that followed. Following the war, Messerschmitt once again assumed the duties of chairman of the board of Messerschmitt, a position he held until he retired in 1973. He died in Munich in September 1978, one of the great engineering geniuses of the twentieth century.

10. The Four-Engine Bomber

Captured in a formation flight that never took place, the Dornier Do 19 and the Junkers Ju 89 heavy-bomber prototypes shown in the illustration represent a lost opportunity that the Germans would later come to regret. In March 1935, Hitler and Goering announced to the world that a new independent branch of the armed forces had been formed in Germany. The new Luftwaffe would operate under the direction of General Keitel, Oberkommando der Wehrmacht (OKW), supreme command of the German army. Generalmajor Walther Wever was named the first chief of the German air staff. Under his sponsorship, Germany had begun development of a four-engine bomber, but the program was canceled shortly after General Wever's death in 1936, and neither airplane was ever produced.

Placement of the Luftwaffe under OKW reaffirmed the German thinking that the air force was essentially an adjunct to ground warfare operations. Its role was to provide reconnaissance for army commanders and tactical support for their operations in the form of bombing and close air support. Many aspects of the Battle of Britain can be traced to this German view of the air force as a fundamentally tactical force. For example, the German failure to design the Bf 109 with more range reflected the thinking that an aircraft whose role it was to support a ground force would not need a very long range. This view of airpower in a ground-support role would also explain the enthusiasm of the Air Ministry for the Ju 87 Stuka dive-bomber in spite of the slow speed of the airplane, which was its ultimate downfall.

General Wever did not share this highly restrictive view of the role of airpower. He was a strategic thinker about both Germany's place in Europe and how an air force should be structured to support that role. From Wever's perspective, Germany was quintessentially the central European power, perfectly positioned to reach east, west, north, and south, as necessary, to influence and compel national events. With the right kind of airplane, Germany could control events anywhere in Europe. Such aircraft could, operating from German soil, strike anywhere in the European continent. The extension of this train of thought led Wever to conclude that the Luftwaffe should have in its inventory a long-range aircraft with large bomb-load capacity—a four-engine bomber. In 1934 Wever had a specification issued to German industry that called for an aircraft capable of flying to the northern part of Scotland or to the Ural Mountains from bases in Germany while carrying a significant bomb load. Both Dornier and Junkers proposed to build prototypes to compete for the contract.

The Dornier Do 19 was the first German four-engine bomber to fly. It was a large airplane—just under 115 feet long—almost as long as today's B-52 bomber. The prototype was powered by four 650-horsepower Bramo radial engines. It had a maximum speed of approximately 200 mph and a range of just under 1000 miles. It had provision for four

two promising, but short-lived, four-engine bomber prototypes: the JU89 (back) and the Do19 (front).

defensive gun positions and was capable of carrying 6600 pounds of bombs. The prototype flew in mid-1936.

Fabrication of the Junkers Ju 89 started one year after the Do 19, and the first of two prototypes flew in December 1936. The Ju 89 had four supercharged DB-600 in-line engines. Its top speed was 242 mph, with range approximately equal to the Do 19. The airplane carried a load of 8800 pounds of bombs.

It was truly remarkable that German industry, so long restricted from technical development, could produce airplanes that were at least as advanced technically as any similar airplanes then being developed by the Allies. Had either of the two been developed and produced for use in the Battle of Britain, it is highly likely that the outcome would have been different. The Ju 89 would have carried a bomb load that was twice that of either the He 111 or the Ju 88, the best of the two engine bombers employed in the Battle of Britain.

In 1936 General Wever was killed in a flying accident. He was succeeded by General Kesselring who had entered the Wehrmacht (the German army) in 1904. Serving initially as an artillery officer, he later served on the general staff during World War I. He switched to the Air Ministry in 1933, and at the time of his selection as chief of staff of the Luftwaffe he was serving as the chief of administration. With his army background, it is not surprising that he subscribed to the view that saw the Luftwaffe as a tactical weapon rather than a strategic force. Goering is reported to have said, "The Fuehrer will not ask how big the bombers are, but how many there are" (Mason 1969). The combination of his own views and the pressures on him to produce quantities of aircraft led Kesselring to conclude that more twin-engine bombers would be a better choice than fewer heavy bombers. He canceled the development of the four-engine bomber.

Erhard Milch, who was serving as the secretary of state for air in the Nazi government, was abroad at the time that the four-engine bomber development was canceled. He had shared Wever's vision and had approved the program, so he was astonished to find that the program had been canceled in his absence. However, Milch's influence had suffered somewhat in the years leading up to 1936 due to his projection that growth of the Luftwaffe to the size needed to be internationally dominant would require more time than Hitler preferred. Hitler, therefore, was less than totally satisfied with the support he was receiving from Milch, a fact that was not lost on those who canceled the four-engine bomber program. In their opinion, producing more two-engine bombers would be viewed more positively by Hitler than would producing fewer four-engine bombers, even if it meant less total bomb-carrying capacity. Milch himself did not actively oppose the decision; he had come to reluctantly accept the accelerated buildup that Hitler demanded and was smart enough to recognize that he would not prevail if he opposed it. The decision to end the four-engine bomber development and to rely on the Dornier, Heinkel, and Junkers two-engine bombers was another key decision that, in retrospect, appears to have been significant in determining the eventual outcome of the Battle of Britain.

11. The Me 110 Multirole Fighter

Even as Willy Messerschmitt was designing the Bf 109, the German Air Ministry had begun to develop requirements for another fighter. In 1934, Messerschmitt received a development contract to design a new kind of aircraft: the multirole fighter. Designing a single aircraft to perform multiple missions was a relatively new concept. Prior to the rapid advances in technologies in the 1930s, airplane designs had been optimized to perform a single mission. Advances in engines and materials led designers to consider aircraft that could perform more diverse missions. This new fighter would be a heavy fighter capable of escorting bomber formations flying deep into enemy territory. While the Bf 109 was a fast, single-engine fighter and extremely maneuverable, it carried only enough fuel (88 gallons) to keep it airborne for a little more than an hour in combat. This new strategic fighter was to have sufficient endurance to escort bombers over longer ranges and

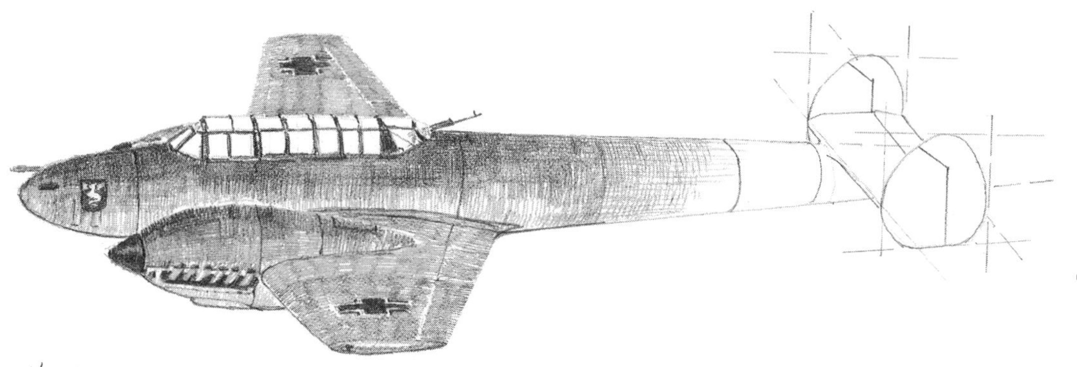

the Bf 110 emerged from Messerschmitt's drawing board as a "multirole" fighter.

the speed to engage enemy fighters en route. To carry the fuel loads needed and to achieve the speeds desired, the Germans opted for a two-engine design. The Me 110, shown in the illustration, emerged from this new approach to aircraft design.

Messerschmitt AG won the contract to produce the new fighter in 1934, and the first prototype flew in May 1936. The Me 110 had a slim fuselage, a tail with twin vertical stabilizers, and low wings. The configuration that flew in the Battle of Britain carried five machine guns, four in the nose and another facing aft, and two 20-mm cannons, also located in the nose. The pilot and the gunner-navigator shared a Plexiglas cockpit. Two 1150-horsepower DB 601A engines powered the airplane to a maximum speed of 350 mph at 23,000 feet. This speed was 20 mph faster than the Hurricane, though a few miles per hour slower than the Spitfire. It had a range of 565 miles and a service ceiling of 32,000 feet. With its powerful twin engines, impressive armament, and high speed, the *Zerstörer*, or Destroyer, as Goering called it, was a favorite of the Nazi propaganda machine. With the publicity given it, the Me 110 was actually more popular as an assignment for young pilots than was the single-engine Bf 109, even though the 109 was the better airplane in terms of performance.

Any system designed to perform multiple missions necessarily represents a series of compromises, and the Me 110 was no exception. Even though the speed of the Me 110 was impressive, the weight of the airplane resulted in less maneuverability than was needed to engage and defeat the agile fighters like the Spitfire and Hurricane. The accounts of the Battle of Britain refer time and again to the defensive circles—Lufbery circles—formed by Me 110s in an attempt to defend themselves against the more agile Hurricanes and Spitfires. The optimum characteristics of aircraft that perform the air-superiority mission are different from those of the airplanes that perform other missions. In attempting to build a multiple-mission airplane, Messerschmitt designed an aircraft that was capable of performing all missions, but could do none of them as well as the airplanes that were optimized for a specific mission. The Me 110 was formidable when it could enter a long dive, shoot at the enemy, then escape; but it was at a huge disadvantage in the climbing, diving, turning environment that was typical of the air war over Britain. It simply could not turn fast enough, nor could it climb rapidly, and the price for those shortcomings was often the destruction of the airplane and crew. Furthermore, although capable of performing as a bomber, the Me 110 did not have the load-carrying capacity necessary to be the credible threat that the best bombers, like the Ju 88, were.

This issue of multirole aircraft is still relevant today. There are modern cases of aircraft that have been designed with multiple missions in mind, or aircraft designed for a specific mission that have subsequently been assigned other roles. Examples include the F-111 aircraft, which was to be used by both the navy and the air force in different missions, and the F-104 aircraft, which was designed as an interceptor and was later used by the German air force in multiple roles, including air-to-ground delivery. The navy eventually withdrew from the F-111 program, declaring the aircraft incapable of meeting its needs. The results achieved by multirole aircraft have tended to be less than outstanding for the same reasons that the Me 110 did not perform well in the Battle of Britain. The compromises and trade-offs made to enable the aircraft to do many things result in the inability to do anything very well. The opposite side of the same coin is the fact that an airplane optimized for a given mission is unlikely to be very good at another mission that demands different attributes. The F-104 was little short of dangerous in the air-to-ground role.

The Me 110 found its niche later in the war, when it was used as a fighter-interceptor against Allied bombers as they attacked Germany. In that application, the *Zerstörer* lived up to its name, as it could take advantage of its diving speed in an environment that demanded less maneuverability than did the mission it was assigned during the Battle of Britain. Approximately 6100 Me 110s were produced before the war ended.

12. Sorley: The Eight-Gun Decision

Even as early as World War I the realization began to dawn on fighter pilots that more firepower in the form of multiple guns could improve the probability of defeating an enemy aircraft in flight. Early in World War I, most aircraft carried a single gun. The two guns mounted on the Nieuport 11 in the illustration represent a nonstandard configuration, a modification no doubt made in recognition that more firepower was needed to increase its effectiveness as a fighter. Although the problem was initially recognized in World War I, it was not until Britain was preparing for the advent of World War II that the problem was subjected to rigorous analysis and aircraft designs altered to ensure that the firepower carried by fighters was adequate to the task that lay before them.

In 1934 the British Air Ministry issued two specifications detailing requirements for the two fighters that would be forever associated with the Battle of Britain. Specifi-

Early effort to increase fire power: adding a second gun to the Nieuport 11.

cation F.36/34 set forth the requirements for the Hurricane, and F.37/34 defined the Spitfire. Both specifications required four machine guns. Squadron Leader Ralph Sorley was in charge of the operational requirements section of the Air Ministry as the specifications were developed. Sorley had studied the new generation of all-metal bombers that were going into service in a number of air forces around the world and was familiar with the performance these aircraft could deliver. A capable analyst and a pilot with operational experience, Sorley estimated that the average pilot could hold an adversary aircraft in its sights for 2 seconds at best. His concern was that the faster bombers being produced would make it difficult to achieve even this brief window when a fighter might actually deliver bullets effectively. Sorley's analysis led him to conclude that the new generation of fighters would need eight machine guns to effectively intercept and down the newer bombers, rather than the four guns that had been the standard.

Sorley took his case to the deputy chief of staff of the RAF and convinced him to change the general requirements for new fighters to reflect an eight-gun array. This by itself, however, did not immediately change the armament configurations for the Hurricane and Spitfire since they were already under development according to existing contracts. By 1935 a new specification (F.10/35) was in development at the Air Ministry that called for six or eight guns in the wings, but offered a reduction in total fuel carried and the deletion of the requirement to carry bombs as trade-offs for the increased weight of the guns. Sorley visited Hawker and Supermarine and spoke with both Camm and Mitchell about bringing their designs into line with the new specification. Both designers agreed that they could accommodate the four additional guns, given the other reductions. Sorley then recommended that either or both of the new aircraft should be ordered into production even if it meant sacrificing the Gloster SS37, the biplane fighter that had won the competition for Britain's newest fighter, which was about to go into production. Both the RAF director of technical development and Air Marshal Hugh Dowding supported Sorley's recommendation. Shortly thereafter, the contracts for the Hurricane and Spitfire aircraft were amended to bring the contract requirements into line with the new specification, and each was then designed to carry eight guns with 300 rounds of ammunition per gun.

13. Specification F.10/35

A specification is a document listing technical requirements that describe the performance of a product or system, the functions that the product is expected to perform, and the connections and contacts (interfaces) that it will have with its surroundings. In the case of military systems, the specification is generally derived from other documents that describe the operational needs that the system will satisfy and the environments in which it will

6 Nov 35. Hurricane's first flight and specification F.10/35 moves from paper to hardware.

operate. When the government buys a new system such as an airplane, the government typically either writes the specification or, at a minimum, approves the specification that a contractor develops. Then, when the government contracts for development or production of the new system, the specification is included in the contract directing the company to design and develop the airplane to perform the functions and meet the performance levels called out in the specification—usually referred to as the *spec.* This is the process used in government contracting today, and it is essentially the same process that was used by the British in contracting with British industry for the aircraft used in the Battle of Britain. The first flight of the prototype Hurricane is shown in the illustration. The initial flight occurred in November 1935, representing the successful culmination of the process that starts with an operational need, is translated into a specification of design requirements, and is completed when a new product emerges to become an active military system.

As Britain approached the 1930s, it was clear that the aircraft operated by the Royal Air Force lagged those being built in the commercial sector, in terms of both the technology employed and the performance achieved. The Schneider Trophy races had made it quite clear that rapid advances were taking place in aeronautical engineering while the RAF continued to fly World War I vintage aircraft. The British government had simply not funded the RAF at levels that would enable the young service to keep up with the advancing technologies. When the decisions finally came to rearm and build a credible defense, the RAF issued Specification F.7/30 and encouraged British industry to submit designs in competition for what would be the first major aircraft production contract in over a decade. The practice was to solicit proposals under a general specification, then to issue a more specific document that refined the requirements of the general spec as necessary to guide the development of a given airplane. The British system for numbering aircraft specifications consisted of a letter to designate the type of airplane described in the spec, the number assigned to the document, and, following a slash, the year in which the spec was issued. Therefore F.7/30 described a fighter and was the number 7 specification issued in 1930. The spec, which was considered a *challenge* specification, called for an aircraft with the following capabilities (Price 1982):

- Not more than 8½ minutes' time to climb to 15,000 feet
- Not less than 195 mph at 15,000 feet
- Service ceiling 28,000 feet
- High maneuverability
- Good pilot visibility
- Four .303 guns and provisions to carry four 20-pound bombs
- Easy and rapid production
- Ease of maintenance

As already described, both Hawker and Supermarine were developing high-performance aircraft that caught the interest of the Air Ministry. The designs of both were initiated under private funding sponsored by the companies involved, but as development progressed it became increasingly apparent that both deserved and needed government support to progress as needed for operational use. Contracts were issued to both companies, each with a specification that was, in effect, an addendum to F.7/30. The spec for the Hurricane was F.36/34 and for the Spitfire, F.37/34. These were very short documents that simply referred to F.7/30 for most requirements, but spelled out any changes that the new aircraft were to incorporate. As far as the Spitfire was concerned, F.37/34 formalized the requirement for the new Rolls-Royce PV-12 engine, but changed little otherwise.

By 1934, the Air Ministry had, as a result of the analytic work performed by Squadron Leader Ralph Sorley, officially concluded that eight guns was the preferred configuration for new fighters. They issued a new general specification, F.5/34, which documented the eight-machine-gun array as the required armament for new fighters. However, the guidance for the design of the Spitfire still referred to a requirement for four guns. Sorley then visited Mitchell to discuss the new requirement. As described earlier in this book, Mitchell agreed that he could fit eight guns into the Spitfire if the requirement for carrying bombs could be deleted and the total fuel capacity reduced.

In 1935, the Air Ministry issued Specification F.10/35, which recognized the advances in aerodynamics that the new generation of fighters represented and also the requirement for increased firepower needed in modern air combat. This was the spec that was to guide the production version of the new Spitfire. F.10/35 called for the fighter to have the following (Price 1982):

- No less than six guns, with eight preferred, mounted outside the propeller disk
- 300 rounds of ammunition per gun if eight guns were provided
- Speeds not less than 310 mph at 15,000 feet
- A service ceiling not less than 30,000 feet

The agreements reached with Mitchell ensured that the Spitfire would be in conformance with the new requirements. Specification F.10/35 was the final design guidance for the Spitfire that flew in the Battle of Britain. The first production prototype Spitfire flew in March 1936.

14. R. J. Mitchell: Developer of the Spitfire

Although the Hurricane represented a transition from the biplane technologies of the 1920s to the modern fighters of the 1940s, the Spitfire was clearly more advanced, breaking new aeronautical engineering ground. The illustration indicates some of the features that give the Spitfire its unique personality. The Merlin engine and the Browning machine guns were common to both of these frontline RAF fighter planes (an excellent supportability design feature in its own right). The airframe design, however, was vastly different from the Hurricane. Mitchell's new fighter reflected new thinking on a number of levels. Based on his experience in designing racing planes, Mitchell chose to optimize performance at the expense of ease of manufacture and support. In spite of the difficulties encountered in building and supporting the airplane, however, the Spitfire will always be remembered as one of the finest aircraft ever built.

No airplane has captured the admiration and excitement of aviation enthusiasts everywhere more than the British Spitfire. In the world of fighter aircraft and their missions there is a hierarchy of glamour. At the top of the pyramid stand those high-performance airplanes responsible for achieving and maintaining air superiority—controlling the skies and making it possible for the other forces and aircraft in a battle to do the job they are designed and prepared to do. These air-superiority fighters are always the premier machines of their time, flying faster and higher than other fighters, and the pilots who fly them rank among the top of their breed. The mission is critical; the men and machines are superb at what they do—like thoroughbred racehorses. The cost of failure in the air-superiority mission is often the loss of the airplane and its pilot . . . and sometimes the failure to achieve a military objective. This was the world of the Spitfire during the Battle of Britain.

Reginald J. Mitchell was born in 1895 at Stoke-on-Trent in Staffordshire, England. He left school at the age of 16 and became an apprentice at a locomotive manufacturing works. He attended night classes at technical colleges, and in 1916, before the age of 22, he was hired by Pemberton Billing, Ltd., a small company that did repair work for the Admiralty and, in addition, produced a few original designs, usually noted more for their innova-

SUPERMARINE SPITFIRE

tion than commercial success. Shortly before the war ended, the company changed its name to Supermarine. By 1919 Mitchell had been elevated to the position of chief designer in recognition of the uncommon ability he had demonstrated in aircraft design. His successes in the Schneider Trophy races cemented his position and reputation as a designer. He remained at Supermarine for the rest of his life, serving later as a director of the company.

The story of Supermarine and the Schneider Trophy has already been told (see "Schneider Trophy Races" earlier in this chapter). The final design, the S.6B, which won the trophy in 1931 and which enabled Britain to retire the trophy permanently, gave some early hints about what form the Spitfire might take, although it was not yet even on the drawing board. Given his success with the Schneider Trophy racers, almost certainly any new Mitchell design would be a low-wing monoplane with a highly streamlined airframe powered by a Rolls-Royce in-line V-12 engine.

The Hawker Hart and Fury had made it clear that new thinking had to be applied to the design of fighter aircraft. In 1930 the British Air Ministry had issued Specification F.7/30, at the time considered an "impossible" specification, as a challenge to the British aircraft industry and to serve as impetus to radically new thinking. Britain was just emerging from its long retrenchment and self-imposed isolationism; Specification F.7/30 was the first opportunity that the Air Ministry had offered industry since the end of World War I. Essentially, every aircraft manufacturer in Britain was interested in winning the competition to see which design best met the specification—and the production contract that would follow. Eight different designs were eventually submitted in response to the new specification.

Supermarine, led by Mitchell, planned to enter the competition. The Supermarine entry was a low-gull-wing design—an all-metal monoplane design powered by a 660-hp Rolls-Royce Goshawk, an evaporatively cooled engine. The airplane was labeled Type 224. The performance of the airplane was a disappointment. With its thick wings, large wheel housings, and an engine that consistently overheated, the Type 224 could muster only 238 mph top speed and took 8 minutes to climb to 15,000 feet. Needless to say, it did not win the production contract.

In 1933, as the Type 224 was being built, Mitchell was operated on for lung cancer. Following the operation he took a trip to the European continent as he was recovering. During that trip, he met and spoke with several German aviators who enthusiastically described the changes and developments that were in progress in Germany, and he observed the growing aviation capability. Convinced that war between Germany and Britain would be the inevitable consequence of the path that Germany had chosen, Mitchell returned to England determined to develop the kind of aircraft that he believed would be needed in such a war. He refused further treatment for his medical condition, and devoted himself to developing the airplane that would come to be known as the Spitfire.

Even while the Type 224 was still in testing, Supermarine entered into discussions with the Air Ministry about building an improved design that would yield the kind of performance that Mitchell's reputation and achievements promised. The new aircraft would be designated the Type 300, and, although based on the Type 224, it would feature retractable landing gear, a cleaned-up wing, and shorter wingspan. A proposal was formally submitted to the Air Ministry.

The Air Ministry was not enthusiastic about the new Supermarine proposal, which offered only marginal performance advantages over the previous design. Mitchell doggedly continued to improve the airplane. A smaller design was developed with thinner wings, but the engine continued to be a problem. Finally, in November 1934, at a meeting of the Vickers Aviation Company board of directors (Vickers had taken over Supermarine), a decision was made to abandon the Goshawk engine. In its place Mitchell was given permission to pursue a design that would accommodate a new engine then in development at Rolls-Royce, the PV-12, which would be called the "Merlin" engine. In the absence of a contract with the Air Ministry, the design work on the new project proceeded under private funding from Vickers.

The decision to abandon the Goshawk engine and modify the Type 300 to use the PV-12 (soon to be the Merlin) engine drew immediate and favorable interest from the Air Ministry. Even though the PV-12 was still in development and was experiencing the normal growing pains associated with new systems, the engine had already demonstrated horsepower in excess of 600 and had as a target 1000 hp. The competition for the aircraft to be built to fulfill the F.7/30 specification was over, but the Air Ministry awarded Supermarine £10,000 to build an improved F.7/30 design. The contract was formalized in January of 1935; the airplane was to be delivered to the government in October of the same year.

The new engine was one-third heavier than the Goshawk; therefore, steps had to be taken to balance out the increased weight in the nose of the aircraft. Mitchell reduced the backsweep of the wing, shifting the center of lift forward. The result was a very slightly sweptback wing, which corresponded quite nicely to a design with an elliptically shaped wing, giving the aircraft the signature shape that the world came to associate with the Spitfire. The elliptical wing offered two principal advantages: it had low induced drag (the drag produced when lift is generated), and it tapered very slowly from the wing root where it joined the fuselage. This slow taper gave ample room for the designers to fold the landing gear into the wing and still be able to accommodate the guns, also located in the wings.

The wing had to be thin to achieve the performance desired, yet still rigid enough to withstand the forces that would be applied during aerial combat. A complex spar system was conceived which enabled the wing to carry the necessary loads. Built like a giant leaf spring, it featured nested girders so that the thickness near the wing root, where loads were very high, was substantial, and the thickness decreased toward the wingtip. In addition, the Spitfire also employed *washout,* a technique that applies a slight twist to the wing such that the angle of attack varies along the length of the wing. In this way, the wing root can be forced to stall while the wingtip is still flying adequately. In the cockpit, the pilot feels the oncoming stall, as the airplane feels as though it is buffeting, but because the outer areas of the wing are still under control, he can maneuver using the ailerons. Pilots often mentioned this feature as one of the outstanding features of the airplane; it could be taken to the limits of the envelope without losing control and falling victim to a pursuing attacker. In addition, Mitchell employed a metal-skinned, monocoque design in which the skin of the aircraft became a part of the load-bearing structure of the aircraft.

Further changes were made to the Spitfire as it neared final design. The armament was increased to eight guns, four in each wing, with 300 rounds of ammunition per gun. To offset the additional weight of the guns, fuel capacity was reduced to 75 gallons. The Merlin engine was altered to include a ducted radiator and the use of ethylene glycol to improve the cooling. In February 1935, the company informed the Air Ministry that it would probably be possible to produce between 360 and 380 Spitfires by March 1939.

The first prototype flew in March 1936. The Air Ministry placed the first order for Spitfires in June, contracting for 310 aircraft.

In the field of systems engineering, the use of integrated teams to design and develop systems is a fundamental concept. The idea is to bring together a group who represent the interests of multiple functional disciplines during design. These might include operational users, engineers, maintenance and supply specialists, and production specialists as well. The purpose of coordinating this diverse group in the design phase is to bring a sense of balance to the process—to ensure that the product can perform as required, can be readily produced once designed, and can be supported once fielded. Failure to consider these issues in the design process almost always means that the design will be difficult to manufacture and that maintenance and support in the field will be difficult and costly.

The transition to production was extremely difficult in the case of the Spitfire. The airplane had been designed based on the experience gained in the development of high-performance seaplanes intended for limited production. Performance was the single criterion upon which every design decision was based. The design itself was very complex—the elliptical wing, with its continuously curved profile and carefully defined washout angles, was difficult to build; the new spar added complexity to the airframe. The Supermarine factory was small and could not handle the production of even the limited 310 aircraft required for the first contract, and therefore production of many of the component assemblies was subcontracted. This demanded very precise control of aircraft configurations and the drawings which described them, but that was not the heritage of Supermarine. In the small production runs typical at Supermarine, workers on the factory floor solved manufacturing problems directly and informally with engineers in the same plant. Subcontracting demanded that multiple parts be produced separately by many subcontractors, then brought together at an assembly point. Communications were accomplished through the medium of engineering drawings. If the drawings were poor, the pieces would not fit. Often they did not. The net result was a design that offered outstanding performance but that was unduly difficult to produce and assemble. Eventually, measures were implemented to bring the system under control and to involve producers who were accustomed to operating in a high-volume manufacturing environment.

It was July 1938 before the RAF received its first production Spitfire, and by the beginning of 1939 a total of 49 had been delivered. By the time the Battle of Britain began, nine squadrons were fully equipped with the Mark I (Mk I) version of the new fighter, and two more were in the process of converting. The Mk I was the version of the Spitfire that was used in the Battle of Britain. A total of 1583 Mk I Spitfires were eventually produced. After delivery of the 175th Mk I, the airplane was fitted with the Merlin III engine, which delivered 1030 horsepower and which featured a crankshaft standardized for interfacing with the de Havilland or Rotol constant-speed propeller. The Mk I existed in two versions: the Mk IA carried eight Browning .303 machine guns, and the Mk IB carried four machine guns and two 20-mm Hispano cannons. The Mk IA had a top speed of 362 mph at 19,000 feet, a range of 395 miles, and a service ceiling of 31,900 feet.

R. J. Mitchell's illness returned in the mid-1930s and by June 1936 he was often too ill to work. He continued to monitor progress on his airplane, however, and he was often observed in his car watching the test flights. Unfortunately, he did not live to see the production version of his fighter. In 1937 he died of cancer at the age of 42. The gift he had given Britain—the Spitfire—will always be recalled as one of the classic airplanes of history. The airplane was so admired by the German pilots who fought in the Battle of Britain that those who were captured after parachuting from damaged aircraft often claimed that Spitfires had shot them down, even if that was not the case. The airplane was so well regarded that defeat by a Spitfire was a face-saving excuse. The Spitfire, and Mitchell, will always be remembered as symbols of the British victory in 1940.

15. 1935–1939 Radar Development

In what might be the only moment of glory in its operational life, a Heyford bomber turns to intercept a radio signal transmitted by the British Broadcasting Company from a tower near the town of Daventry. Developed between the wars, the Heyford was nearing the end of its useful life when it was selected to participate in one of the earliest radar experiments. In order to prove conclusively that reflected radio signals could indicate the presence of aircraft in flight, a test similar to that illustrated here was conducted in 1935. While a team of scientists hovered over an oscilloscope, the aircraft was flown to deliberately intercept the transmitted signal. When the beam was interrupted, the signal was measurably displaced, indicating the position of the aircraft relative to the transmitter. This phenomenon would lead to development of the capability to detect both the range and the altitude of approaching aircraft, two critical requirements for an air defense system.

British concerns that their defenses against approaching aircraft were inadequate continued after World War I into the mid-1930s. As late as July 1934, in air exercises conducted by the RAF, it had been amply demonstrated that a bomber could easily penetrate British airspace without being detected. The government had attempted to solve the problem using huge, concrete acoustic mirrors intended to capture and use the sounds made by passing aircraft to locate them, but these had proven completely ineffective. At the suggestion of A. P. Rowe, an Air Ministry scientist, they then turned to a group of outside scientists to investigate a number of potential solutions that had been suggested and to recommend an approach for further development. The committee was dubbed the Committee for the Scientific Survey of Air Defense and was headed by Henry Tizard, rector of the Imperial College of Science and Technology.

A Heyford bomber turns "final" to intercept a BBC signal, and usher in the Radar era; Daventry, 1935.

Tizard brought Robert Watson-Watt onboard, and the committee immediately set about the task of mapping out a coherent program to develop a realistic and scientifically credible solution to the problem. The first order of business was to reject unrealistic approaches that could not be supported by physics and the technologies then available. Among these was the suggestion that a "death ray" might be developed to down approaching aircraft; another that had substantial backing was the recommended use of infrared detectors. The death ray was dismissed as a schoolboy fantasy; the infrared approach was also rejected as impractical. One month later, Watson-Watt produced a paper demonstrating that radio waves might be used to detect airborne aircraft. This suggestion followed from earlier work that had demonstrated the measurement of the ionosphere using reflected radio waves. Watson-Watt suggested that the committee investigate to determine whether radio waves might be transmitted and reflected back to a detector so that the position of the aircraft could be determined. The experiment illustrated in the drawing followed.

Dowding, then head of the research and development branch of the Air Ministry, used the success of the experiment to get funding approved to pursue and develop the concept. A research station was built at Orfordness on the Suffolk coast, transmitters were erected, and trials using RAF aircraft began during the summer of 1935. The technology was referred to officially as *radio direction finding* (RDF) to mislead any curious investigations. It continued to be called RDF officially until 1943, when the name was changed to *radar* (radio detection and ranging) to harmonize with the name used by the Americans.

Within weeks the team was demonstrating that aircraft could be tracked at distances up to 40 miles.

Within nine months after the first meeting of Tizard's committee, the Air Council recommended the construction of a chain of stations that would cover the south coast from Southampton, some 70 miles southwest of London, to Newcastle, on the east coast and far to the north. The chain started with three stations. The first was established at Bawdsey Manor, northeast of London. Bawdsey became the experimental station where new concepts were tested and proven before being integrated into the system of stations. Substantial research was required to find the correct tower height, the appropriate wavelengths, and the direction-finding techniques that would enable the system to detect and determine both the azimuth and the altitude of incoming aircraft. Bawdsey demonstrated the potential of the new technology, and by May 1937 the first three stations had been constructed. In air exercises conducted in August of 1937, the stations demonstrated detection of aircraft at distances up to 100 miles, and trained operators were able to correctly identify formation raids. A further experiment was conducted wherein the detection and tracking capabilities of the new radar stations were linked to the filter room and command and control apparatus at Biggin Hill. The filter room had responsibility for sorting out friend from foe and passing relevant information to Group Operations, where decisions were made regarding which squadrons to alert and launch in order to intercept inbound formations.

The staffing at Bawdsey consisted of both civilian scientists and military operations specialists. This combination of unique expertise and points of view enabled the newly emerging system to rapidly develop in a way that ensured that the technology would be correctly focused to solve real military needs. The prototypes for the filter room and operations room that played such an important part in the command and control of fighters in the air were developed at Bawdsey, thanks in no small part to this unique civil-military partnership.

Among the last problems to be solved were those that dealt with identification by radar controllers of friendly aircraft once airborne. When groups of German and British fighters met in the air, all of them showed up as returns on the controllers' radar scopes. The controllers needed some means of identifying which were British and which were the enemy. The first solution, called "Pipsqueak," enabled controllers to identify a formation of fighters as friendly. Using Pipsqueak, one fighter in each flight would tune his transmitter to continually broadcast a signal which could be received by the ground direction-finding stations. The ground stations could then identify British fighters, plot the progress of the flight, and direct the flight toward the positions of attacking bombers that were sensed on radar. This was one of the first attempts to positively identify and control friendly aircraft in flight. While Pipsqueak solved one problem—directing and positioning a flight of friendly aircraft so that they could find and attack enemy formations—it did not solve the problem of distinguishing between individual aircraft once an engagement had occurred. In air combat, the engagement of fighters and opposing fighters quickly results in a freewheeling melee in which fighters violently maneuver as they attack and are attacked. Viewed on a radarscope and without some means of identifying one's own forces, it is impossible to know which blips represent friendly aircraft and which are enemies. The solution to this problem was called "identification friend-or-foe" (generally known as IFF), which is a technique that provides a distinctive appearance to the return signal given by aircraft tuned to a given channel so that a controller viewing the radar can differentiate between the signals received. With this addition, the air identification issues associated with the air defense system were solved.

By the time the Battle of Britain had begun the Chain Home (CH) system, as it became known, included 20 stations, each of which had a 350-foot transmitter tower and a 240-foot receiving tower. By that time the range of the radar had improved to the point that the stations could be farther apart than originally planned, and coverage over the south coast was extended all the way to Land's End, at the southwestern tip of England. Earlier experimentation had indicated that the system was ineffective when aircraft approached at altitudes less than 3000 feet, so a supplementary system, called Chain Home Low (CHL), was developed to close that gap. There were 30 CHL stations, which, together with the 20 CH stations, completed the air defense system that would prove to be as important to the air defense of Britain as the Hurricane or Spitfire.

16. Junkers and Ju 88

The bomber fleet that the citizens of London would come to dread consisted of three primary aircraft: the Junkers Ju 88, the Dornier Do 17, and the Heinkel He 111. Night after night this deadly trio rained destruction on the city. Of the three, Hugo Junkers's Ju 88 was clearly the most capable weapons system. It was the best of the German bombers during World War II and was far superior to the Ju 87 Stuka, although that airplane, also designed and built by Junkers, is perhaps better known.

Hugo Junkers was a scientist by training and a university professor by vocation. He started investigating the design of aircraft while at the university, and he became one of the pioneers in the development of metal aircraft. He built a series of all-metal monoplanes, including the Junkers Ju 52/3, which was a mainstay of Lufthansa's international civil air fleet during the early 1930s. One of Junkers's protégés in the 1920s was Erhard Milch, who became head of Lufthansa and then later became the individual charged with rebuilding the Luftwaffe for the Nazi party. While Milch was becoming ever more powerful in the Nazi hierarchy, Junkers became ever more publicly critical of the Nazis. In a

A Ju88, Germany's best two-engine bomber, breaks ground on an early test flight.

monumental act of perfidy, Milch used the power of his position to force Junkers to turn ownership of his corporation over to the German state in 1934. Junkers died six months later.

The Ju 88 was produced by the Junkers company after the death of Hugo Junkers. The design itself was the result of a competition held by the Luftwaffe, which was typical of the German selection process. The Ju 88 went from drawing board to first flight in a single year, 1936. It was a cantilever low-wing, twin-engine monoplane with all-metal construction. Originally powered by Daimler-Benz 600 engines and armed with three machine guns, the production version of the aircraft was capable of reaching 286 mph at 16,000 feet while carrying a 4000-pound bomb load. It was an exceptional aircraft, not only in terms of performance, but also in terms of handling qualities and versatility. Later models of the Ju 88 employed the Jumo 211B 1200-horsepower engines in order to make the DB 600 engines available for use in fighters.

Production of the Ju 88 in sufficient numbers was impeded by a top-level decision to make a dive-bomber out of the airplane after design and initial production were completed. This required that the wings and frame be strengthened to handle the increased aerodynamic loads, and it further required installation of speed brakes to control the speed during descents. The airplane went into operational service in 1939. Although never a dive-bomber in the sense of the Ju 87 Stuka, the Ju 88 was capable of bombing from a long, descending flight path that raised its speeds to levels that made it very difficult for even the best British fighters to catch it if their intercept angles were not precisely correct. The Ju 88 is generally acknowledged to have been comparable in every sense to the British Mosquito bomber, which was an exceptional aircraft in its own right.

17. Mk II Gunsight

With the advent of 300-mile-per-hour aircraft in the 1930s, designers needed to take another look at the airplane as a complete weapon system. The opportunity to inflict lethal damage on a fast-moving target while flying in a 300-mph shooter was limited at best. At least two technical improvements were needed—more firepower and more accurate aiming. The Mark II reflector gunsight addressed the aiming problem. As shown in the illustration, the gunsight was mounted front and center, replacing the World War I–vintage ring-and-bead sights for those pilots fortunate enough to have the new technology.

To fully appreciate the problem faced by a fighter pilot engaged in air-to-air combat, first imagine yourself as a bird hunter. As the birds fly overhead, you attempt to shoot them down by aiming at a lead point along their flight path and somewhat in front of them. You hope that your pellets arrive at a point in the sky at the same time the birds do; if so, you get a kill. As a hunter will tell you, this is not easy, especially when the birds are close to you and moving fast. But now imagine that, to complicate the task, you are also moving through the air at speeds comparable to the birds, so not only must you consider the movement of the birds as you aim, but also the effects of your own movement. This is precisely the problem faced in air-to-air gunnery.

When aircraft first began to engage in aerial gunnery, the only help the pilot had available was the ring-and-bead gunsight. As the name implies, this sight consisted of an iron ring close to the pilot's eye in combination with a bead on the far end of the barrel. The pilot would—just like the bird hunter with a shotgun—attempt to sight along the barrel, centering the bead at the desired aim point while mentally calculating the necessary deflection (lead angle) depending on the dynamics of the situation encountered. In reality, there were only a few gifted pilots who did this well, and their names became well known—von Richthofen, Galland, and Malan to name several. Most pilots tried to solve the problem by taking as many variables as possible out of the equation—whenever possible they just drove up behind the enemy and fired point-blank.

As the Battle of Britain loomed, a partial solution to the aiming problem was found in the optical reflector gunsight. The reflector gunsight had optics that projected the image of a *pipper*—a lighted circle with horizontal lines—onto a piece of reflecting glass that was positioned just in front of the pilot's face as he looked forward through the windscreen. The pipper was focused at infinity so that the pilot could keep the target and his gunsight in focus simultaneously rather than switching focus from the target to the gunsight, as had been the case with the ring and bead. Furthermore, the reflected pipper did not obscure the target as the ring-and-bead gunsight often did. The sight had two collars at the base; one allowed the pilot to adjust for the range at which his guns were harmonized, and the other allowed him to set the horizontal lines for the wingspan of the tar-

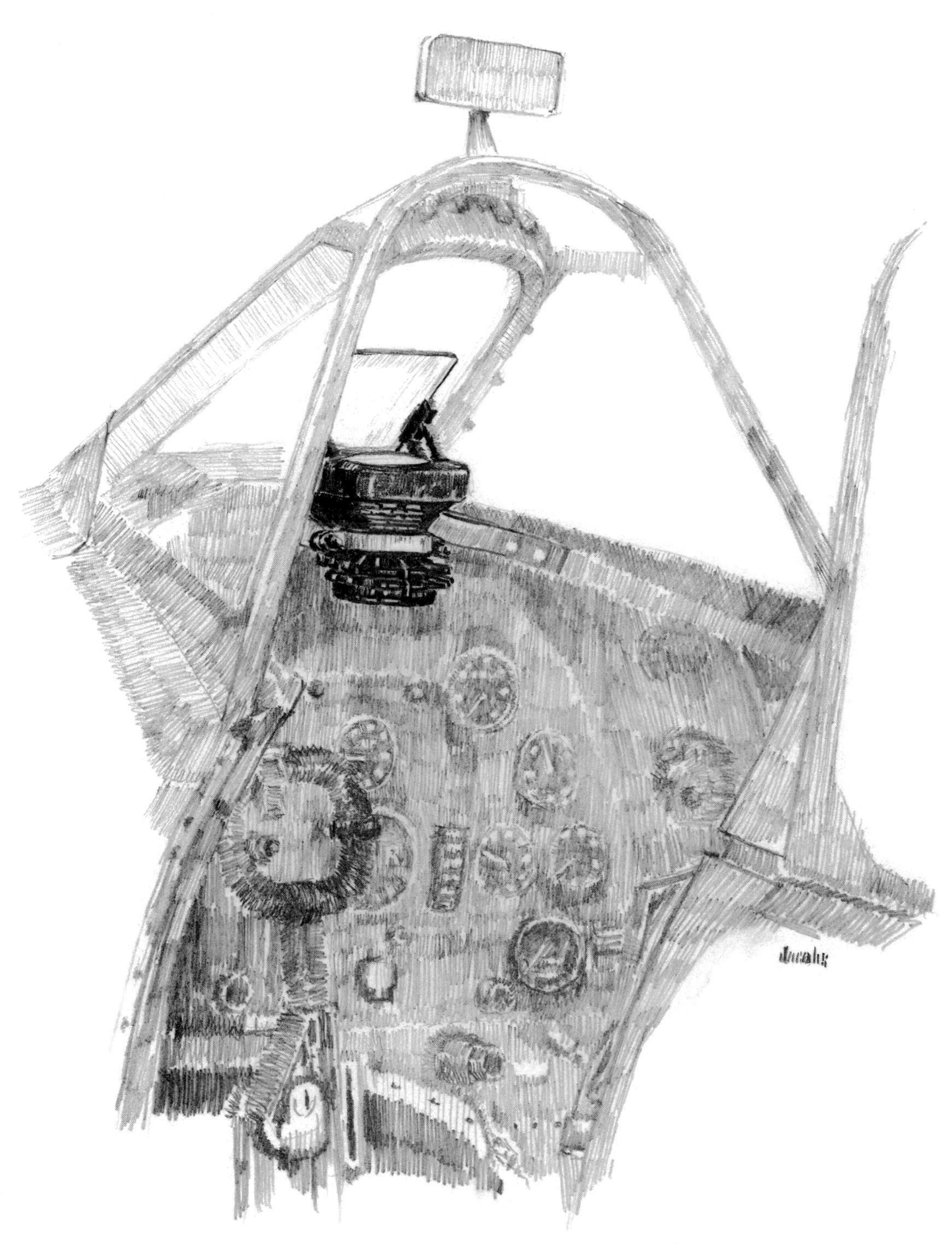

Mark II Reflector gunsight in a Spitfire cockpit.

gets he expected to encounter. When tracking an enemy aircraft, the pilot would approach until the wingspan of the target appeared to be approximately equal to the gap between the horizontal lines. This would indicate that he was at the proper range to fire. Then, by maneuvering to bring the pipper to bear on the target or at the appropriate lead point for deflection (off-angle) shooting, the pilot was provided a rough firing solution. While the reflector gunsight was not as accurate as the computing sights used today, it was a substantial improvement over the ring-and-bead sights of World War I.

The idea had been patented in 1900 and had been further developed for use in aircraft by a German firm, Oigee of Berlin. The Scottish firm of Barr and Stroud developed the forerunner of the sight that eventually became the Mk II. Research on reflector sights continued until 1937 when the sight was patented. After design problems involving the reflecting glass were ironed out, the RAF accepted the sight and placed an initial order for 1600 units with Barr and Stroud. The technology was held at high levels of secrecy, and the financial restrictions of the 1930s meant that few reached the operational fleets, so little was known of the new developments leading up to 1940. Interestingly, the first squadron to be operationally fitted with the new sight was Squadron 65, which was flying Gloster Gladiators at the time. The reflector sight was to become the standard for use on British fighters, but, as the Battle of Britain approached, funding and supply problems prevented them from being available in sufficient numbers to outfit most frontline Hurricanes and Spitfires. The majority of the fighter pilots in the Battle of Britain went to war with the ring-and-bead sight.

Gunsights today have progressed and improved on these early sights. Today's pilot has a *head-up display,* or HUD, which combines radar and flight computers that sense the target and then compute a gun-firing solution that accounts not only for the movement of the target, but also for the movement of the shooter. The HUD provides all of the necessary instrument readings—airspeed, altitude, radar range, and closure rate—so the pilot need never take his eye off the target while tracking it for the kill. If the pilot can keep the target within a target designator indicator long enough for the computer to solve the problem, a hit is nearly certain . . . but it is still true that a pilot rarely gets more than a couple of seconds to get off a good shot before his adversary will have maneuvered out of position. The more things change, the more we are reminded that the basics remain the same.

18. Spitfire Production Problems

In February 1936 Vickers Aviation informed the Air Ministry that, if both the contractor and government testing of the Spitfire were completed by April 1936 and a production order placed by May 1, then the company "should be able to start production . . . in September 1937, at a rate of five per week. In this event it should be possible to turn out

Sept 26th He 111's from KG55 destroy the Woolston Spitfire factory. the

shadow factory at Castle Bromwich is able to take up the slack.

a total of between 360 and 380 by 31 March 1939" (Price 1982). This optimistic forecast was to prove more difficult to achieve than either the company or the Air Ministry imagined.

In the first place, testing exposed a number of problems that had to be solved before the Air Ministry would commit to a production contract. Testing of the prototype production Spitfire, serial number K 5054, actually continued into October 1936. However, in spite of the fact that government testing was just getting under way, the Air Ministry placed an order for 310 airframes. A production specification was issued (F.16/36) which described the changes needed based on testing of the prototype. The major differences were requirements to stiffen the wing structure, to raise the maximum speed to 450 mph indicated airspeed, and to increase the fuel load to 84 gallons. It became apparent almost immediately that Supermarine had little chance of delivering at the rates or schedules originally projected.

In April 1936, as the Spitfire was moving toward production, the government announced a new policy intended to increase the production capacity for aircraft beyond that which existed within the industry itself. This scheme involved using the automobile industry, with its labor pool, skills, and knowledge of mass production, to supplement airframe and engine output through the use of "shadow factories." Arrangements were made with Austin, Rover, Daimler, and others to manufacture components of the Mercury radial engine. Both Fairey and Blenheim aircraft were manufactured by Austin and Rootes Group, respectively, and Rolls-Royce established its own shadow factories at Crewe and Glasgow for the manufacture of Merlin engines. Other factories produced components and assemblies such as propellers and munitions. By the end of 1938 there were over 2 million workers involved in producing aircraft assemblies in plants outside established aircraft manufacturing facilities. These shadow factories proved to be a boon to the British, both in terms of the added production capability that they represented and by the fact that they represented a dispersal of production among different, geographically separated sites. This ensured that production would not be halted by the loss of any single facility to bombing, whereas production that is concentrated in a few locations raises this risk considerably. The illustration shows a bombing raid conducted on September 26, 1940, with the intent of knocking out the primary Spitfire production plant at Woolston. By that time, the British had established a viable production facility at Castle Bromwich and they were able to maintain production with little lost time due to the raid. Achieving this state of distributed production capacity for the Spitfire had not been an easy task.

At the time it received the production contract for the Spitfire, Supermarine employed approximately 500 people and was busily engaged in producing other aircraft already on contract for the RAF. The company had small facilities appropriate for the small orders that it had typically handled. Supermarine had neither the facilities nor the workers to produce a modern fighter in volume. The Air Ministry encouraged Supermarine to subcontract a substantial portion of the work involved. Though the company did engage another company to construct tail sections, management was reluctant to allow more of the hard-won work to go to other companies. Furthermore, Supermarine was in the process of developing the technical data that would describe all the changes required in the new specification and was in poor position to provide potential subs with the necessary drawings that would permit them to construct subassemblies. By early 1937, the Air Ministry had become overtly concerned about the production status of the airplane.

A series of exchanges between the Air Ministry and Sir Robert McLean, who headed Supermarine, finally resulted in McLean's agreement to subcontract as much as 80 percent of the Spitfire production. This agreement was accompanied by a reduced production rate, but it quickly became apparent that Supermarine could not meet even the reduced requirement. The company was behind on creating the necessary drawings to permit subcontracting, and the matériel ordering system was chaotic to the point that parts and materials were often not delivered as needed to enable production. Furthermore, when parts were delivered, they often did not fit. This discipline of managing the form, fit, and function of systems in production is referred to as *configuration management.* Supermarine possessed little of the knowledge and discipline needed to do this well. It was July 1938 before the RAF received its first production Spitfire.

Even as production slowly increased, it was becoming increasingly apparent that the arrangement that had Supermarine handling the management and assembly of Spitfires was never going to produce aircraft at the rates necessary for the war effort, and the fact that war was imminent was becoming more obvious by the day. The British, casting about for a source of knowledge and skills needed to produce complex systems at high volumes, turned once again to the automobile industry. The government financed the construction of a huge plant at Castle Bromwich, which would be equipped and operated by Morris Motors Ltd. Construction began in the summer of 1938. Lord Nuffield, a production manager who had made his reputation as a production engineer, would manage the operation.

Nuffield was determined that the factory would use extensive tooling and jigs so that relatively unskilled labor could turn out aircraft the same way cars were manufactured. The difference was that, in the production of cars, designs were finalized and few changes were allowed once production was begun, thus the heavy use of jigs and tooling was a viable production strategy. However, the design of the Spitfire was in constant change as the RAF sought to take advantage of the latest technologies to improve performance and improve survivability, causing the jigs and tooling to be changed often. Nuffield, a traditionalist and an idealist in a highly volatile design environment, refused to produce aircraft until he had a firm design baseline; consequently, although he was producing a number of parts and assemblies, he was producing no complete aircraft.

The matter came to a head in May of 1940, shortly after Lord Beaverbrook was appointed as minister of Aircraft Production. Beaverbrook telephoned Castle Bromwich, demanding to know why no aircraft had yet been produced. Nuffield responded with a vigorous defense of his management and the capabilities of the plant, adding that Beaverbrook could either have Spitfires or modifications in designs, but not both. Then, playing his trump card, Nuffield closed saying, "Perhaps you would like for me to give up control of Spitfire production?" Beaverbrook, seizing the moment, responded, "Very generous of you, Nuffield; I accept." And he rang off (Price 1982).

Beaverbrook then took control of the factory and handed responsibility for production to Vickers. Pulling key management and staff from within their factories and coordinating the production efforts of Supermarine with Castle Bromwich, they created a hybrid production system which used a combination of semiskilled workers to produce those items that were stable in design and skilled labor to produce those items that required great flexibility and rapid change. In July, Castle Bromwich turned out 23 aircraft, and in September the plant produced 56. Production quickly ramped up to levels that permitted the equipping of fighter squadrons in time for use in the Battle of Britain.

Hawker had few of the problems in producing the Hurricane that Supermarine experienced with the Spitfire. There are various reasons for this. In the first place, the Hurricane was a considerably less complex aircraft than the Spitfire, employing none of the advanced technologies that Mitchell had designed into his airplane. Furthermore, Camm had designed the Hurricane to be produced on the production line, where relatively high volumes of aircraft were already being produced. The Hurricane was designed for producibility. In contrast, the Spitfire evolved out of the high-performance seaplane experience within Supermarine, where aircraft were handmade and produced in units of one or two instead of hundreds. The company had no experience in either physically manufacturing aircraft in large numbers or in the management of high-volume production. In spite of the problems encountered, however, production of both aircraft proved to be adequate for the needs of the RAF during the summer of 1940; it was the availability of trained pilots that would prove to be the far higher risk to the defensive effort.

19. Drop Tanks

A member of the new Luftwaffe's Condor Legion makes a steep turn while patrolling the skies over Spain in 1937. The pilot is most likely unhappy to be flying a "dog" like the Heinkel He 51 in a ground-support role while his fellow fighter pilots are flying the high-performance Bf 109 and performing air-superiority missions. Although he does not realize it, this pilot has an advantage that within four short years his Bf 109 mates would dearly love to have—the auxiliary fuel tank. The idea of fitting external fuel tanks on fighters to extend their range and endurance was first tested by the Luftwaffe in Spain.

The combat range of the Bf 109 was severely limited due to lack of fuel. This problem caused concern for the Luftwaffe on a number of levels. During the early phases of

An He 51 with drop tanks in Spain, 1936 ... old airframe, advanced concept.

the Battle of Britain, it meant that Bf 109s had to be flown toward the English coasts in massive shifts so that the attack could be pressed without giving the RAF periods of respite and recovery. It also meant that Goering would be forced to concentrate his fighter squadrons on the Pas de Calais, as close as possible to the British mainland, in order to squeeze the last mile out of the aircraft acting as bomber escorts. Finally, during the London phase, it meant that fighters escorting bombers en route to the city often had to turn around and head for home before the bombers reached their targets, leaving the bombers to the mercy of the British fighters.

The failure of the Luftwaffe to use external drop tanks during the Battle of Britain remains a mystery. They had developed the technology and used it operationally in Spain in both the He 51s and the Bf 109s. The Germans had even designed a fuel tank, the *Dackelbauch* (dachshund belly), and used it on the Me 110 escort aircraft based in Norway. Furthermore, they fitted later models of the Bf 109 to accommodate the tanks. General Adolf Galland, writing after the war, stated that the use of tanks could have extended the range of the Bf 109s by 125 to 200 miles, a distance that he argued would have been decisive (Heaton 1997). After the war, Field Marshal Erhard Milch, who organized Lufthansa, the German state airline, and later rose to the post of minister for aviation, wrote that the failure to use drop tanks was one of six mistakes that turned the tide of the war against Germany (Air University 2001). Milch attributed the failure to the military, stating that the pilots refused to use the tanks due to their poor design. The fact is, the Bf 109E models that fought in the Battle of Britain were never configured with the lines and connections to permit them to carry drop tanks. Once again, it was a failure of leadership that allowed this design shortcoming to become a fatal flaw; it did not have to be that way.

20. Ju 87 Wins Dive-Bomber Competition

As Germany rebuilt her armed forces, the doctrine that guided the use of airpower generally assumed that aircraft would be used in a tactical role to support ground forces. At that time, bombs were often delivered from aircraft flying straight and level. Bombs delivered in this manner were usually as dangerous to the friendly forces as they were to the enemy. The Ju 87, shown in the illustration, was Hugo Junkers's response to Ernst Udet's desire to arm the Luftwaffe with dive-bombers that were capable of performing precision tactical bombing.

Having seen dive-bombing demonstrations in the United States, Udet bought two Curtiss Hawk IIs and demonstrated them to members of the German Air Ministry in late 1933. The Air Ministry followed those demonstrations with a new requirement for a dive-bomber and called for a fly-off competition between contractors interested in producing the new aircraft. In 1936, four companies submitted prototype designs—Arado, Blohm and Voss, Heinkel, and Junkers. The Heinkel He 118 and the Junkers Ju 87 quickly established their superiority, and the contest was really between those two. Udet, who held the position of chief of the development section, was flying the He 118 when a malfunction caused him to bail out. The decision was made; the Ju 87 would be Germany's new dive-bomber.

Called the *Stuka*—short for *Sturzkampfflugzeug* (dive-bomber)—the Ju 87 began active service with the Luftwaffe in 1937. The airplane saw service in the Spanish civil war, where it was very effective in attacks against point targets. The airplane was powered by the Jumo 211 A-1 engine. The armament consisted of two machine guns mounted in the wings and a single machine gun in the rear cockpit. The landing gear were fixed (not retractable) and the gull-wing design gave it a distinctive appearance in flight. The simplicity of the design made maintenance and repair quite easy, and the Stuka was often turned around for as many as six missions in a single day during the Spanish civil war. The aircraft was extremely effective as a psychological weapon in the early phases of World War II. The Luftwaffe mounted sirens on the landing gear so that the Stuka seemed to lit-

The JU 87 was designed exclusively for dive bombing.

erally scream as it attacked. It became the symbol of blitzkrieg seen around the world in newsreels as it rolled in and screamed to earth. The Stuka carried one 1100-pound bomb under the fuselage and four small 110-pounders under the wings.

Forcing the nose down into a dive angle that was often nearly 90 degrees to the ground, extending the dive brakes to keep the speed from building to the point that it would rip the wings off the airplane, releasing a bomb so that it does not collide with the airplane, and then pulling out of this screaming dive using *g* forces that load the wings up to six times the loads encountered in straight and level flight—these are demanding requirements for any airframe. Most airframes are simply not strong enough to handle the loads encountered in dive-bombing. The stresses on the pilot are great as well. Altitude may change thousands of feet in seconds; in fact, many Stuka pilots had their eardrums pierced and tubes inserted to allow the changing air pressures to equalize without ear blocks. The same gravitational forces (*g* forces) that stress the airframe also impact the pilot. While pulling out of a high-angle dive, the pilot must fight to remain conscious as the blood in the head is pulled down by the force of gravity, and the arms and legs feel six times their normal weight as the pilot tries to move them against the tremendous forces. The Ju 87 performed the mission for which it was designed quite well. It was a stable dive-bombing platform and could deliver its load with a high degree of accuracy. But however well conceived it may have been for supporting tactical maneuvers within the European continental theater, it was not well designed for employment in the Battle of Britain. It had relatively short range (370 miles) and was far too slow (232 mph at 13,500 feet) to defend itself against the British fighters. These performance shortfalls were compounded by the inherent vulnerabilities of the dive-bombing mission. It was withdrawn from the Battle of Britain after only a few weeks of heavy losses. Goering and Udet had hoped the Stuka would be able to perform as a fighter-bomber, but in the final analysis it survived more as a symbol than as a significant contributor to the German war effort.

21. Gun and Cannon Development

In spite of the many bullet holes in the vertical tail of the downed He 111 in the illustration, the fact is that it was brought down not by gunfire but as a result of a midair collision. Attacked by a British Hurricane, the bomber took numerous hits from the Browning .303 guns, but the damage done was not sufficient to down the airplane. The lethality of the British fighter armament was a continuing issue during the Battle of Britain. British ammunition tended to be very light in comparison to that fired by the German cannons, and too often German aircraft escaped even after having been hit many times.

The issues of how many pieces and what type of armament to mount on an aircraft and where to mount them are vitally important—should the aircraft carry machine guns,

He 111 downed by collision after extensive gunfire couldn't do the job.

or should it carry cannons, or perhaps a mix of the two? Should the guns be in the wings, or on the fuselage? Aside from the lethality of the guns themselves, the answers to these questions impact the design of the aircraft and the tactics that pilots use. Both the Germans and the British wrestled with these issues as they converged on the final designs of the aircraft they would send into the Battle of Britain.

In Germany, Messerschmitt experimented with various combinations of guns and cannons. He considered mounting four .50-caliber machine guns, two in the fuselage and one in each wing; he tried two machine guns in the fuselage augmented by a 20-mm cannon mounted in each wing. Some of the Bf 109s in the Battle of Britain had a single fuselage-mounted cannon with four machine guns in the wings; however, the combination that most British fighter pilots faced was two 7.9-mm machine guns mounted on the nose and two 20-mm cannons, one in each wing. The cannon offered the twin advantages of increased range and heavier projectiles in comparison to the machine gun.

The Germans employed a modified Swiss Oerlikon cannon that had originally been used as a ground-based antiaircraft weapon. They modified the breech to increase the rate of fire, shortened the barrel, and reduced the powder charge carried in the shells to reduce the weight. It was a good air-to-air weapon even though the modifications slowed the impact velocity of the shells, making the cannons less effective than they might otherwise have been. In spite of the fact that the muzzle velocity of the cannons was slower than that of the machine guns, the shell was 14 times heavier than the machine-gun rounds used by the British. While the modified cannon had a fast rate of fire compared to antiaircraft weapons, the Oerlikon was still slow by air-to-air standards; it could fire only 17 shells during the typical 2-second window that the pilot had available to shoot down another fighter (Deighton 1979). Furthermore, each cannon carried only 60 shells, so a 2-second burst would use 25 percent of the ammunition available. These shortcomings notwithstanding, the German cannon combinations were generally acknowledged to be superior to the British eight-machine-gun arrays.

For their part, the British had started the design of the Spitfire and Hurricane with a requirement that they carry four machine guns, but Squadron Leader Ralph Sorley's analysis had led to incorporation of eight Browning machine guns (.303 caliber) as the standard for both. These guns were derivatives of the American Colt .30-caliber gun, manufactured under license to Browning. The British bored the .30 barrel out to .303 caliber in order to accommodate the ammunition they used. Each gun carried 300 rounds of ammunition. The issue in air combat is how much lead or steel can be placed on the target and with what velocity it hits the target. In spite of the fact that the British carried eight guns, the two gun–two cannon combination used in most Bf 109s could deliver 18 pounds of projectile weight to the 13 pounds that the British could deliver. In 1940 the RAF performed tests comparing the performance of several different armament configurations. Eight different combinations of guns and cannons were tested, and the recorded results indicated that the two-cannon combination was the most effective, while the eight-gun array favored by the British in the Battle of Britain was fourth in terms of effectiveness. The British eventually incorporated cannons into the armament used on the Spitfire, and a few of these cannon-armed versions saw action before the Battle of Britain ended.

Both sides had difficulties associated with their armament. The German cannons suffered from low muzzle velocities that reduced their lethality, a situation that was compounded if the cannons were fired from extended ranges. The British machine guns, on the other hand, suffered from the light weight of the shells they used. In one recorded incident, six Spitfires fired 7000 shells at a Do 17 bomber and still failed to bring it down

(Hough and Richards 1990). The lightness of their weapons forced British fighter pilots to bring their aircraft as near as they dared to opposing aircraft in order to ensure that their bullets hit with maximum effect. The British also faced other problems. When they climbed out through the clouds to higher altitudes, they found that the condensation from the clouds often froze in the barrels and breeches, causing jams at altitude when the guns were most needed. The initial fix was to place gauze or tape over the barrels to prevent the moisture from settling into the guns; the long-term fix was to heat the barrels electrically, drawing power from the engines.

The armament that an air-to-air fighter carries is in many ways the last link in the chain that determines whether it is a success as a total weapon system. The command and control element must ensure that the aircraft is directed toward the correct targets and is positioned to attack; the airplane then must be capable of performance that enables it to maneuver successfully in preparation to fire on the enemy; but unless the aircraft is also able to fire a lethal projectile that brings the target down, none of the other elements will have accomplished the ultimate objective of the system. The issue is *total system performance,* and the armament subsystem is a vital part of the whole fighter weapon system.

22. Barrage Balloons

Imagine yourself as a German bomber pilot. You are approaching London on a low-altitude run when you spot several huge balloons floating above you. Suddenly you realize that the balloons have heavy steel cables hanging below them. You know that if you hit one of these cables with a wing or an engine, you are going to crash. Adding power, you pull up violently, hoping to clear the balloons and fly above them. As you pull up to the higher altitude your aircraft is suddenly bathed in the white light of a searchlight; your increased altitude has now made it possible for the antiaircraft guns to see and track you. The shells begin to explode in midair around you. You begin to think that if you just survive this mission it will be a miracle. Barrage balloons like those shown in the illustration were routinely used during and after the Battle of Britain to force German bombers to higher altitudes, where aiming was more difficult and tracking them with antiaircraft fire was easier.

The concept of using barrage balloons (or balloon barrages, as the British called them) was introduced during World War I when the Germans were attacking Britain with the Gotha bombers. Balloons were floated at altitudes up to 10,000 feet. They carried steel cables that hung below, and there were also steel cables that joined the balloons to each other. The primary effect of the balloons was not to destroy aircraft, although a few were brought down by the balloons, but, rather, to force the aircraft to fly at predictable altitudes. Accurate bombing was more difficult at the higher altitudes; furthermore, antiaircraft guns could be more effective when the altitudes were more predictable. The con-

30 Group's balloons denying the Luftwaffe low level airspace over London, 1940.

cept was so successful that the British increased the use of them considerably during the Battle of Britain, although they concluded that the cables that connected balloons to each other substantially increased the complexity with little gain in effectiveness, so they ceased that practice.

As the British prepared to defend against the inevitable air attacks that would come in the Battle of Britain, barrage balloons were an integral part of their plans; 450 balloons were to be stationed around London, with additional units stationed at key ports and other major industrial centers. By 1938, the Balloon Command had been formed under the command of Air Vice Marshal O. T. Boyd. The command consisted of two balloon groups, and authorization had been granted authority to deploy 1450 balloons at key points, including the city of London. During World War II, the balloons were generally flown at about 5000 feet; the primary purpose, as in World War I, was to force aircraft to higher altitudes, which made bombing less precise and further exposed the attackers to antiaircraft defenses. The balloons also discouraged the use of dive-bombing techniques.

Anything that distracts attacking pilots and makes the mission more difficult and dangerous to fly contributes to the defense. The British tested the concept of balloon barrages in World War I, found it effective, and adapted it with minor changes to their defense strategy in World War II. Like many old ideas that are suddenly new again, the balloons were a part of the fabric of the defense system that prevailed against the seemingly overwhelming odds that the country faced in the summer of 1940.

23. Heinkel and the He 111

The He 111 shown here had a glazed, streamlined, and offset cockpit to improve visibility for the pilot. The nonsymmetrical appearance of the airplane and the throbbing of its unsynchronized twin engines made the He 111 a recognizable symbol of the German bombing campaign to many British citizens.

In 1922, Ernst Heinkel established the Heinkel-Flugzeugwerke at Warnemünde on the Baltic Sea. His contributions to aircraft design were significant in building the Luftwaffe's growing strength in the years leading up to World War II. Among his more notable designs were the He 51, an escort-fighter that flew in the Spanish civil war and was the first aircraft to carry external drop tanks; the He 59, the seaplane that carried the Red Cross symbol but which was attacked by the British, who believed it was also doubling as a maritime reconnaissance aircraft; the He 115, recognized as one of the best seaplanes of its time and used by the Germans for various roles, including minelaying; and the He 111, one of the three bombers that conducted missions over England during the Battle of Britain. Heinkel's criticism of Hitler's regime led to the loss of his company to the Nazi state in 1942, but control of the company was returned to him after the war, and he continued to be active in aeronautical engineering until his death in 1959.

the He III cockpit represented advanced airframe technology: glazed, streamlined, and offset for visibility.

The He III was one of those aircraft that was ostensibly built for nonmilitary purposes, but which was clearly intended for military missions. Designed as a fast mail carrier and passenger aircraft for Lufthansa in the 1930s, the prototype had provisions for three gun positions and a 2200-bomb load. Initially employed as a bomber in Spain, the aircraft performed well enough against the fighters encountered in that theater to lead the Germans to be overly optimistic about its probable performance in the Battle of Britain. Beginning with models produced in 1938, the nose section was redesigned to include an offset cockpit and extensive glazing with windows to improve pilot visibility. The airplane was powered by twin Daimler-Benz DB 601A engines which delivered 1100 horsepower each. It was relatively slow, with a maximum speed of 247 mph at 16,400 feet. It could carry 4410 pounds of bombs and had a maximum range of 1224 miles. To compensate for the slow speed of the airplane, the Germans added up to 600 pounds of armor to protect the crews—which slowed it even further.

The He III did not fare well against the fighters it encountered in the Battle of Britain. It was simply too slow, too low, and too lightly armed to survive against the likes of the Spitfire and Hurricane. By September 1940, the He 111 had been relegated to night-only bombing in southern England. Germany produced more He 111s than any other bomber, and the airplane, by virtue of its sound, appearance, and numbers, came to symbolize the German bomber attacks in the eyes of many Britons.

24. Dornier Bombers

Casting a considerably larger shadow as it breaks ground in the illustration than it did as an operational bomber during the Battle of Britain, the Dornier Do 17 was originally designed by Claudius Dornier as a fast mail carrier and passenger airplane. Dornier began his career as a metallurgical engineer. He worked for a short time with the Graf Zeppelin Company, but when Zeppelin declined to diversify into aircraft development, Dornier left

Casting a small shadow for the Luftwaffe, a DO17 breaks ground in 1938.

to form his own company. He developed some of the earliest large, all-metal aircraft, among them the Do X, a giant 12-engine aircraft that flew from Germany to New York in 1931. Dornier was forced to cease producing aircraft as a result of restrictions levied after World War II, although he remained active as a consultant in aeronautical engineering until his death in Switzerland in 1969.

The Do 17 made its first flight in 1934. Distinctive in its appearance, it had a long, thin fuselage with twin vertical stabilizers on the outer ends of the horizontal tail section. It was on occasion referred to as the "Flying Pencil." Poorly designed as a passenger carrier, the airplane was mothballed and was in storage when the Luftwaffe rediscovered it in 1935 and decided to convert it to a fast bomber. New prototypes were produced, and the airplane went into service in 1937. The Do 17, like the He 111, was sent to Spain, where it performed well enough against the opposing fighters to convince the German high command that it could be counted on as a bomber in World War II.

In 1937, the Do 17 experienced a moment of glory when it was fitted with a Daimler-Benz 600 A engine for the Zurich Air Show and proved faster at 248 mph than most of the fighters at the event. But the Dornier bomber, like other bombers, was powered by engines that were less than the best. There were too few of the DB 600 and 601 engines available, and these were reserved for the Bf 109 fighters. The Do 17Z, the most produced variant, had two Bramo 323 nine-cylinder engines with a two-stage supercharger that delivered power in the 1000-horsepower range. It was capable of airspeeds up to 265 mph at 16,400 feet, and had a range of 745 miles with a service ceiling of 26,400 feet. With a bomb load of only 2200 pounds and no armor, it was the least well known of the three bombers that made up the German assault force. The airplane carried up to eight freestanding machine guns and was used both as a bomber and as a long-range reconnaissance aircraft.

Dornier produced several other aircraft of note in addition to the Do 17. The Do 18, also originally designed as a mail carrier, was a seaplane used for maritime reconnaissance; only a hundred or so were produced. Reconnaissance was a continuing theme at Dornier. The company also built the Do 215, which was used for high-altitude reconnaissance. The other Dornier design that merits a historical footnote is the Do 19, which was one of the prototype four-engine bombers that General Walther Wever had built before he died in the air accident that led to the cancellation of that initiative.

25. Boulton-Paul Defiant

The drawing illustrates one of several critical weaknesses of design, as the pilot of a Boulton-Paul Defiant maneuvers into the six o'clock position intending to attack and kill an Me 110, while his gunner, oblivious of the pilot's intent, is focused on a different tar-

Fundamental Defiant problem. Pilot and gunner engaging separate targets simultaneously.

get. The Defiant was designed to isolate the pilot and gunner from each other so that each might devote his total attention to his assigned tasks. The problem, as shown in the illustration, was that in the high-speed, violently maneuvering environment of air combat, shooting and flying must be carefully integrated to take advantage of those rare instants in time when a target is vulnerable. The Defiant was ineffective as a fighter and, worse, was dangerous for the crews who flew it.

By the 1930s, as an offshoot of their work on bombers, the British company Boulton and Paul had developed an effective quad-gun array mounted in a hydraulically powered turret that could turn 360 degrees while firing. In World War I, Britain had success with a two-seat fighter, the Bristol, in which the gunner occupied the second seat. The concept was that a gunner who did not have to split his attention between flying the aircraft and shooting might be a much more effective gunner. The Air Ministry, combining the concept of the newly developed turret with the previous experience, issued a specification in 1935 that required a two-seat fighter with revolving turret-mounted guns and speed in the 300-knot range. Boulton and Paul designed and built the Defiant in response to the specification. The airplane had nice lines, with the turret mounted behind the cockpit. It had good handling qualities and, at the time of its deployment in 1939, was hailed as the "fighter with the sting in its tail." It was powered by the Merlin III, had a range of 465 miles, and possessed maximum speeds of 304 mph at 17,000 feet.

The primary missions intended for the Defiant were to be defensive patrol and bomber interception. Things did not proceed as planned, however. Production delays resulted in only a few deliveries before the Battle of Britain began, and with the advent of the German attacks, the airplane was forced into a fighter interceptor role similar to that assigned to the Hurricanes and Spitfires. The strict division of responsibilities between pilot and gunner proved disastrous. With the turret mounted immediately behind the cockpit in the Defiant, the guns could not be fired directly forward; they were restricted to an arc that excluded the 16 degrees directly along the line of flight of the aircraft. With the gunner and the pilot physically separated, the gunner was further distracted by having to direct the pilot by radio. The pilot had to imagine the gunner's line of fire in order to position the aircraft to shoot—something like trying to aim a gun by viewing it through a rearview mirror. To make matters worse, the guns could not be depressed to angles below the horizontal, so the Defiant had to maneuver to positions below the enemy before it could fire on them. The quad guns were successful when the Defiant was able to maneuver and attack bombers, but in air-to-air engagements against fighters it was a death trap. The Defiant had a few initial successes, primarily because it caught a few attacking Bf 109s unaware as they approached from the rear. However, once the Germans had figured out the configuration, the Defiant proved insufficiently maneuverable to be effective and too slow to defend itself. To compound matters, the turret was difficult for the gunner to enter, which slowed the responsiveness of the Defiant when it was in ground-alert status.

The turret was also very difficult to escape from once airborne, and the gunner had little hope of bailing out when the aircraft was hit in the air. It suffered disastrous losses to the Bf 109s and was soon consigned to a night fighter role, but it was little more successful in that mission than it had been in the interceptor role. In summary, the Defiant made few significant contributions to the British war effort. The design was based on a mistaken premise from the outset, and the fundamental flaw that separated the flying function from the associated gunnery doomed the airplane from the start.

26. U.S. Technology

Among the many remarkable aspects of the Battle of Britain, nothing is more remarkable than the ease with which technology flowed from country to country to enable the combatants to develop the capabilities each needed. The illustration of the Hurricane makes the point that this exchange of technologies, particularly technologies that originated in the United States, played a substantial part in making it the weapon system that it was. The Merlin engine evolved from the Liberty and Curtiss V-12 engines that originated in the United States; the 100-octane aviation gasoline that enabled the Merlin to achieve its full power potential was supplied by the U.S. Esso Corporation; and the .303 machine gun used by British fighters was derived from the U.S. Colt .30-caliber machine gun, produced under license by Browning. There can be little doubt that the British fighters would not have been as successful as they were in the absence of technologies that originated in the United States. The United States also manufactured fighters for the British government; some of the first P-51 fighters produced by the United States were built for Britain and flew British colors when they entered combat.

One of the key technologies that the United States made available to Britain was 100-octane aviation gasoline. From 1938 to 1940, the British busied themselves improving the Merlin engines by improving the supercharger. By increasing the boost pressure within the engine for short periods they could increase the horsepower from 990 to over 1300 horsepower. But this could be done only if an improved fuel were used. The standard 87-octane aviation fuel produced unacceptable levels of detonation at the higher pressures. Esso and Shell were the companies that had the capability to produce 100-octane fuel; however, Shell could not produce it in the quantities required, so the RAF turned to Esso as their primary source. The 100-octane fuel had been secretly in use by the U.S. Army Air Corps since 1938. The use of the higher-octane fuel at the higher boost pressures produced a significant increase in climbing performance. The sudden increase in British fighter performance came as a surprise to the Germans until they were able to analyze a downed Spitfire and discovered that the fuel was of considerably higher octane than the standard aviation gas. Initially there were some problems clearing Esso to sell 100-octane fuel to Britain because of the American Neutrality Act, but Roosevelt

three "Yank" technologies... 100-octane fuel, Colt guns, and Curtiss influence on the Merlin.

worked an arrangement that allowed the sale to proceed in time for its use in the Battle of Britain (Hough and Richards 1990).

The Browning machine gun used on all British fighters in the battle also owes much to the United States. The original plans for the Hurricane had two guns in the fuselage and two in the wings, but as the Air Ministry requirement changed to eight guns, Camm decided to place all eight guns in the wings. This demanded that the guns be both light and reliable. An additional requirement was that the RAF guns be capable of firing the same ammunition that the British army used. The solution came in the form of an agreement between Colt, of the United States, and Browning, of Great Britain, that licensed to Browning the right to modify and manufacture the Colt .30 machine gun as needed for use in RAF applications. The Colt gun was bored out to .303 caliber, and extensive modifications were made to the breech to accommodate the peculiarities of the British cordite propellant. The resulting Browning machine gun was highly reliable and very light. A full set of eight guns with 2500 rounds of ammunition weighed just over 400 pounds.

The flow of technology was hardly a one-way street, however. As discussed in other sections of this book, the British incendiary-round technology was made available to the United States armed forces for a total charge of $1. This technology became a mainstay for the U.S. air forces, for both fighters and bombers, in all theaters for the remainder of World War II. Furthermore, the British Merlin engine was far superior to the American Allison engine that originally powered the P-51, and it was the Merlin that finally enabled the P-51 to provide adequate fighter escorts to the American bombers as they targeted Germany once the campaign against the German homeland began in earnest.

Possibly some of the most curious events of the war also involve the transfer of technologies across national borders during the years leading up to the Battle of Britain. Some of the transfers worked to the mutual advantage of countries that would be adversaries within just a few short years. For example, an American, Alfred Gassner, was a co-designer of the Junkers Ju 88, Germany's best bomber at the time of the Battle of Britain, and a German, Edgar Schmued, who had worked at both the Fokker and the Messerschmitt companies, played a major part in designing the American P-51 Mustang aircraft. Perhaps the most interesting and surprising of technology transfers was the arrangement that allowed the Germans to use the Kestrel engines, forerunner to the Merlin, in the competitive fly-off that led to the selection of the Bf 109 as Germany's frontline fighter. This is particularly surprising, since the fly-off occurred in 1935, well after the British had begun to rearm for war, which by then they recognized as almost certain. But before pointing fingers, one might as well ask why the variable-pitch propeller technology that was used on the Bf 109 was licensed to Germany by the U.S. firm Hamilton-Standard. Clearly, both Britain and the United States approved the transfer of critical technologies to Germany, even while it was well understood at national policy levels that the probabilities were high that both countries would eventually have to fight against the German systems they were, in effect, helping to develop.

27. Barking Creek

The event pictured in the illustration, with two Spitfires attacking a flight of Hurricanes, occurred as Britain prepared for the coming war with Germany; it resulted in the loss of two airplanes and the death of one British pilot. Known as the Battle of Barking Creek, it was the impetus for the development and installation of identification-friend-or-foe (IFF) transponders in all British fighters before the war began.

On September 6, 1939, a searchlight battery reported to 11 Group command and control that they had spotted a flight of aircraft suspected to be hostile. Upon hearing that they had inbound hostile aircraft, 11 Group ordered that six airplanes be launched to intercept the inbound aircraft. Hurricanes of 56 Squadron were alerted and took off to investigate. The radar station at Canewdon then picked up the flight of Hurricanes and reported them as inbound from the seaward side—thus, in effect, confirming the visual sighting of the searchlight battery. There has been considerable discussion about whether this could have been a misinterpretation by the radar crew, or whether it was a momentary technical failure of the radar itself. Supporters of both sides have debated the issue, but it is unlikely that the argument will ever be absolutely resolved. The fact is that the radar station detected the Hurricanes and took them to be hostile. To compound the situation, the sector controllers, in their zeal to ensure that the intercept mission would be successful, had launched 14 aircraft, rather than the 6 requested by 11 Group. Having judged the flight to be hostile and finding the flight to be larger than initial reports suggested, the controllers then launched additional Hurricanes and Spitfires and vectored them into the area of the "hostile" aircraft. Visibility was not good, and the combination of sun and haze made identification difficult, at best. A section of Spitfires from 74 Squadron intercepted a section of Hurricanes from 151 Squadron and, mistaking them for Bf 109s, proceeded to shoot down two of them, killing one of the pilots.

Among those who were aware of the incident, the Battle of Barking Creek became the subject of wry and cynical songs in the local music halls. To the RAF, however, it was no laughing matter. It highlighted the imperative need for identification and control of individual aircraft in the air. Pipsqueak had given controllers the means to track flights of RAF fighters and vector them toward groups of inbound aircraft, but once the air-to-air battle was engaged, the sky, and the radar scope, became a dynamic whirling, diving mass of engagements between individual aircraft. Without some means of radar identification of individual aircraft, the controllers had no way of knowing whether a given blip on their screen was an RAF or a German airplane.

Radar growing pains at Barking Creek: poor identification, poor control, friendly casualties.

The answer was the development and installation of what became known as *IFF* (identification friend or foe) in RAF aircraft. The IFF was a transponder, which was tuned to a designated frequency. When "queried"—that is, when a radar signal was transmitted to the aircraft carrying the IFF transponder—an altered signal would be returned to the radar receiver such that the controller would see a somewhat elongated blip, clearly distinguishing the aircraft with IFF from those without. By mid-July of 1940, all operational aircraft in the RAF had been fitted with IFF transponders. This was truly the final step in the build-out of the Chain Home radar system. It now extended from initial detection, to launch and control of RAF fighters, through the interception of enemy fighters. This system of positive control enabled the British to use their limited resources with the utmost of efficiency as they confronted the numerically superior German forces.

The combat forces of the United States today emphasize the role of communications technologies in their war plans. Commanders recognize that with good intelligence and positive control of the forces at their command, they can bring the right mix of forces to bear at a designated place and time on the battlefield. Command and control becomes as important to victory in this scenario as any specific weapon or combat unit. The British truly operated from this very same perspective in 1940, when, with the development of IFF, they completed the Chain Home system that enabled the Hurricanes and Spitfires to be placed at the most opportune locations, to be launched at precisely the right moment to achieve intercepts, and to be directed to the locations where the enemy would be found. In this sense, the Chain Home system was as important as the weapons themselves; without the command and control system, the weapons could never have performed as effectively as they did.

28. Radio-Beam Flying

As Germany turned increasingly to night bombing attacks, the issue of finding targets while flying through a dark and often cloudy sky became a significant concern. The drawing illustrates the approach that Germany took to solve the problem, showing an He 111 bomber navigating toward a target, guided by the *X-Gerät* electronic beam system. A glance at a large-scale map of Great Britain will confirm that finding London, even during the blackout, was not the problem; the Thames Estuary points like an arrow toward the heart of London. Earth and water appear significantly different from the air even on dark nights, so German bombers would have had relatively little difficulty finding the city. And, of course, once a few bombs were dropped, subsequent bombers could use the fires and smoke from the earlier attacks to assist them in setting their aim points. The problem of finding targets really arose when the targets were the smaller, less recognizable cities and industrial sites.

To assist their bombers in hitting these hard-to-find targets, the Germans developed an ingenious system based on transmitted electronic beams that guided the pilot to the

Electrons pointing the way: an He111 with X-Gerät system providing guidance to the target.

target area, and even signaled him when he was in the immediate vicinity of the target. The initial version of this navigation system, called *Knickebein* by the Germans, consisted essentially of a set of signals to guide the pilot to the target, and a second signal transmitted from a site geographically separated from the course guidance signal that told the pilot when the plane was in the target area. A transmitter in Germany or France provided a continuous signal as long as the pilot was on course toward the target. If the airplane drifted left of the designated course, the pilot would begin to hear Morse code dots instead of the continuous signal; if right of the course, the pilot would hear dashes. By broadcasting a second signal that intercepted the course guidance signal, the bomber could be informed that it was at the designated target. In this way, even under conditions of complete darkness and cloud cover, bombs could be placed in the vicinity of the desired target. It has been estimated that the accuracy of the *Knickebein* system was sufficient to drop a bomb within 300 yards of a given target. While this may sound like a substantial miss, one must recall that a group of bombers would attack any given target, and if bombs fell, on average, within a 300-yard circle around the intended target, the dispersion of the bombs would be sufficient to do considerable damage to a town or a factory.

The point has been made that little is static in warfare, and this is particularly true where technology is the basis for competition. It did not take long for the British to determine the nature of the *Knickebein* system and to devise means for countering it (see following subsection, Foreign Systems Exploitation). The Germans naturally followed with an improved navigation-bombing system, which they called *X-Gerät. X-Gerät* was similar in concept to *Knickebein,* but more sophisticated and considerably more accurate. Like *Knickebein,* there was a course guidance signal, but there were three intercept signals, rather than the single beam of *Knickebein.* The first signal intercepted warned the crew that they were approaching the target; the second told the bombardier to punch a time-clock mechanism; and the third told the bombardier to punch the clock a second time. The system then released the bombs automatically using a timing device. The timing device compensated for wind and ground-speed effects, and the accuracies achieved were in the 120-yard circular error range. To make the electronic bombing system even more effective, the Germans used *X-Gerät* with a specially equipped and trained pathfinder unit called *Kampfgruppe* 100. This unit was trained to operate with *X-Gerät,* and they used bomb loads comprised primarily of incendiary bombs so that the bombers that followed would see the fires and know where the target was without the need for all attacking aircraft to carry the electronic system.

It is a testimony to the engineers and scientists on both sides of the conflict that they were able to devise systems like *Knickebein* and *X-Gerät,* and that they were able, in both cases, to rapidly find counters. This whole area of the use of electronics in both offensive and defensive roles has continued to grow in importance in modern warfare. In many ways, as complex and modern as the technology involved is, the nature of the contest remains very similar to that in these earlier efforts in which initiatives were launched, countermeasures were developed, and new measures were then developed—all within relatively short spans of time. The events of the early 1940s heralded this new type of electronic warfare and showed the direction and the pace that it would take from the start.

29. Foreign Systems Exploitation

A downed or captured aircraft represents a considerable opportunity to find and exploit information that has military value. The *Knickebein* electronic beam navigation system developed to guide German bombers to targets obscured by night and weather conditions was initially compromised when the British found evidence of it onboard the downed He 111 pictured in the illustration.

Knowledge about enemy capabilities and systems can be invaluable during a military conflict. Suppose, for example, that one side could capture an enemy fighter or bomber and become familiar with the way that airplane handles, how fast it turns or climbs, and how well the pilot can see out of the airplane. Using that information, new tactics could be developed specifically to exploit the weaknesses inherent in the enemy's design. Even a destroyed aircraft can provide information about onboard systems—the way particular systems operate or the tactics they employ. Investigation of an enemy's systems and exploitation of the information developed is not new; it was a natural part of both sides' efforts during the Battle of Britain, and it continues to be a normal part of intelligence gathering today.

One of the more interesting examples of the exploitation of intelligence gathered from enemy systems was the British development of countermeasures to defeat the *Knickebein* navigational system. The British were aware that the Germans were using some sort of radio beams to guide bombers to their targets during night raids over England. In October 1939 an He 111 was shot down north of Edinburgh. Investigation by Dr. R. V. Jones, a physicist attached to the Air Ministry Directorate of Intelligence, led to the discovery that the Lorenz navigation system on the airplane had been modified to make it significantly more accurate than anything that the British had developed to date. Analysis of the aircraft logbook further revealed frequencies used for the navigation transmissions and locations of the transmitting stations. Additional corroboration came when the British interrogated a German Luftwaffe prisoner of war and the prisoner told them about the system of radio navigation that could guide an airplane from the Continent to London with a divergence of less than 1 mile.

In June 1940 a flight of Anson trainers was established and given the explicit mission to intercept and locate the transmitted *Knickebein* signals. With the information at

an He 111 crash landing in Scotland yielded significant information about German technology.

hand they were quickly able to locate the signal at 31.5 MHz. The next order of business was to develop measures to counter the system. The Air Ministry solicited the assistance of Dr. Robert Cockburn of the Telecommunications Research Establishment in developing a means of distorting or otherwise jamming the German system. Dr. Cockburn discovered that he could modify electrodiathermy equipment to eliminate the dots and dashes that told pilots they had drifted off course. These sets were distributed to various points throughout the country with instructions for their use; in addition, the RAF began to transmit false continuous tones using their own Lorenz equipment. The result so distorted and confused the signals emanating from the Continent that the system was rendered essentially useless. The *Knickebein* system was soon replaced by the improved *X-Gerät* system and naturally the British effort began immediately to counter that.

There were other cases of exploitation of enemy systems by both sides. In February 1940 the British forced another Heinkel He 111 to land on English soil. The airplane was so little damaged that the British were able to repair and then fly it. They systematically evaluated the airplane, identifying all the strengths and weaknesses possible. By the time the Battle of Britain began, the British pilots and others involved in the defensive effort knew chapter and verse about the He 111, which is no doubt partially responsible for the fact that the airplane was no more effective in the battle than it was.

Of course, the British were not the only ones capable of gleaning excellent intelligence from enemy systems. The Americans had made their top secret 100-octane aviation gas available to the RAF just in time for use in the Battle of Britain. It gave the fighters that were using the Merlin engines a sudden—and significant—boost in power. In late August of 1940, a Spitfire was forced to land on the Continent. It was immediately subjected to careful evaluation, during which the Germans discovered that the fuel used was green in color and of much higher octane than the standard 87 octane used in their Bf 109s. This, no doubt, gave impetus to fuel development activities in Germany.

Today, interest in obtaining the equipment used by opponents and exploiting the intelligence derived from those systems is no less compelling than it was in Europe in 1940. As both sides did then, the United States has made a routine practice of obtaining and flying the systems it may face in combat whenever possible in order to better develop the means of countering them in combat. The U.S. Air Force routinely trains its fighter pilots using aircraft that simulate the performance of opposing systems and employ the tactics that pilots are likely to encounter in the air. The United States was quite rightly concerned about the information that might have been deduced by enemy air forces when an F-117 stealth fighter was shot down while on a mission during the Kosovo campaign in March of 1999. The collection of intelligence derived from the physical evaluation of foreign systems has been, and will continue to be, a normal activity of nations that may someday meet one another in combat. This was the case both before and during the Battle of Britain, and it remains so today.

30. Aircraft Colors and Markings

Aircraft coloring and marking schemes have two fundamental purposes—concealment and identification. Coloring schemes are designed to camouflage the airplane so that it is difficult for enemy pilots to see and attack it; markings, on the other hand, are intended to identify the country and unit to which the aircraft is assigned so that pilots do not fire on friendly aircraft and can find their teammates in the air. The two pictures illustrate the fundamental concepts associated with aircraft coloring and markings. The Germans and British took very similar approaches to coloring their aircraft, but their systems for marking the airplanes for purposes of identification were quite different. Interestingly, the fact that the two took similar approaches to coloring made it even more important that the marking schemes employed be effective in differentiating friend from foe.

It is a generally recognized principle in close air-to-air combat that the pilot who first sees the other has a huge advantage, especially if the airplanes are of comparable performance. The Spitfire and the Bf 109 were very nearly evenly matched, so the first to spot the other would often get the first shot. The coloring schemes used by fighter aircraft, both in the Battle of Britain and today, are intended to prevent the enemy from securing this advantage. Coloring, to be effective, demands that the aircraft appear to fade into the background against which it is flying. It is the same problem that all fighting systems face—how to hide an object in plain sight. In the case of an airplane, the background against which it appears depends on the location of the viewer relative to the airplane. If the attacker is above the target, the background against which the airplane appears is the surface of the earth, typically either vegetation or open water, and one would want the coloring of the airplane to appear similar to that surface. If the attacker is below, the background against which the target is viewed is the sky. In the Battle of Britain, both Germany and Britain prepared their airplanes to fly over the English countryside, consequently the schemes both countries chose were almost the same. The British generally used a mixture of tan and dark green or tan and black on the top of the aircraft and a light blue or cream color on the bottom. During the early stages of the Battle of Britain, some aircraft used a combination of black and light colors on the underside, with one black wing and the other painted light blue or cream.

The Germans, on the other hand, generally used a coloring scheme that featured green on dark green, or simply dark green on the top and a light gray or blue on the bottom. In some cases they painted the nose of fighters, or some portion of it, yellow to make identification of friend and foe easier, and bombers often carried white strips either on the wingtips or on the vertical stabilizer to assist in formation assembly. As might be expected, aircraft that operated primarily as night fighters and bombers were often painted black; in any case, the light colors of the undersides of the airplanes would be changed to black to obscure them when viewed against the night sky.

The most important function of aircraft marking is to prevent friendly fire from downing an airplane. The aircraft of every nation are uniquely marked for that purpose. In the case of Britain, national markings consisted of the well-known bull's-eye roundel symbol in various forms, but always with a red circle at the center. A blue circle with the red center was typically painted on the top of the wings. Concentric circles of blue and white with the red center were typically painted on the bottoms of the wings, and circles of yellow, blue, and white with the red center marked the sides of the fuselage. The German national symbol was the Balkan cross (✚) of black on a white background, which was painted on the tops and bottoms of the wings and on the sides of the fuselage. If the coloring schemes and national markings used by both countries were relatively straightforward and similar, the approaches each took to unit identification markings were generally different and, especially in the case of the Germans, quite complex.

The *Geschwader* was the largest operational field unit in the German Luftwaffe hierarchy, typically consisting of 90 to 120 aircraft. These large units were then broken into *Gruppen* of 30 to 40 aircraft, each of which included three or four *Staffeln* of 10 to 12 airplanes. There were various types of *Geschwader*—fighters, bombers, Stukas, and others. The marking used in the *Jagdgeschwader* (Bf 109 units) typically included a colored number on the fuselage to the left of the Balkan cross that identified both the individual aircraft and the *Staffel* to which it was assigned. A symbol was painted to the right of the cross to identify the *Gruppe*. In addition, other symbols were often used to identify the rank and position of the pilot; for example, a double chevron identified the pilot as the commander of a *Gruppe*. The *Jagdgeschwader* was identified by emblems and symbols painted on the fuselage of the aircraft. In addition to unique aircraft markings, Germany also often colored the noses of their Bf 109s a distinct yellow to assist pilots in identifying their own aircraft.

Bombers, heavy fighters, night fighters, and other units used a somewhat different approach to markings. They employed a four-character system, two to the left of the Balkan cross symbol that identified the *Geschwader* to which the aircraft was assigned, and two right of the cross, the first of which identified the specific aircraft. It was generally either colored or outlined in the color corresponding to the *Staffel* to which the aircraft was assigned. The fourth character associated the *Staffel* with the *Gruppe* to which it was assigned. The German system was very complex, yet quite precise in tracing the hierarchical unit structure to which the airplane belonged.

British markings were much less complex. The typical British aircraft carried two characters to the left of the roundel and one to the right. The two characters left of the roundel identified the unit to which the aircraft was assigned. These codes were very much like the German codes—assigned according to a coded listing and somewhat arbitrary. In fact, a single squadron often used multiple codes, and a given code was often used by more than one squadron—all in all, quite confusing. However, it must be recalled that much of

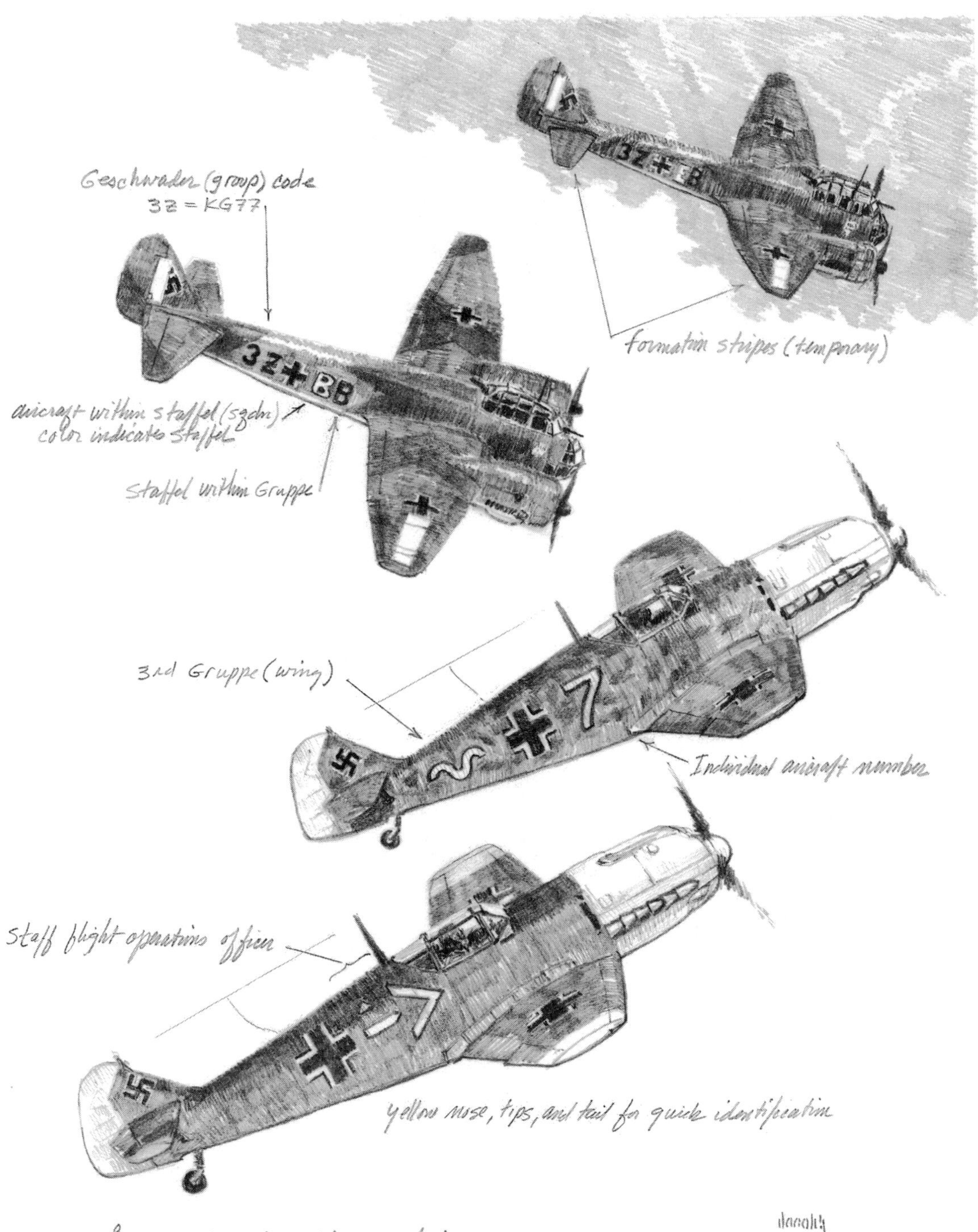
Geschwader (group) code
3Z = KG77
Formation stripes (temporary)
aircraft within staffel (sqdn)
color indicates staffel
Staffel within Gruppe
3rd Gruppe (wing)
Individual aircraft number
Staff flight operations officer
yellow nose, tips, and tail for quick identification
Luftwaffe aircraft markings

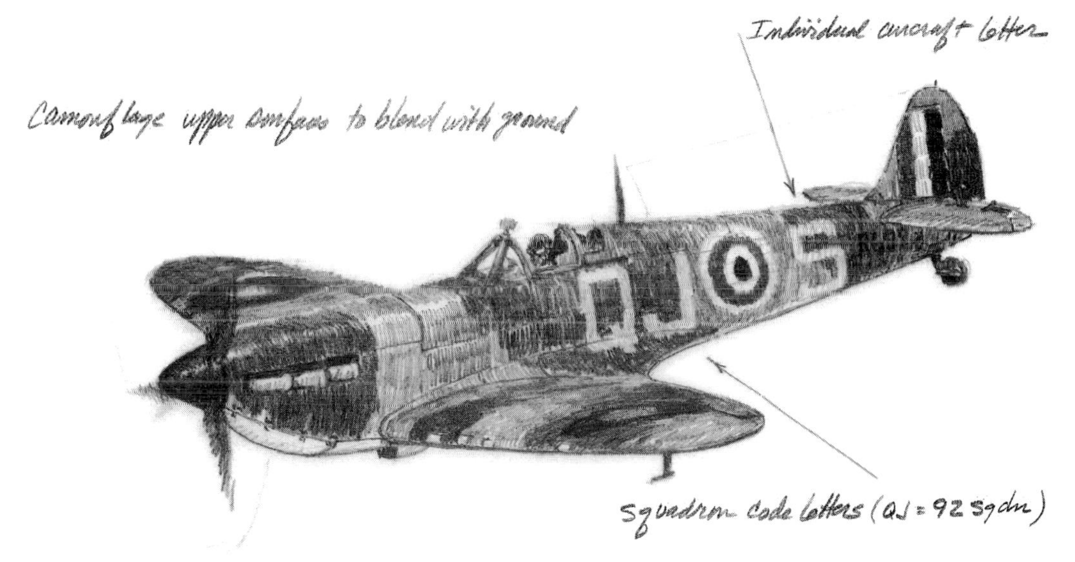

RAF aircraft markings

the identification of British aircraft in the air was handled through the command and control system that included Pipsqueak and IFF as well as aircraft-to-aircraft communications. The character right of the roundel identified the specific aircraft and also was often used as the radio call sign of the aircraft.

Aircraft coloring and marking is as much a part of the preparation for combat today as it was in the Battle of Britain. Some changes have taken place, but the objectives are the same. Today's air-superiority fighters are usually painted a light gray or blue color to enable them to hide against the sky, whereas the aircraft that will operate at lower altitudes will generally have a camouflage paint scheme that varies with the type of terrain they expect to encounter. Most aircraft carry greens and browns very similar to those that the British and Germans used to mimic the vegetation of the European countryside, although some carry tans and browns for desert combat. Some things change very little with time.

31. DeWilde Incendiary Bullet

An incendiary round offers a pilot two very distinct advantages. First, the incendiary round is designed to penetrate the skin of the target and then ignite, so it tends to have substantially more destructive capacity than a round that carries only a steel or lead projectile. The second advantage is that the pilot who fires the round can see the ignition if he hits the target; therefore, the incendiary provides a pilot the means to make real-time corrections to his aim point while he is firing. The Bf 109 shown in the illustration, fatally damaged with relatively few bullets, illustrates the increased effectiveness that an incendiary round can produce. The history of the development of the incendiary round used in the Battle of Britain is one of the more intriguing tales to come out of the war.

In 1938, a Swiss inventor, Paul Rene DeWilde, and his partner, Anton Kaufman, demonstrated their new incendiary round to the British Military Purchasing Commission. The Air Ministry had indicated an interest in buying the rights to produce the round in anticipation of the coming war. Independently, and in spite of official opposition to his efforts, C. Aubrey Dixon, a small-arms expert who worked in the design department of the Royal Arsenal at Woolwich, London, was also working on the development of a new design for an incendiary round. In fact, official opposition to his efforts was adamant to the point that Dixon's budget had been officially limited to "no more than £10 to pursue any single new invention." However, with the unofficial assistance of departmental chemist J. S. Dick, Dixon produced a prototype round that represented a significant improvement over that offered by DeWilde.

The design featured a chemical mixture with a small steel anvil in the nose of the bullet. A steel ball was behind the nose assembly, and when the bullet struck a target, the inertia of the ball would cause it to fly forward, striking the anvil and igniting the chemical mixture. Dixon included a soft nose that collapsed on impact, creating a larger impact

Fire from incendiary bullets could make a few hits go a long way.

area and encouraging ignition as well. Not only was the Dixon design effective, but it offered the significant advantage of being easily produced in volume, whereas the DeWilde design was so difficult to produce that it required a certain amount of hands-on labor to construct the round.

To keep the design secret and to obscure the source, the British paid DeWilde a substantial sum to buy his patent. They incorporated some superficial features of the DeWilde round into Dixon's design and officially referred to the incendiary as the "DeWilde round." Then they produced Dixon's design by the hundreds of thousands. Dowding commented that the round was extremely popular with pilots during the Battle of Britain and was considered far superior to any incendiary rounds that they had used previously (Dowding 1946). The round was so effective that it was initially restricted to use only by Fighter Command; Churchill expressed concern that its use by other commands might result in the prosecution of downed aircrew for war crimes. This restriction was lifted in January 1941.

Dixon later sold his design to the United States for $1 as a part of the "Reverse Lend Lease" program. Over 3 billion Dixon rounds were manufactured in the United States for use in World War II. General Hap Arnold stated at one point that his crews would just as soon fly without their machine guns as without the incendiary ammunition. Dixon, an army captain when he designed the round, was eventually promoted to the rank of colonel. He and Dick shared a prize of £3000 for their invention. Dixon was awarded the Order of the British Empire in 1945 for his contribution to the war effort. The citation in the *London Gazette* of June 14, 1945, stated, "Colonel Dixon . . . designed the incendiary small-arms bullet used by RAF fighters during the Battle of Britain [and] onwards."

Dixon retired from the army in 1950 with the nominal rank of brigadier. He was recalled to serve as chief inspector of armaments with the Army of Pakistan until his final retirement in 1959. He died at the age of 83 in 1984, a quiet hero of the Battle of Britain in every sense of the word.

32. Harmonize Guns

The Hurricane and Spitfire both carried their guns in the wings of the aircraft. The downside of mounting the guns in the wings rather than along the centerline of the airplane is that they had to be "harmonized." Harmonization involves making slight adjustments to each gun mounting so the bullets tend to converge at a predetermined point in front of the airplane, similar to the way the headlights of a car are adjusted to illuminate a point on the road in front of the car. The illustration shows a Hurricane's guns being harmonized on the ground prior to release for combat. This could be very impressive to watch as clouds of gun gas and smoke rose, accompanied by the tremendous noise of the guns and brass cartridges spewing from the ejectors under the wings.

Harmonizing a Hurricane's eight Brownings at an airfield gun butt.

As the Battle of Britain began, concerns arose regarding the distance at which the guns aboard British fighters were harmonized. The regulation standard for British fighter aircraft as the battle began was 650 yards. This meant that a pilot needed to place his aircraft 650 yards away from an enemy aircraft to maximize the number of bullets that would hit the target if his aim were true. The problem was that there is a certain amount of normal dispersion, or scatter, associated with gunfire. Even if the pilot placed his aircraft at the correct distance and had the target in his sight, the bullets were dispersed in the area of the target. Furthermore, the problem was compounded for British pilots in the Battle of Britain because they were using relatively light ammunition (.303 caliber), which lost a great deal of energy at the standard distance. By way of comparison, the German cannon projectile fired from the Oerlikon cannon was 14 times as heavy as the British .303 round. In too many cases, direct hits by British pilots did not bring down German aircraft. Mason describes an incident in which three Spitfires fired 5600 rounds at a Do 17 and still failed to bring it down (Mason 1969). To deliver their bullets with maximum effect, top British fighter pilots like Sailor Malan and Al Deere found that they needed to be much closer to the target. Soon the de facto standard had been shortened to 250 yards, and pilots requested that their armament officers reharmonize their guns for close-in combat. The practice was eventually recognized as the standard for the RAF, and all fighters were adjusted to the new standard distance. This practice would continue until Britain began to arm its fighters with cannons. The larger shells with more explosive charge and heavier projectiles made kills from longer distances possible, but the cannon-equipped fighters appeared too late to make a significant impact during the Battle of Britain.

33. Enigma and Ultra

ē·nig' ma *n.* a riddle; a baffling matter.
ul·tra *adj.* going beyond others.

Messages between military units are necessarily encoded or enciphered to prevent those who might intercept the communication from understanding either the subject or the details of such messages. Since official messages may contain information about the disposition of troops and equipment and plans and orders for military action, countries spend a great deal of time and effort developing means to safeguard their own information and to intercept and understand the official communications of other countries. This is, if anything, even more true today than it was in the years leading up to the Battle of Britain.

The German cipher machine, known as "Enigma," was developed well before World War II. The purpose of the machine was to encipher Morse code messages between stations throughout Germany. Similar in appearance to a typewriter, the design of the machine was initially based on the use of three rotors that would systematically substitute a different letter for any letter of the alphabet selected for transmission. Words were typically broken into groups of five letters, so the phrase "BATTLE OF BRITAIN" might look something like "NXDFG AXIYH KTCSZ". The key to the machine was changed daily, so that not only would the substituted letters change constantly, but also the ciphers would not be repeated from one day to the next. A recent public television program on the Enigma estimated that it would take a modern supercomputer a year to decipher the code even today (Public Broadcasting System 2001.)

Work on cracking the Enigma ciphers had begun early in the 1930s. In fact, a disgruntled and poverty-stricken German soldier had offered to sell 300 drawings of the machine to the French in 1931. When the French declined, the soldier offered the drawings to the British, who also showed no interest. The drawings were finally sold to the Poles, and it was they who, in fact, initially deduced the internal workings of the machine. The British formed the Government Code and Cipher School (GC&CS) in 1938 and recruited personnel specifically to break the Enigma code under the code name "Ultra." Working with the Poles, GC&CS, led by the likes of the genius Alan Turing, inventor of the computer, set about the task of cracking the cipher so that daily transmissions could be read. Combining the brilliance of the code breakers and some very logical analyses of human behavior, CG&CS finally cracked the code on May 22, 1940.

The fact that Britain had access to the daily communications transmitted via the Enigma has raised some interesting and controversial issues, several of which have concerned the extent to which Air Chief Marshal Sir Hugh Dowding had access to that information. Some have argued that Dowding did, in fact, have access to critical information and that possession of the information led him in some cases *not* to take action that might have benefited the war effort because to do so would have revealed to the Germans that the Enigma transmissions were being read. Others have made the case that Dowding was not initially among those given access to Ultra intelligence data and that it was only in late October 1940 that he began to receive the decrypted information—near the end of the Battle of Britain and well after the battle was won. The proponents of this view argue that Dowding often received information either too late to act or in too fragmented a form to be operationally useful.

It is likely that British Intelligence had not fully organized itself during the Battle of Britain to collect, disseminate, and use the intelligence that the Ultra made available. Having just cracked the code in May 1940, there was too little experience and the database too sparse at the time the battle got under way to support a conclusion that the intelligence gathered would have been very useful during the July to October 1940 time frame. The analytical methods and the procedures necessary to exploit the information would not have matured sufficiently in that short a time span. Ultra was useful in gathering intelligence about organizational structures and orders of battle, but it is unlikely that significant events were, or could have been, altered during the battle as a result of breaking the cipher. On the other hand, as procedures matured and the experience base grew, Ultra intelligence became quite significant to the combat effort by 1941 and was a significant factor for the remainder of the war.

The same issues that are raised and debated about the availability and use of intelligence remain as pertinent today as they were during the 1940s. Even in these times of space-based satellites and communications downlinks that transmit data at the speed of light, arguments about procedures, roles, and responsibilities continue. Intelligence gathering remains an often tedious business of gathering many threads of information and then trying to weave them into an understandable tapestry. Just as in those earlier years, the issues of who gets intelligence, when, and in what form remain the subject of bitter debates. Commanders today often feel that they do not get the intelligence in as timely a fashion as they need it to take decisive action. The timeline shown in the illustration makes the point that a defender is always at a disadvantage, because the steps necessary to

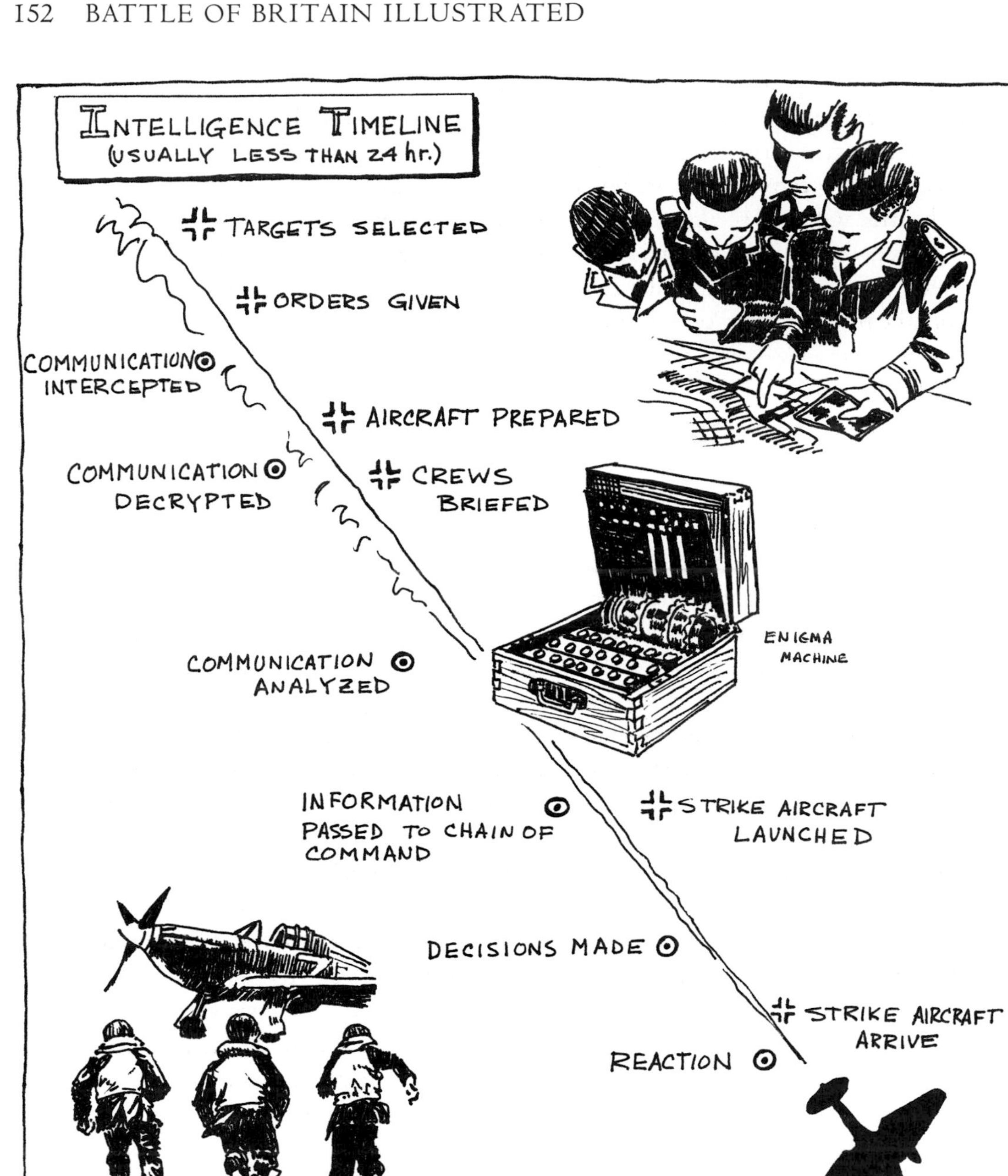

INTELLIGENCE INTERCEPT PROBLEM: TIMELINE IS TOO SHORT FOR DEFENDERS.

produce useful intelligence often result in the intelligence getting to those who can act on it with little time margin for action. Some argue that the system would be improved if raw intelligence data were given directly to operational commanders, and others argue just as strenuously that the commanders are not as well equipped as the intelligence agencies to analyze the data and draw accurate conclusions. There is every reason to expect that these debates will be with us indefinitely. Commanders will always want more information and want it more quickly; the agencies that collect and disseminate the information will always assert that the need for security and careful analysis argue against speed as the sole criterion for judging good intelligence.

34. Aircraft Modifications and Upgrades

The drawing indicates a few of the key modifications that both sides made to upgrade their premier fighters during the Battle of Britain. Although a complex fighter may spend years in development, it often spends many more years—even decades—in operational service, so the task of modifying existing designs continues for the life of the airplane. Once an airplane is designed, manufactured, and placed into service, the job of keeping it up-to-date and capable of performing its assigned missions begins. As new technologies emerge that offer improved performance or as enemy capabilities are changed and improved, the current design must be modified and upgraded to keep it competitive with its adversaries. In many ways, this task is even more difficult than the initial development of the airplane, because the engineer works in a more constrained environment where every change considered must be carefully evaluated to ensure that it improves the existing design.

In the relatively short period between the time when the Spitfire and Hurricane were put into operational service and the Battle of Britain began, the British rushed to make critical improvements that would give their fighters an edge in the war they knew would soon begin. Some of these have already been mentioned. The Merlin engine was modified to boost pressure inside the cylinders, an upgrade made possible through the use of 100-octane aviation gasoline. In June 1940 work began to fit the Spitfires and Hurricanes with de Havilland and Rotol constant-speed propellers, increasing both the altitude capability and the rate-of-climb performance of both. In addition, thick, bulletproof windscreens were added to the fighters and armor plate installed behind the pilot's seat, both of which improved the survivability of the airplanes and crews. It was also during this period that the British began to paint the undersides of their fighters the light blue favored by the Luftwaffe, which made them less visible from below. During the days before the more intense fighting began, British pilots began to insist on having their guns harmonized at 250 yards, rather than the 650 yards originally set.

As the battle raged, other modifications were made based on observations and experiences gained in combat, including changes to the engines and fuel systems. Different

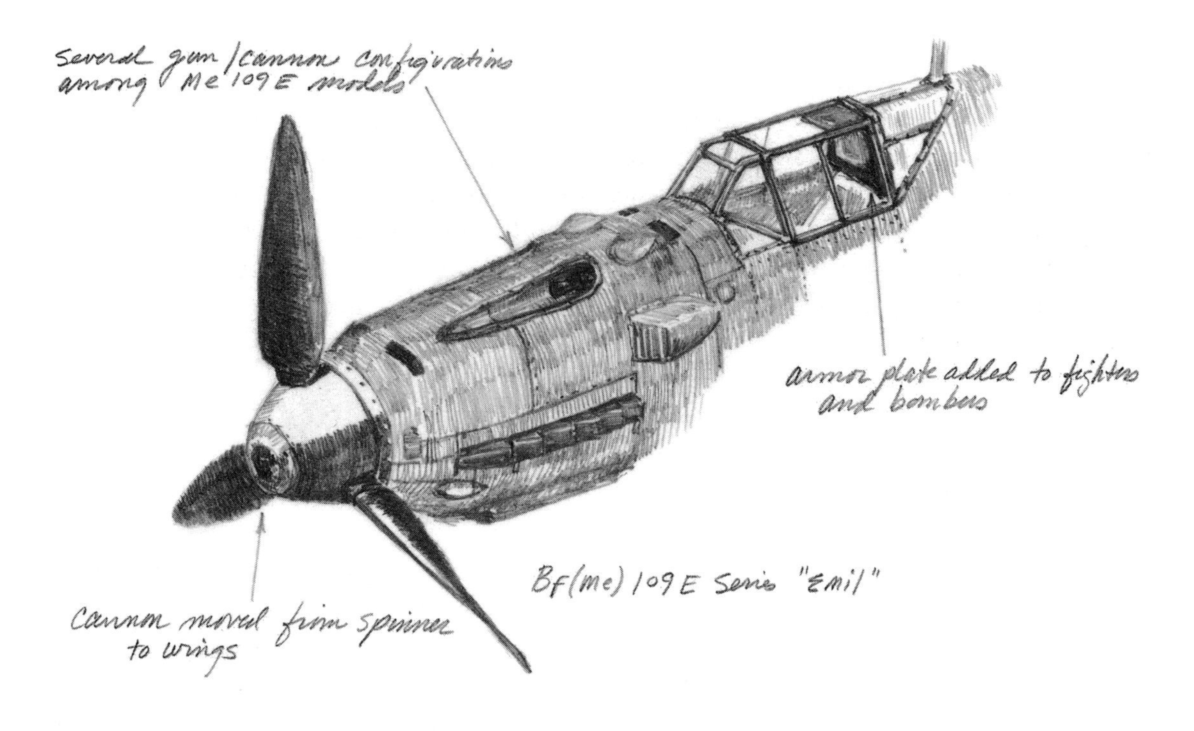

RAF and Luftwaffe aircraft upgrades to improve performance.

coolants were used, better priming systems and improved magnetos installed, spark plugs upgraded, and carburetors modified—to name some of the simpler modifications. A simple yet vital addition was the installation of rearview mirrors on the canopy bow of all fighters so that with a glance upward pilots could tell whether enemy fighters were tracking them. One upgrade that illustrates the difficulties of introducing new technology into existing designs was the addition of the 20-mm Hispano cannon to the Spitfire. The British recognized the superiority of the Oerlikon cannons carried by the Bf 109s compared to their machine guns, and by 1939 they had begun to mount the Hispano cannon in the Spitfire. In the spring of 1940 the cannon was judged sufficiently engineered to be used in operational fighters, and 19 Squadron was transitioned to the cannon-equipped Mk IB Spitfire. The cannons proved so unreliable that the squadron commander requested the squadron be reissued their machine gun–equipped Mk IA Spitfires. The modification that made the cannon the routine armament carried by the Spitfire was not fully completed until after the Battle of Britain was largely over.

The Germans also had an active program to upgrade and improve their combat aircraft. When the Bf 109E model was produced—the model that was most active in the Battle of Britain—the engine was upgraded from the Jumo 210 to the DB 601A. The cooling requirements for the new, more powerful engine were such that the entire cooling system had to be redesigned, including redistribution of the weight aft of the center of gravity to balance the increased weight of the new engine in the nose section. In addition, as noted in other sections of this book, Messerschmitt experimented with many different armament configurations on the Bf 109 in a constant search for the combination of cannons and guns that provided the most firepower consistent with the constraints imposed by the design, particularly the wings. As the British fighters continued to resist and effectively counter the fighters that escorted the bomber armadas, the German bombers, and even the fighters, began to carry additional armor plating to protect both the crews and the vulnerable points on the aircraft, particularly the engines. This armor was often positioned to counter attacks from the rear, and this in turn gave rise to the use of head-on frontal attacks by some British fighter squadrons.

Indeed, both sides in the Battle of Britain had an active policy of modifying and upgrading their systems to keep them effective and competitive in the air war. No air force, whether in 1940 or today, can survive long in the absence of this kind of activity. If there is a significant difference between then and now, it is that today the pace of technology is much more rapid than it was then. Even as a modern airplane is being designed and developed, the technology employed begins to age, often becoming obsolete before the system is actually fielded. In this dynamic environment, the challenge becomes not only to find the new technologies to substitute for those that have become obsolete, but also to design the system so that new technology can be incorporated easily and rapidly, thus enabling it to continue to be as effective as possible even as changes are made to the original designs.

35. Seaplane Technology

We often associate seaplanes with the Heinkel He 59 and the controversial air-sea rescue missions that have been the subject of considerable dispute and acrimony since 1940, but the fact is that Germany produced a group of excellent seaplanes that had capabilities spanning a number of operational missions. Shown here laying mines, the Heinkel He 115 was an excellent design, and the aircraft was often used to perform minelaying, reconnaissance, and bombing missions in British waters.

The Treaty of Versailles, among its many restrictions, denied Germany access to engine technology and forbade them from developing aeroengines. The result, naturally, was that the performance and reliability of German engines lagged behind the levels of the Allied nations well into the 1930s. As Germany began to rebuild its air capability, one avenue available to them was the development of air transport. As they exploited this opening, Lufthansa became one of the world's leading airlines by the early 1930s. However, as their interest in intercontinental air transport grew, the Germans found that the lack of reliability of their aircraft engines was a cause for concern. Out of this concern grew the idea that ocean crossings should, whenever possible, be made by aircraft that were capable of landing on the water in an emergency—seaplanes. It was a matter of controlling a recognized risk. Both Heinkel and Dornier made significant contributions to seaplane technology in the years leading up to the Battle of Britain. The seaplanes that German aircraft designers built were quite large and often slow—good transport aircraft, but at a distinct disadvantage in air combat. In spite of this, these seaplanes were central to the excellent air-sea rescue capability that Germany mounted during the Battle of Britain.

Perhaps the most interesting—and controversial—of the German seaplanes was the Heinkel He 59. Painted white and carrying the red cross recognized on both sides as the symbol for the noncombatant rescue and medical evacuation function, the airplane became the subject of heated debate between Germany and Britain during the Channel phase of the Battle of Britain. The He 59 was a large, biwing airplane with a wingspan of 78 feet, length over 57 feet, and standing over 23 feet high. It was powered by two BMW engines, but was capable of only 137 mph at sea level. It had a range of just over 1000 miles and carried its fuel inside the floats that extended below the airframe. During the period when they were concentrating their attacks on British shipping, the Germans began to use the He 59 to rescue downed airmen in the English Channel. The British, convinced that the Germans were using the He 59 to perform combat reconnaissance missions to locate British ship convoys, accused the Germans of taking advantage of the conventions that normally allowed Red Cross vehicles to operate in a combat area without being fired upon. They refused to observe the conventions where the He 59s were concerned and, on July 14, ordered British fighters to fire on them when they were observed near operational areas or British capital ships and territory. Of course, one might ask the question—where

an He 115 from KFG 906 deploying an LMB aerial mine.

else would one likely find downed airmen *except* near combat areas and near British shipping? Nevertheless, the British began to routinely fire on the airplanes. The Germans publicized the issue prominently, but subsequently repainted the aircraft using a camouflage pattern and armed it with machine guns. The He 59 was a fine seaplane and continued to be used through the war in various roles, including armed reconnaissance, minelaying, and bombing, either with conventional bombs or with torpedoes.

Heinkel produced another excellent seaplane for Germany: the He 115. This airplane was used in many roles throughout the war and was judged to be one of the finest seaplanes produced by any nation on either side during the war. The He 115 was a monoplane with the wing mounted at midfuselage. Like the He 59, it was a large vehicle, standing on two large floats. It was powered by two BMW nine-cylinder engines, which propelled it at speeds of over 200 mph with a maximum range of over 2000 miles. The He 115 was used in minelaying missions, especially in coastal waters, as well as in bombing and torpedo missions. It was capable of carrying torpedoes or mines in addition to two 550-pound bombs, or it could carry five 550-pound bombs. Armament consisted of two 7.9-mm machine guns.

Dornier also made substantial contributions to the development of seaplanes. The Do 24 flying boat was an excellent example, cited by some as the best air-sea rescue seaplane of all those used in World War II. It must have been a good airplane; it went into service in 1937 and continued to be used by various nations until it was finally retired in 1967. The Do 24 floated on the fuselage with floats extending from either side at the waterline. With an 88-foot wingspan, it was the last large operational seaplane in Europe. The Do 24 had three Bramo 323 nine-cylinder engines mounted in a high-wing configuration. It was capable of airspeeds in excess of 180 mph, with a range in the vicinity of 3000 miles. It carried a machine gun in a turret located in the nose and a 20-mm cannon located in a dorsal turret between the cockpit and tail.

In addition, Dornier built the Do 18, an armed maritime reconnaissance aircraft originally designed as a mail carrier. Similar in dimensions to the He 59, the Do 18 was a high-monowing design. It was powered by two Junkers Jumo 205 engines that were both mounted above the cockpit and on the centerline of the aircraft in a push-pull configuration. The aircraft floated on the bottom of the fuselage and on two winglike floats that extended from either side of the fuselage. Armament consisted of one machine gun and one 20-mm cannon. Maximum speed was 165 mph, and the range was just over 2000 miles. A machine gun turret was mounted on the top of the fuselage in 1939, but the airplane was never produced in significant numbers. A total of only about 100 of all types were eventually built.

Arado also produced a seaplane, the Ar 196. A true maritime aircraft, the Ar 196 was assigned to German warships as onboard reconnaissance systems. Relatively small, as befits an aircraft that occupies deck space on a ship, the airplane was only 40 feet long and 14 feet high. With a single nine-cylinder BMW engine, its appearance was somewhat similar to a Stuka with floats when viewed from the side. The Ar 196 was well armed, carrying two 7.9-mm machine guns and two 20-mm cannons in the wings. It could carry two small (110-pound) bombs on pylons under the wings. The maximum speed was over 190 mph; the range, 670 miles; and the service ceiling, 23,000 feet.

The British did not pursue the development of seaplane technology in any substantial sense during the 1930s and, consequently, did not have any of significance during the Battle of Britain. This is somewhat surprising, given that Britain had won the Schneider Trophy on the strength of Supermarine's seaplane designs, but the British strategy was to fight sufficiently close to the homeland that her pilots would normally be able to reach British soil even when their aircraft were damaged. Britain's resources and focus were entirely on producing the fighters that would be the key to the air defense. There is no question that Britain would have saved pilot's lives had they developed seaplanes, and there is no question that the shortage of pilots became one of Dowding's greatest concerns as the battle wore on. However, to have a capability in combat, one must begin the development well before the fighting begins or do without. In the case of seaplanes, Britain did without.

36. Photo Reconnaissance

The drawing shows two uses of photography in aerial combat. First is the classic use of film to perform the aerial reconnaissance mission, shown here by the Ju 86 performing a high-altitude aerial reconnaissance mission. The second is the use of gun camera film to record the view along the line of fire of fighter aircraft. Both have important roles in warfare; the first is critical to planning bombing campaigns, and the second is important to confirm pilot reports and to assess the probable effectiveness of fighter engagements.

Reconnaissance is a necessary and integral part of a bombing campaign. The planners—those who choose the targets and decide what kind of bombs to use, what type airplanes and how many shall be used to strike a target, and what routes shall be flown to approach and depart the target area—need reconnaissance to tell them what state the target is in, what weather conditions exist over the target area, and how it is defended. Following a strike, information is required about the effectiveness of the attack: Was the strike successful? To what extent? Are further strikes needed to complete the destruction of the target? Is the target being repaired so that strikes will have to be repeated at some later date? Without this kind of information, a bombing campaign is truly being conducted blindly and it will not be effective in the long run. Targets will be assumed destroyed when they are not; some targets will be repeatedly attacked even after they have essentially been destroyed; and aircraft and crews will be lost because they were uninformed about potentially hazardous defenses en route.

The Germans conducted an active air reconnaissance program as a part of their bombing campaign against Britain during 1940 and 1941. No day went by without

Confirmed enemy (not friendly) engaged

Confirmed shots fired

Two uses for aerial photography: photo recon and documenting attacks

GUN CAMERA FILM

confirmed target damaged or "probable"

confirmed kill (crew bails out)

Jacobs

flights by single bombers flying over the British coast to perform the dangerous air reconnaissance task. The Luftwaffe had seized on the idea of using bombers as reconnaissance aircraft during late 1939. Dornier had developed an export version of the Do 17, called the Do 215. Before these planes could be exported, they were seized and modified for the reconnaissance mission by mounting three cameras on each. Two or three *Staffeln* were equipped with these photo-reconnaissance bombers. In addition, the standard Do 17Z was used regularly to gather intelligence information. Recognizing the vital role that they played in planning the bombing campaign, the British often intercepted and shot down these reconnaissance aircraft. In fact, losses became so great after a couple of months into the Battle of Britain that the Luftwaffe began to consider means by which they could gather the needed information out of range of the guns of the British fighters.

The solution was to develop high-altitude photo-reconnaissance aircraft that would fly above the normal operating ranges of the Hurricanes and Spitfires. They modified the Junkers Ju 86, a twin-engine bomber that resembled the Do 17, with its long, thin fuselage and twin tail boom, fitting it with supercharged diesel engines that permitted it to operate above 30,000 feet. At these altitudes, they were out of reach of British antiaircraft artillery and fighter interceptors. Of course, to effectively conduct photo reconnaissance at altitudes above 30,000 feet demands optics that produce pictures with sufficient resolution to make them useful for targeting and damage assessment. However, even the German optics, generally recognized as among the world's best, were not up to the task. They simply could not provide the level of detail required to make the pictures useful for intelligence purposes.

Britain also experimented with photo reconnaissance in the period leading up to and during the Battle of Britain in order to gain information about German capabilities. When France fell, Britain lost access to continental bases for reconnaissance. They then equipped the Spitfire with cameras, and as early as 1939 these were used to gather information about German force dispositions—how many, what type, and where they were located. For this mission, the guns, ammunition, gunsight, and other armament-related equipment were removed from the aircraft and two aerial cameras installed. Carrying three times the normal fuel load gave the airplanes a range of over 1500 miles at altitudes above 30,000 feet. This concept of the lone fighter-reconnaissance aircraft flying "alone, unarmed, and unafraid" was a radical departure from the norm, which, until that time, depended primarily on the bomber for reconnaissance. Today, both are routinely used, as are specially designed aircraft such as the SR-71 and, of course, space-based satellites.

Gun cameras are important in confirming claims and reports made by pilots recently returned from fighter engagements. Gun cameras are trained and focused to record the results whenever the pilot pulls the trigger that activates his guns. Prior to the use of gun cameras, the results of aerial combat were generally recorded during intelligence debriefings after combat sorties. The delays involved and the lapses in memory caused by the emotion of the kill-or-be-killed atmosphere, compounded by the confusion normal to combat, where multiple airplanes may be shooting at the same target, often led to misinformation about the number of kills and who should be credited with them. The Battle of Britain is rife with reports of massive numbers of kills having been claimed by both sides that later had to be revised downward because of these factors. Gun cameras went far toward solving these problems. They could confirm that the target was an enemy aircraft and that shots were fired and damage occurred, and they could also be used to arbitrate claims regarding aircraft kills. The film records automatically when the guns are fired, and it is downloaded and developed when the aircraft returns from a mission. The camera doesn't forget and doesn't exaggerate—it just records what the pilot saw.

37. Airfield Defense

During the phase of the battle when the Royal Air Force and its airfields were the primary targets of the Luftwaffe attacks, the British undertook a number of innovations aimed at improving the survivability of their airfields, improving the efficiency and effectiveness of ground operations on the airfields, and, as a last resort, bringing the airfields back into action after they had been attacked.

While barrage balloons were mostly used in the defense of population centers, they were at times employed in the defense of key airfields and other assets. As already described, the function of the balloons was primarily to force bombers to higher altitudes, thus creating unplanned maneuvers while the bomber was aiming at the target and exposing the bomber to antiaircraft fire as well. The British also developed the parachute-and-cable (PAC) rocket system that could be quickly and cheaply deployed and used at any airfield. The PAC rocket would fire a parachute with trailing cable to altitudes of about 1000 feet. These were fired just as bombers approached the airfield and were intended to engage the aircraft in flight while it passed overhead. The illustration records the events of August 18, 1940, when PAC rockets downed two Dornier Do 17s during an attack on the airfield at Kenley.

Refueling, rearming, and servicing an aircraft in order to get it ready to fly between missions is referred to as "turning the airplane around." Quick turnaround reduces the exposure of the airplane and crew on the ground and enables the airplane to fly second, third, and even fourth missions within a single day. To improve turnaround time demands not only thoughtful aircraft design, but also requires that the ground-support equipment and procedures be designed with that purpose in mind. British ground crews developed various clever vehicles that could bring fuel, ammunition, oil, oxygen, and starter cartridges to the aircraft to reduce turnaround times. An experienced ground crew could turn a fighter around in as little as 15 minutes if no maintenance was required. As soon as an aircraft landed, an assessment was made of its condition. If the airplane was in generally

A Do17 is brought down by a "Parachute and Cable" system.

good condition, it would be serviced on the field. If engine damage was indicated, it was sent to a hangar where colocated civilian Rolls-Royce engine specialists would either immediately fix it or change the engine. In addition, Lord Beaverbrook developed a special factory facility to which pilots could fly damaged aircraft directly from combat missions for immediate attention and more complex repair operations.

The RAF also developed rapid repair capabilities for the airfields in the event that the defenses were breached. Crater-filling brigades were organized at most airfields, sometimes composed of civilians, but often made up of the enlisted men who performed the many support functions required at an operating fighter base. In a number of cases the same men who were responsible for maintaining and repairing the fighters were also detailed to repair the fields so that the fighters could operate. This double duty not only created extraordinarily long workdays, but also slowed both the repair of the aircraft and that of the airfields themselves. Churchill, during an inspection of the airfield at Manston, which was repeatedly bombed during the airfield phase of the program, recommended that large civilian companies be formed to assist with airfield repair. In his concept, these companies would be well equipped and would be highly mobile so that they could travel to multiple sites as needed. He even went so far as to suggest that airfield repair itself be camouflaged to give the impression that the fields were still out of action in order to deflect future attacks.

In war it is important to develop good weapons, and, as demonstrated during the Battle of Britain, it is at least as important to develop the command and control systems that enable those weapons to be concentrated and employed effectively. Sometimes lost in the enthusiasm to develop these more glamorous components of the nation's defense are the focus and attention needed to ensure that "the kingdom is not lost for the want of a nail," (i.e., that the seemingly smaller, less important aspects of the total system are not ignored or overlooked). Airfield operations and defenses are typical of these details that support and make possible the performance of the other parts of the total system. Britain demonstrated the total systems approach to their own defense that enabled them to prevail in that summer of 1940; their example is instructive to all.

38. Air-Sea Rescue

The state of their engine technology forced the Germans to include seaplanes in their inventory as a safety measure. Germany produced a number of very good seaplane designs during the 1930s, among them the Heinkel He 59 and the excellent Dornier seaplanes, Do 24 and Do 18. Consideration of the risks associated with overwater air travel naturally led Germany to develop many tools and techniques for rescuing survivors of air mishaps at sea. It should come as no surprise that Germany had, by 1940, developed an excellent air-sea rescue capability. They had the airplanes designed to do the job, and they

Seaplane technology applications: Rescue, minelaying, reconnaissance.

Rescue of pilots who could return to combat was fair game.

had already thought through many of the issues associated with developing a credible air-sea rescue service. The fact that the coming battle would be largely fought in the air and would occur over the Channel and the British homeland added urgency to the German need for air-sea rescue. British pilots might bail out and land on their home soil, ready to return to combat almost immediately. German crews, by contrast, would have to be picked up and brought back to the European continent; otherwise, they would die or be captured and, in either case, be lost to the German cause.

In 1939, as Germany prepared for the assault on Britain, the war games conducted by the General Staff indicated that an air-sea rescue capability would be a significant and necessary adjunct to the bomber and fighter forces that would carry the battle to Britain. This was a practical recognition of the consequences of attacking the British homeland and flying their fighters to the limits of their range and endurance. Air-sea rescue would be necessary, and the existence of such a capability would be an important boost to the morale of the aircrews engaged in the campaign. In August 1939, the first German air-sea rescue unit, equipped with Heinkel He 59s, became operational at Norderney, an island base in the North Sea.

One of the most difficult tasks in rescue is to find a downed pilot, especially in the absence of radio signals. A search-and-rescue pilot, looking out of the cockpit of an airplane, might easily see 50 square miles of ocean or more, depending on altitude. A pilot in the water in these conditions is like the proverbial needle in the haystack. The Germans did an excellent job of providing pilots with the means to increase the likelihood that rescuers would see them from the air. Pilots flew with inflatable vests and dinghies; they wore highly visible yellow skullcaps and carried dye capsules that would color the water around them. Pilots were also equipped with flare guns that could be used to attract the attention of rescue aircraft flying nearby. In addition, the Germans distributed floating buoys along the routes that the bombers took so that downed pilots could climb onto them and await rescue. They also equipped a number of seagoing rescue ships that patrolled the combat areas to pick up aircrew members as they parachuted or ditched in the Channel.

To complement these well-thought-out preparations, the Germans organized an air-sea rescue service staffed by both military and civilian pilots whose role was to fly into the combat area, locate downed crews, land on the water, and rescue them, as shown in the illustration. This was, by any definition of the word, an extremely dangerous mission. Once a rescue aircraft had landed in the water, it was often difficult to get the crew onboard, especially if they were wounded. The Dornier seaplanes that floated low in the water on the fuselage made this task easier, although sitting low in the water made it more difficult to see the downed crewmember if the waves were significant. Takeoff with a loaded airplane in the water was also extremely difficult, especially if the wind and waves were working against one. The German rescue aircraft were, by most accounts, sitting ducks for British fighter pilots, who were encouraged to treat them as combatants and shoot them down whenever they were found in the combat area, whether airborne or on the water. The bravery of the German air-sea rescue crews and the selfless manner in which they conducted their assigned missions was admirable.

In contrast to the well-organized German air-sea rescue operation, the British did not provide a similarly organized and equipped capability for their aircrews—at least at the time of the Battle of Britain. Instead, the British depended on a relatively few rescue launches, augmented by the random assistance provided by fishing vessels and the like when they happened to see downed crewmembers and were able to reach them. Perhaps the belief was that pilots would be able to reach British airspace before bailing out; perhaps it was simply an issue of priorities, and they had focused exclusively on fighter and bomber production to the exclusion of other configurations. Whatever the cause, the failure to provide air-sea rescue had its price. Approximately 220 British crewmembers died in the waters of the English Channel during the first 20 days of the Battle of Britain. Even in summer a person is unlikely to survive in those cold waters for more than a couple of hours. Furthermore, the British had none of the automatically inflatable vests, flare pistols, and sea dyes common on the German side. Although no data is available on the numbers of men who might have been saved had there been an organized air-sea rescue capability, there can be little doubt that Britain later regretted her failure to have provided for that before the fighting began. By September, when pilots were Dowding's scarcest resource, the need for this kind of service must have become glaringly obvious. Later in the war Britain developed a substantial air-sea rescue capability, but not before the Battle of Britain was over.

Modern fighter pilots are supported by air-sea rescue that is clearly designed on the German model. They carry automatically inflatable vests with automatic radio beacons that send signals to assist searchers in finding them. They have inflatable dinghies, flares, and food and water to sustain them. The world's air forces all have air vehicles that can perform the search-and-rescue role, although in most cases today these are helicopters rather than seaplanes. In addition, launches and ships designed specifically to handle any sea condition that might exist in a rescue situation are an integral part of the rescue organization. Today's armed forces worldwide recognize both the practical value and the moral imperative of taking whatever action is necessary to recover the crews who fight their wars.

39. Night Fighter Challenges

Today it is sometimes difficult to imagine the degree of difficulty an interceptor pilot must have experienced in finding enemy aircraft at night during the Battle of Britain. Modern pilots have the benefit of years of technology development on their side; the combination of highly evolved command and control systems in combination with excellent airborne radar and guided weapons simplify the problem—although it still remains far from simple. Imagine the situation that faced the RAF in 1940. German bombers

23 July 40. A Blenheim makes the first successful night intercept with airborne radar.

inbound to English targets would be scattered across the darkened skies. The bombers carried no running lights and the airplanes were painted the dark greens and blacks intended to make them as difficult to see as possible. This was compounded by the presence of clouds and rain that periodically swept across England, particularly during the later stages of the battle, when night operations increasingly became the norm. To the average fighter pilot, it must have seemed sheer happenstance to see one of the dark bombers against the dark sky—and luckier still to shoot one of them down.

To remedy the situation, Britain began to develop airborne radar, building on the knowledge and experience gained through development of the Chain Home system. A perennial problem with bringing ground-based technologies into the aerial environment is that, even though the scientific principles remain the same, problems are invariably encountered as designers attempt to deal with issues such as size and weight limitations and electrical interference with other onboard systems (today referred to as *electromagnetic interference,* or EMI). These issues had all complicated and slowed the integration of radar into airborne platforms in 1940. Furthermore, the problem was not made easier by the fact that Britain, in a number of cases, assigned to the night intercept mission those aircraft that had demonstrated an inability to compete with top-line German fighters in the daytime. The Bristol Blenheim and the Boulton-Paul Defiant aircraft were both examples of this policy. The result was that Britain's night fighters often lacked agility, were relatively slow, or both.

The task of bringing the airborne radar through the last stages of development and up to operational status fell to the Bristol Blenheims assigned to the Fighter Interceptor Unit at Tangmere, a primary RAF fighter base. On the night of July 23, the Chain Home system detected Do 17s inbound toward London. The course and altitude information was passed to Tangmere, and the controller then carefully steered a Blenheim to an intercept position. Radar contact was established, after which the pilot maneuvered until the Do 17 was silhouetted against a bright moon. The pilot then aimed and fired, destroying the Dornier and simultaneously logging the first aerial kill in history accomplished using airborne radar. The drawing illustrates this event, showing the Blenheim in its night fighter role. Interceptors guided by airborne radar downed a total of six German aircraft during the Battle of Britain. The interceptors were, in general, too slow to catch their targets unless they were fortuitously positioned, and the technology was, at the time, too erratic for dependable operation. Some urged Dowding to increase the use of day fighters in the night intercept role, but Dowding persisted in the belief that improved night interceptor designs and further work on the airborne intercept radar were the keys to the solution of the problem. Later, as the Bristol Beaufighter came on-line and technical advances were made in airborne radar technology, results began to improve substantially. This technology would come to be regarded as indispensable and a key enabler of intercept operations that today can be effectively conducted at all times of day and in all kinds of weather.

CHAPTER 3

THE TACTICAL TIMELINE

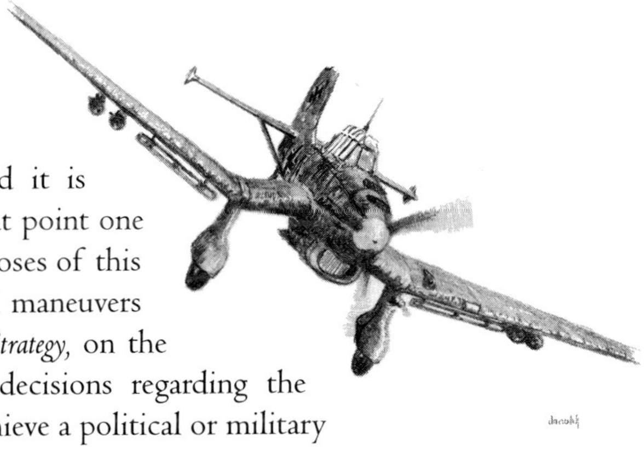

TIMELINE OVERVIEW

Tactics and strategy often overlap, and it is nearly impossible to say precisely at what point one stops and the other begins. For the purposes of this book, *tactics* include the formations and maneuvers used to engage the enemy in combat. *Strategy,* on the other hand, involves the higher-level decisions regarding the employment of national resources to achieve a political or military objective.

As in the case of technology, the evolution of tactics and strategy also displays unique characteristics. For example, specific tactics are often attributed to a single, perceptive innovator. Furthermore, key tactical developments often endure for long periods of time; tactical air forces today still employ the "finger four" formation that was developed by the German pilot, Werner Moelders, during the Spanish civil war. The development of new technology is, to the contrary, a progressive process that constantly seeks to replace current technology with new technology. Another pattern associated with tactical development is that it occurs most rapidly and significantly during periods of combat, as those involved in the life-or-death contest of man versus man or machine versus machine look for ways to gain advantage or ensure their own survival. Technology development, on the other hand, is usually slower to develop and is likely to evolve most significantly during periods of preparation for war.

Let us examine the evolution of strategy and tactics as they progressed from the later years of World War I to the Battle of Britain. As we do, look for the patterns of development and for the relationships that exist between tactics and the other views—historical and technological—that we have already examined. We will begin with one of the earliest individual contributors, Oswald Boelcke, who is generally credited with developing the first set of formal guidelines for pilots engaged in air-to-air combat.

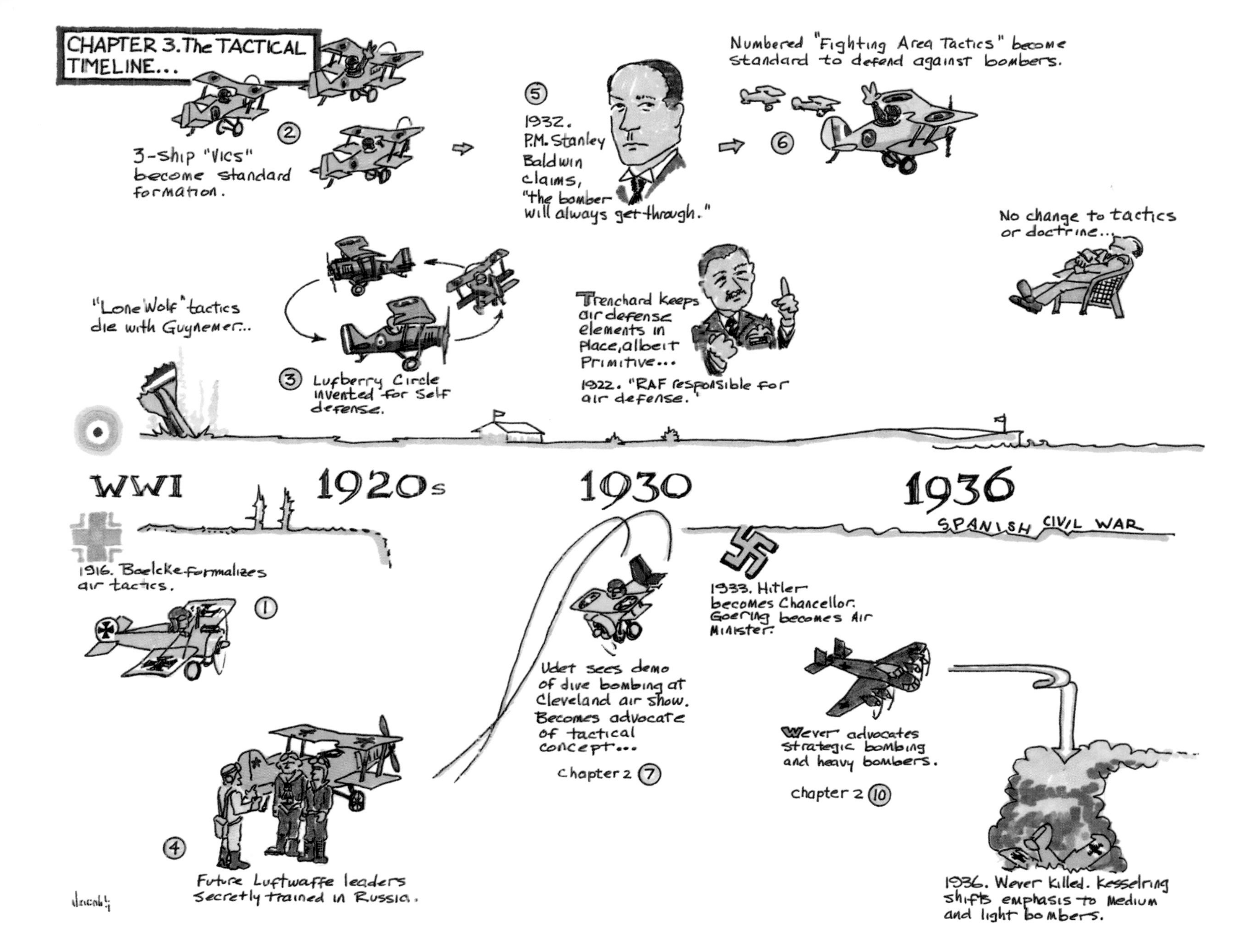

CHAPTER 3. The TACTICAL TIMELINE...
3-ship "Vics" become standard formation.
②
1932. P.M. Stanley Baldwin claims, "the bomber will always get through."
⑤
Numbered "Fighting Area Tactics" become standard to defend against bombers.
⑥
No change to tactics or doctrine...
"Lone Wolf" tactics die with Guynemer...
③ Lufberry Circle invented for self defense.
Trenchard keeps air defense elements in place, albeit primitive...
1922. "RAF responsible for air defense."
WWI
1920s
1930
1936
SPANISH CIVIL WAR
1916. Boelcke formalizes air tactics.
①
Udet sees demo of dive bombing at Cleveland air show. Becomes advocate of tactical concept...
chapter 2 ⑦
1933. Hitler becomes Chancellor. Goering becomes Air Minister.
Wever advocates strategic bombing and heavy bombers.
chapter 2 ⑩
④
Future Luftwaffe leaders secretly trained in Russia.
1936. Wever killed. Kesselring shifts emphasis to medium and light bombers.

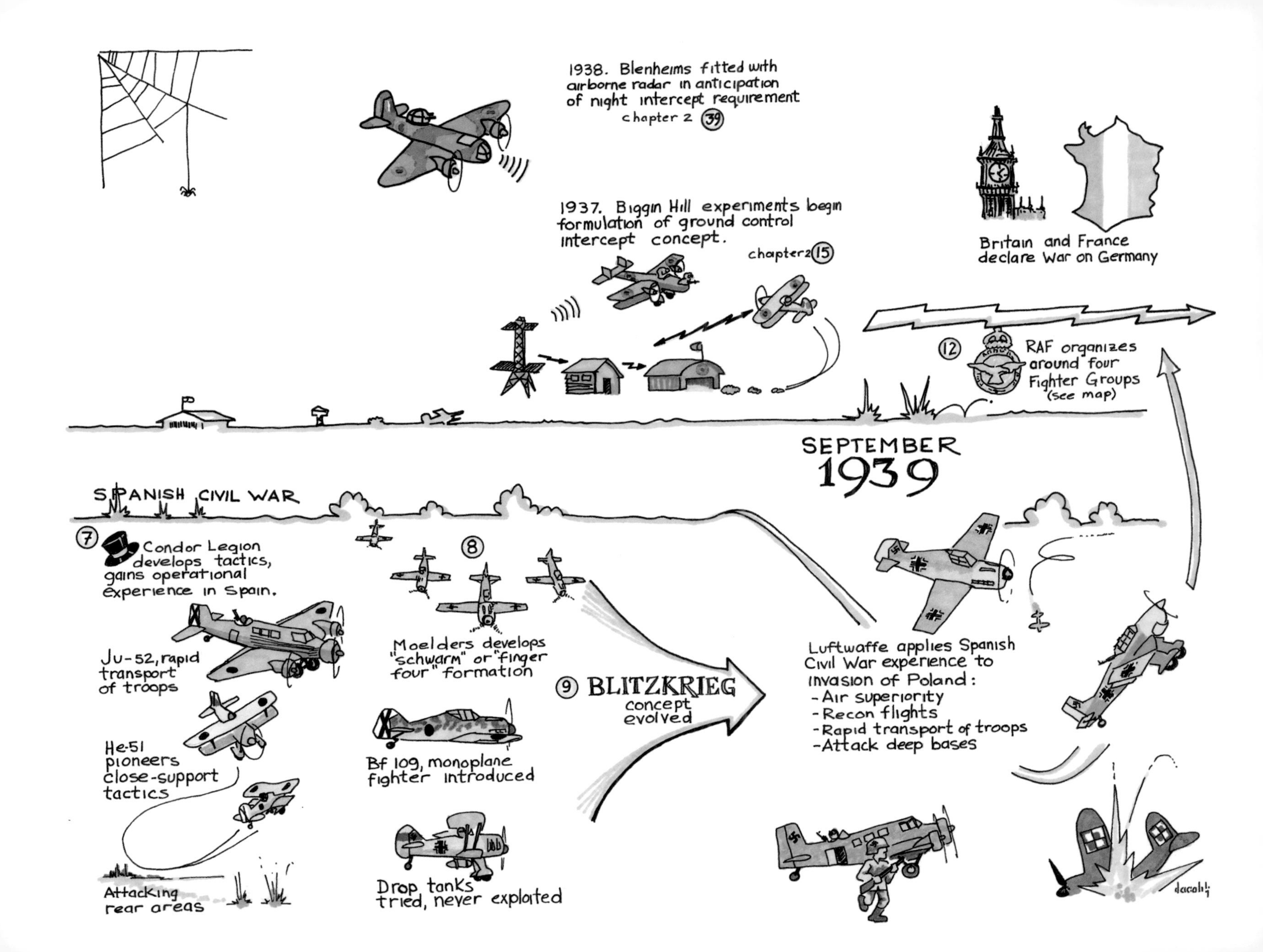
1938. Blenheims fitted with airborne radar in anticipation of night intercept requirement
chapter 2 (39)
1937. Biggin Hill experiments begin formulation of ground control intercept concept.
chapter2 (15)
Britain and France declare War on Germany
(12) RAF organizes around four Fighter Groups (see map)
SEPTEMBER
1939
SPANISH CIVIL WAR
(7) Condor Legion develops tactics, gains operational experience in Spain.
Ju-52, rapid transport of troops
He-51 pioneers close-support tactics
Attacking rear areas
(8) Moelders develops "schwarm" or "finger four" formation
Bf 109, monoplane fighter introduced
Drop tanks tried, never exploited
(9) BLITZKRIEG
concept evolved
Luftwaffe applies Spanish Civil War experience to invasion of Poland:
-Air superiority
-Recon flights
-Rapid transport of troops
-Attack deep bases
dacoli

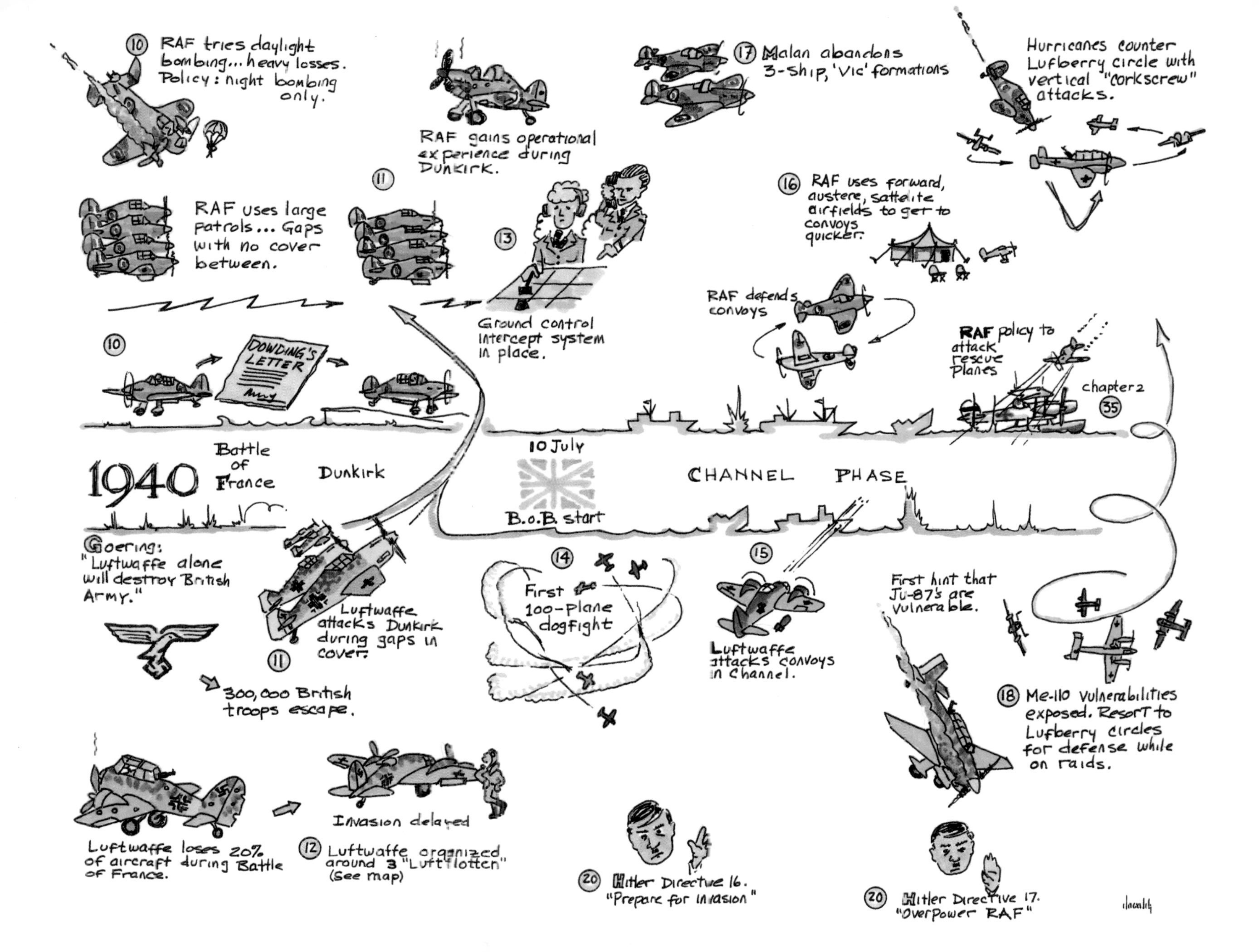

10 RAF tries daylight bombing... heavy losses. Policy: night bombing only.
RAF gains operational experience during Dunkirk.
17 Malan abandons 3-ship, 'Vic' formations
Hurricanes counter Lufberry circle with vertical "corkscrew" attacks.
RAF uses large patrols... Gaps with no cover between.
11
13
16 RAF uses forward, austere, sattelite airfields to get to convoys quicker.
Ground control intercept system in place.
RAF defends convoys
10
DOWDING'S LETTER
RAF policy to attack rescue planes
chapter 2
35
1940
Battle of France
Dunkirk
10 July
B.o.B. start
CHANNEL PHASE
Goering: "Luftwaffe alone will destroy British Army."
Luftwaffe attacks Dunkirk during gaps in cover.
11
14
First 100-plane dogfight
15
Luftwaffe attacks convoys in Channel.
First hint that JU-87's are vulnerable.
300,000 British troops escape.
18 Me-110 vulnerabilities exposed. Resort to Lufberry circles for defense while on raids.
Invasion delayed
Luftwaffe loses 20% of aircraft during Battle of France.
12 Luftwaffe organized around 3 "Luftflotten" (See map)
20 Hitler Directive 16. "Prepare for invasion"
20 Hitler Directive 17. "Overpower RAF"

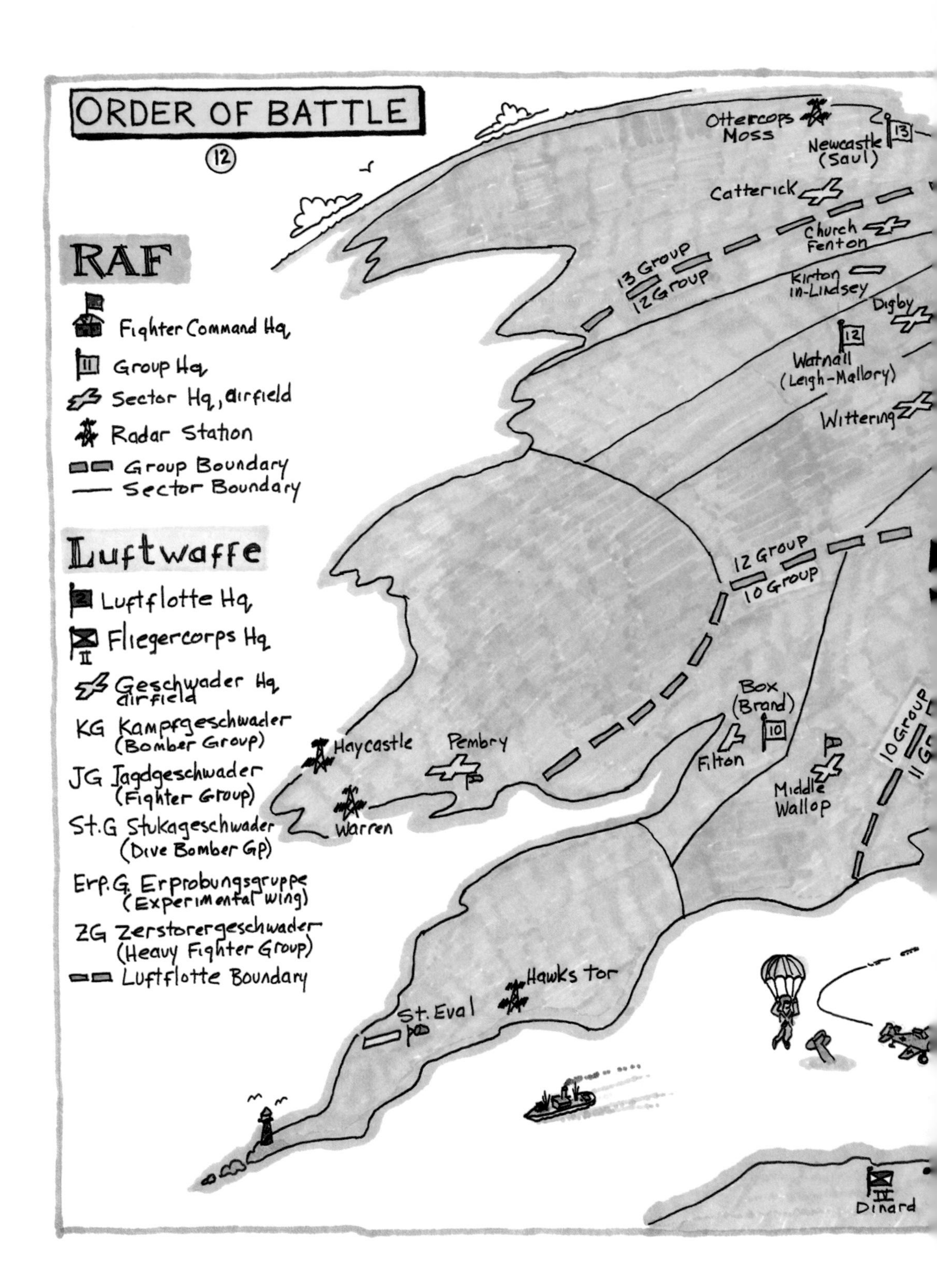

ORDER OF BATTLE
12
RAF
Fighter Command Hq.
Group Hq.
Sector Hq., airfield
Radar Station
Group Boundary
Sector Boundary
Luftwaffe
Luftflotte Hq.
Fliegercorps Hq
Geschwader Hq. airfield
KG Kampfgeschwader (Bomber Group)
JG Jagdgeschwader (Fighter Group)
St.G Stukageschwader (Dive Bomber Gp)
Erp.G. Erprobungsgruppe (Experimental wing)
ZG Zerstorergeschwader (Heavy Fighter Group)
Luftflotte Boundary
Ottercops Moss
Newcastle (Saul)
13
Catterick
Church Fenton
13 Group
12 Group
Kirton in-Lindsey
Digby
12
Watnall (Leigh-Mallory)
Wittering
12 Group
10 Group
Box (Brand)
10
Filton
Middle Wallop
10 Group
Haycastle
Pembry
Warren
Hawks tor
St. Eval
Dinard

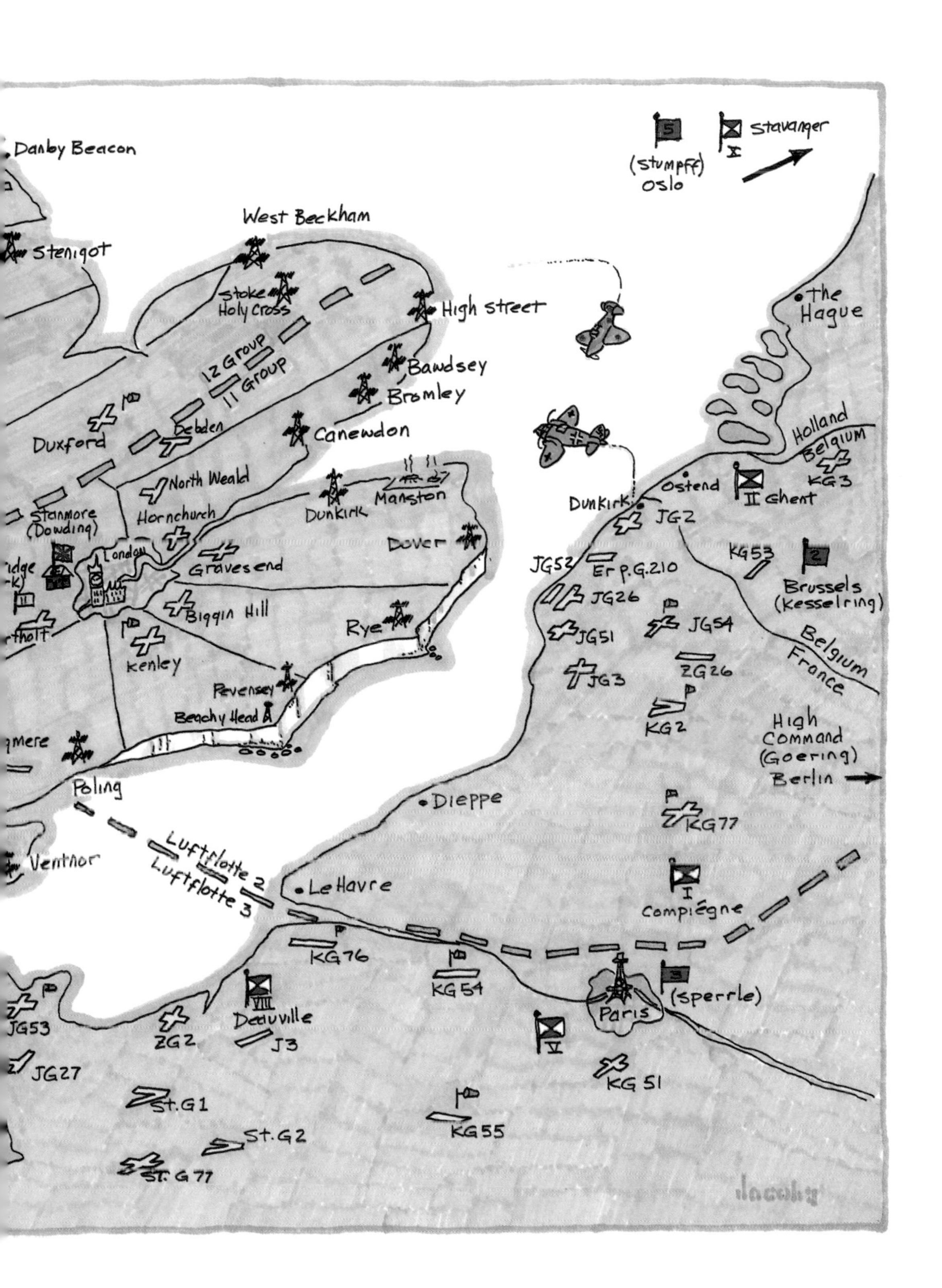

Danby Beacon
Stavanger
(Stumpff)
Oslo
West Beckham
Stenigot
Stoke Holy Cross
High street
The Hague
12 Group
11 Group
Bawdsey
Bromley
Duxford
Debden
Canewdon
Holland
Belgium
KG3
North Weald
Manston
Ostend
Ghent
Dunkirk
Stanmore (Dowding)
Hornchurch
Dunkirk
JG2
Dover
London
Gravesend
JG52
Er p.G.210
KG53
Brussels (Kesselring)
JG26
Biggin Hill
Rye
JG51
JG54
Belgium
France
Kenley
ZG26
JG3
Pevensey
Beachy Head
KG2
High Command (Goering) Berlin
Poling
Dieppe
KG77
Ventnor
Luftflotte 2
Luftflotte 3
Le Havre
Compiégne
KG76
KG54
(Sperrle)
Deauville
Paris
JG53
ZG2
J3
JG27
KG 51
St.G1
St.G2
KG55
St. G 77

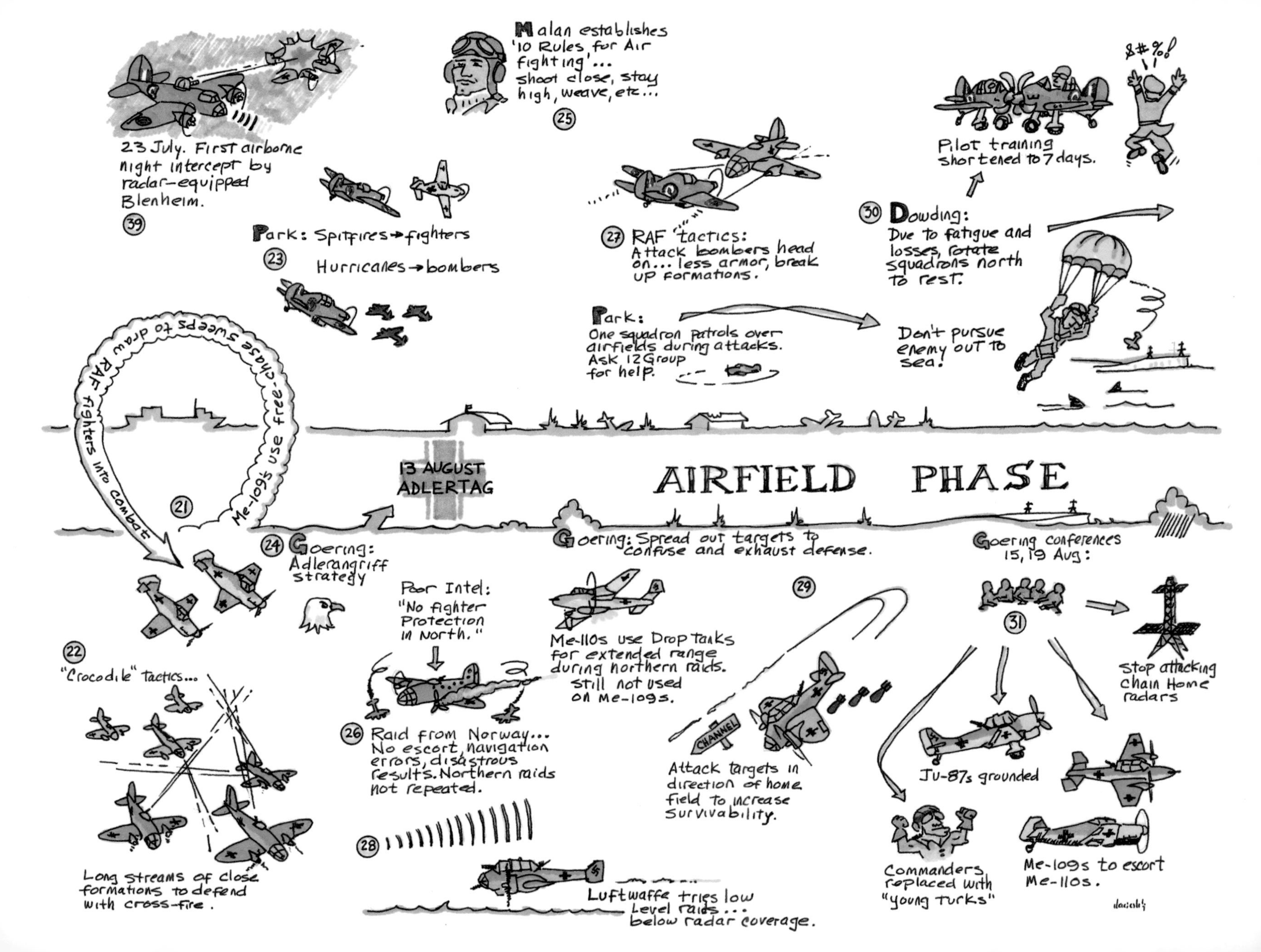
AIRFIELD PHASE
13 AUGUST
ADLERTAG
23 July. First airborne night intercept by radar-equipped Blenheim.
39
Malan establishes '10 Rules for Air fighting'... shoot close, stay high, weave, etc...
25
Park: Spitfires → fighters
Hurricanes → bombers
23
27 RAF "tactics: Attack bombers head on... less armor, break up formations.
Pilot training shortened to 7 days.
&#%!
30 Dowding: Due to fatigue and losses, rotate squadrons north to rest.
Park: One squadron patrols over airfields during attacks. Ask 12 Group for help.
Don't pursue enemy out to sea!
21
Me-109s use free-chase sweeps to draw RAF fighters into combat
24 Goering: Adlerangriff strategy
Goering: Spread out targets to confuse and exhaust defense.
Goering conferences 15, 19 Aug:
31
Poor Intel: "No fighter Protection in North."
Me-110s use Drop tanks for extended range during northern raids. Still not used on Me-109s.
29
Stop attacking Chain Home radars
22
"Crocodile" tactics...
Long streams of close formations to defend with cross-fire.
26 Raid from Norway... No escort, navigation errors, disastrous results. Northern raids not repeated.
CHANNEL
Attack targets in direction of home field to increase survivability.
JU-87s grounded
28
Luftwaffe tries low level raids... below radar coverage.
Commanders replaced with "young turks"
Me-109s to escort Me-110s.

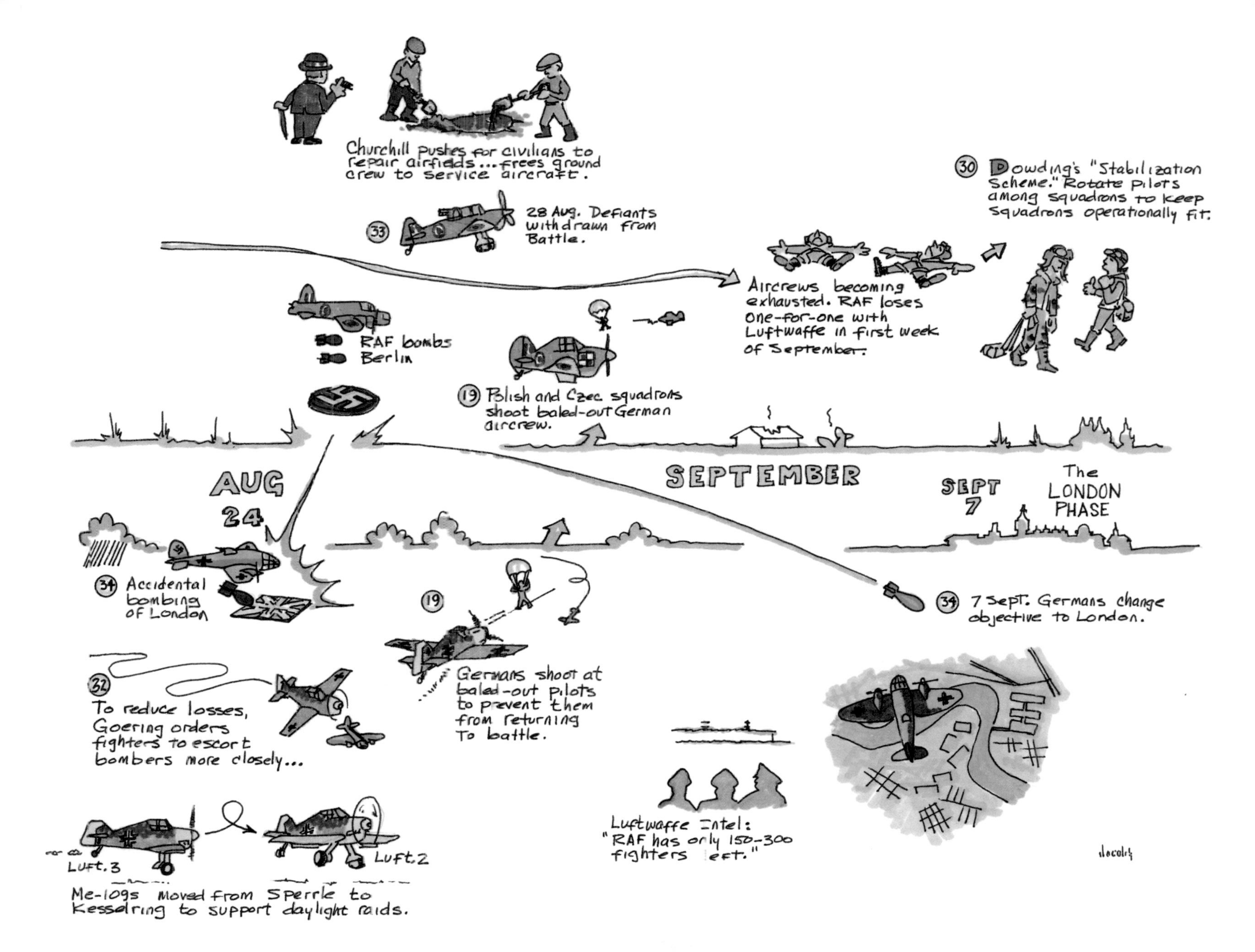

Churchill pushes for civilians to repair airfields...frees ground crew to service aircraft.
(33) 28 Aug. Defiants withdrawn from Battle.
(30) Dowding's "Stabilization Scheme." Rotate pilots among squadrons to keep squadrons operationally fit.
Aircrews becoming exhausted. RAF loses one-for-one with Luftwaffe in first week of September.
RAF bombs Berlin
(19) Polish and Czec squadrons shoot baled-out German aircrew.
AUG 24
SEPTEMBER
SEPT 7
The LONDON PHASE
(34) Accidental bombing of London
(19) Germans shoot at baled-out pilots to prevent them from returning to battle.
(34) 7 Sept. Germans change objective to London.
(32) To reduce losses, Goering orders fighters to escort bombers more closely...
Luft. 3
Luft. 2
Me-109s moved from Sperrle to Kesselring to support daylight raids.
Luftwaffe Intel: "RAF has only 150-300 fighters left."

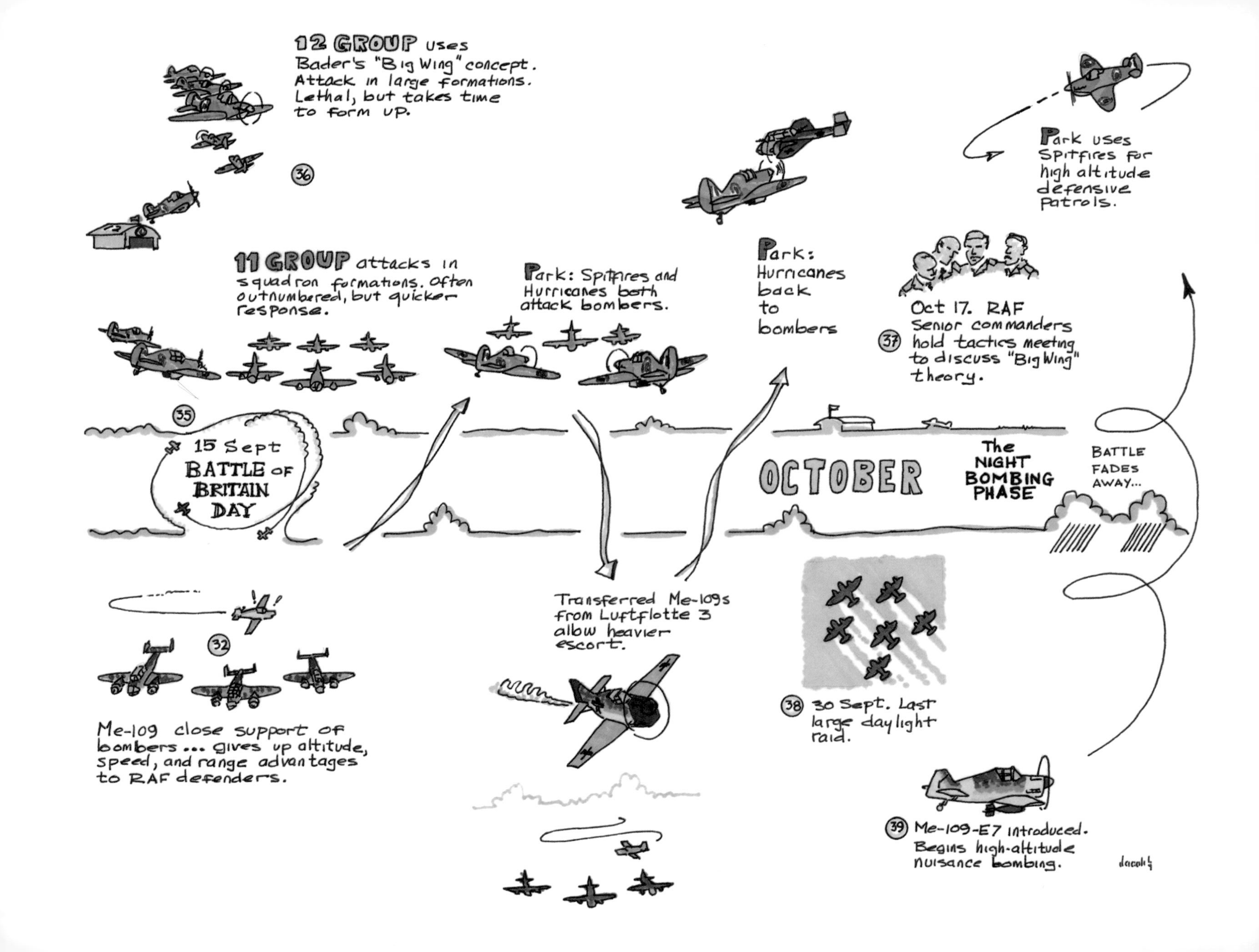

12 GROUP uses Bader's "Big Wing" concept. Attack in large formations. Lethal, but takes time to form up.
36
11 GROUP attacks in squadron formations. Often outnumbered, but quicker response.
Park: Spitfires and Hurricanes both attack bombers.
Park: Hurricanes back to bombers
Park uses Spitfires for high altitude defensive patrols.
Oct 17. RAF Senior commanders hold tactics meeting to discuss "Big Wing" theory.
37
35
15 Sept BATTLE OF BRITAIN DAY
OCTOBER
The NIGHT BOMBING PHASE
BATTLE FADES AWAY...
Transferred Me-109s from Luftflotte 3 allow heavier escort.
32
Me-109 close support of bombers... gives up altitude, speed, and range advantages to RAF defenders.
38
30 Sept. Last large daylight raid.
39
Me-109-E7 introduced. Begins high-altitude nuisance bombing.

A self-described loner, Oswald Boelcke, was recognized early in World War I as one of Germany's master tacticians. He won the Iron Cross when he was 23 years old and later won the *Orden Pour le Mérite*, Prussia's highest award for bravery, with a squadron mate, Max Immelmann, in 1916. Boelcke was renowned for his exacting and precise approach to flying. In an attempt to provide his pilots with a set of guidelines to help them in aerial combat, he enunciated what he termed "Dicta Boelcke"—Boelcke's rules for air combat. This was the first recognized effort to formalize and standardize air combat tactics. His rules were simple, and they captured the key attributes of successful air-to-air engagements—seizing on the advantage of surprise, aggressive offensive maneuvers, and group cooperation. Boelcke's rules were excellent guidance in World War I, and for the most part they remain excellent advice for the fighter pilot of today. This is an outstanding example of tactics attributed to a single individual that have endured to the present.

In spite of his own rules, Boelcke and other outstanding fighter pilots of World War I tended to favor solo hunting. The favored tactics of the era featured lone pilots flying out to look for targets of opportunity. One of the legendary engagements between fighter pilots was between Ernst Udet, of Germany, who would eventually log 62 air combat kills, and Georges Guynemer, of France. Well-known for his flying ability as well as for his aircraft, which he named *Vieux Charles*, Guynemer had participated in over 600 air-to-air engagements; he had survived being shot down seven times; and he had 53 kills at the time of his death. He and Udet circled and maneuvered for over 8 minutes—a lifetime at the edge of destruction—when Udet's guns jammed. Udet later reported that as soon as Guynemer realized his opponent was helpless, he performed an inverted pass, waved, and departed. Guynemer was last seen attacking another aircraft near Ypres, Belgium, in September 1917; he never returned. The death of Guynemer marked the end of the lone-wolf hunter era. After this, fighter tactics would adhere more closely to the Dicta Boelcke. Pilots would fly and attack in groups—formations—that would provide mutual support and covering fire while engaged in the dangerous business of aerial combat.

Groups of airplanes not only bring more firepower to an engagement, they offer mutual support against surprise attack. By the last years of World War I, the air forces of the combatant nations had all moved away from the solo tactics of the earlier years and toward use of groups of airplanes in the air. The British, like other nations, wrestled with the problem of how best to organize groups of aircraft in flight. In the final analysis, the combination of their own organizational structure and the need to use hand signals to communicate between airplanes in a formation led them to adopt the tactical formation called the *vic*. A vic was a formation of three aircraft flying in tight, wingtip formation. It featured a flight lead with a wingman on either wing. The three-aircraft vic became the standard formation for the RAF as it entered the 1920s.

With the advent of the multiple-airplane formation as the preferred tactical unit, pilots began to experiment with group tactics to overcome some of the weaknesses inherent in solo tactics. Raoul Lufbery, who held both French and American citizenship, volunteered and flew for the Lafayette Escadrille, a fighter unit of Americans flying for France in World War I. Known as an excellent teacher and tactician, Lufbery became an ace, logging five enemy kills after only three months of flying with the Escadrille. He is primarily remembered for the tactical maneuver attributed to him, the *Lufbery circle.* This formation involves several aircraft flying together in a circle so that each aircraft guards the rear quadrant of the airplane in front. Although the maneuver has inherent weaknesses as an air-to-air combat tactic, it is significant as one of the first efforts to recognize the need for mutual fire support among fighters and to adapt air combat maneuvers to address that need. The Lufbery circle continued to be used into World War II, and one could argue that the circling maneuvers used today to position tactical aircraft during dive-bombing missions have roots that date back to this World War I maneuver.

The Treaty of Versailles notwithstanding, General Hans von Seeckt, as chief of the German army command and head of the *Reichswehr,* set about the task of rebuilding a credible German air combat capability in the early 1920s. Because the treaty clearly prohibited such activity, von Seeckt proceeded to build a clandestine air force. Agreements were reached with Russia giving the Germans the opportunity to establish a training operation for military flying combined with an aircraft test facility at Lipetsk, a base located 220 miles southeast of Moscow. Between 1924 and 1932 the Germans trained over 200 pilots and observers at Lipetsk. This was also the training ground for those who would lead the Luftwaffe into the Battle of Britain and World War II. New aircraft were tested at this site and tactics and skills were honed that would enable a highly professional Luftwaffe to suddenly emerge into the daylight in the mid-1930s, seemingly from nowhere.

In Great Britain, General Sir Hugh Trenchard was the chief of staff of the Royal Air Force during the 1920s. A concerted move was afoot, particularly by the Admiralty, to have the young service absorbed back into the army and the navy. The RAF was fighting for its very existence. Trenchard's contributions were crucial in the context of the Battle of Britain—which at this point was still years in the future. Under Trenchard's stewardship, the continued existence of the RAF as a separate service was endorsed by the Salisbury Committee in June of 1923. Then, in a further tribute to the way Trenchard had managed the fledgling service (successfully handling missions assigned while minimizing costs), Prime Minister Stanley Baldwin ordered the RAF to take a primary role in the Home Air Defense Force. He further ordered that the air force was to be increased from 18 squadrons to 52. These two things—affirming the RAF as a separate service on equal footing with the army and the navy and then assigning the RAF responsibility for home defense—were, in retrospect, critical to the British victory in the Battle of Britain.

It often seems that branches of the armed forces spend an inordinate amount of time and energy battling over organizational details and the roles and missions assigned to each. It is a fair question to ask whether this is necessary. The differences between the way the Germans and the British organized in the years leading up to World War II is instructive in answering this question. In Britain, a significant national debate took place over the issue of whether the RAF should survive as an independent military force. The RAF had been formed from assets that had previously belonged to the army and the navy; the RAF had assumed responsibility for missions that had been assigned to those forces. After the war, the Admiralty, in particular, argued for absorption of the air force back into the senior services. The decision by the government to maintain an independent air force led to the development of a centrally controlled defensive force that would develop the integrated capability to effectively oppose the Luftwaffe. Had the RAF not survived as a separate service, it would almost certainly have been decentralized and required to protect and defend the army and navy units to which the air units would have been attached. Germany, to the contrary, organized their air forces as an adjunct to the army. The idea that the air force existed primarily to support ground operations significantly affected the development of German air power. For example, an air force whose mission was first and foremost to support ground operations would tend to favor many tactical aircraft that could be dispersed and controlled by the various ground commanders in preference to concentration of development in fewer strategic bombers. Such an air force would not be overly concerned with the development of long range fighters, since the assumption would be that these aircraft would generally be supporting short-range operations normal to ground-support missions. Furthermore, this type of air force would be very interested in improved tactical bombing accuracy since, in a ground-support role, bombing must be precise to be effective without damaging one's own ground troops.

Ernst Udet, a World War I ace who had continued to fly as a barnstormer and racer in events all over the world after the war, was invited to fly in the U.S. National Air Races in Cleveland in 1931, where he became aware of the capabilities of the Curtiss Gulfhawk when he observed a dive-bombing demonstration. Until that time, German bombing—even tactical bombing—was typically accomplished from level flight, and it was often inaccurate to the point of representing a danger as much to German troops as to the enemy targets. Udet was therefore much impressed with the accuracies demonstrated by the dive-bombers. As a former combat pilot, he not only had the ability to understand the tactical significance of what he saw, but he was also a man with friends in high places. Supported by his friend and fellow combatant from World War I, Hermann Goering, Udet returned to the United States in 1933 to fly the dive-bomber. He purchased two Curtiss Hawk II aircraft, which were returned to Germany and demonstrated to the German air staff. This demonstration eventually led to the development of the Stuka dive-bomber, the aircraft that most symbolized the blitzkrieg style of warfare that Germany unleashed on Europe in the late 1930s.

Implicit in the theories of air warfare enunciated by Giulio Douhet in the early 1920s was the assumption that no defense existed against the bomber, a widely accepted assumption on both sides of the English Channel. In 1932 Stanley Baldwin, the prime minister of England, while engaged in a debate before the House of Commons on the subject of disarmament, made the following statement, "I think it is well also for the man in the street to realize that there is no power on earth that can protect him from being bombed. Whatever people tell him, the bomber will always get through. The only defense is offense. . . ." This strong belief in the invincibility of the bomber, which was common among both politicians and military professionals, had a strong influence on British fighter tactics.

By 1925, with the RAF established as an independent military service, a new command was created to handle air defense. An area facing to the south of England was designated as Fighting Area, the geographic area where the RAF expected to engage and fight an attacking air force. The development of aerial combat tactics within Fighting Area evolved from the observations and practices of World War I in combination with the prevailing thinking of the day that bombers would be unescorted in combat. The British had adopted the tight, three-aircraft vee, or vic, as the fundamental aerial formation employed. The fighters developed a series of carefully orchestrated maneuvers that would provide each aircraft with an opportunity to fire on the target—presumed to be the unescorted bomber. These close-formation maneuvers were designated "fighting area tactics." They were numbered, standardized, and mandated for all units to use. Even though they were to prove inferior to the German aerial formations and their use cost the lives of numerous British pilots, they nevertheless remained the standard for British fighter engagements until after the Battle of Britain had begun.

The first chief of staff of the Luftwaffe was General Walther Wever. Unlike Goering and his fellow generals (with the possible exception of Milch), Wever was a man with a strategic vision of air warfare. He saw air power as a centrally controlled force which might strike anywhere in Europe from the homeland rather than as a resource to be commanded by army generals. General Wever initiated the development of a four-engine bomber. Unfortunately for Germany, Wever was killed in an aircraft accident in 1936, leaving the disposition of the bomber program in the hands of Kesselring, who was not a supporter of the strategic employment of air power. Rather, he supported the model that favored two-engine bombers to support shorter-range tactical operations needed by army commanders. The four-engine bomber program was soon abandoned, a decision that would later prove fortuitous for the British.

The Spanish civil war began in 1936. Germany, and especially the Luftwaffe, was quick to take advantage of the opportunity to test the machines and tactics that they had been developing in secret since the early 1920s. The *Legion Kondor* (Condor Legion), a German unit composed of both air and ground forces, was sent under cover to fight on the side of Franco. During this conflict, many of the techniques and tactics later employed in the Battle of Britain—the rapid movement of troops by air transport, the concepts of rear interdiction and aerial reconnaissance, the organization of ground support activities—were developed or refined. New formations for organizing large numbers of bombers and tactical combat formations and maneuvers for fighters were developed. New aircraft were introduced and tested under combat conditions while the doctrine that would govern their employment was formulated. The operational experience gained in the Spanish civil war served the Luftwaffe extremely well; by the time the Battle of Britain started, it is generally agreed that the Luftwaffe was the best organized and most experienced air force in the world.

As might be expected, the lessons learned by the Luftwaffe were not always the right ones. Some of the observations and conclusions led to missed opportunities that had significant impacts on the outcome in the Battle of Britain. The failure of the Germans to exploit their development of external fuel tanks was mentioned in an earlier section of this book. These tanks were developed and tried on the He 51 fighters in Spain, but the Bf 109, which was severely range-limited, was never fitted for drop tanks during the battle of Britain. German bombers had little difficulty outrunning enemy interceptor aircraft during the Spanish civil war, which led to an erroneous conclusion that they would be able to likewise outrun British fighters. They did not reckon on the new generation of fighter aircraft, even though they had developed and were flying one themselves—the Bf 109. The Germans also found that they could operate in Spain without highly sophisticated communications, so the development of air-to-ground and air-to-air communications did not receive priority attention as Germany prepared for war with Britain. While the benefits that accrued to the Luftwaffe from the Spanish civil war experience were substantial, the cost and miscalculations that arose from that experience were also significant.

One of the most significant tactical innovations that came out of the Spanish civil war was the development of the *Schwarm* or "finger four" tactical combat formation. The formation is called the "finger four" because, if viewed from above, the aircraft in the formation would appear to be positioned relative to each other just as are the fingertips on the human hand when the fingers are extended. The formation was described and documented by Werner Moelders, one of the top Luftwaffe aces during World War II. The finger four was a fluid formation that enabled pilots to truly work cooperatively in the air, providing each other mutual protection as the formation engaged enemy aircraft. It was a formation that permitted the participants to maneuver, taking maximum advantage of the situation as it developed while using the eyes of all pilots to avoid surprise. It was everything that the British vic-3 formation was not—flexible, effective, and lethal—a fact that the British would realize only after the Battle of Britain had begun. This is yet another case of a tactical innovation that is largely attributed to a single individual and which lasts for years. In fact, modern fighter pilots in every country in the world learn to fly a variation of the finger four as a routine part of their training in air-combat maneuvers.

During the late 1930s, two significant technical events occurred in Britain that had significant tactical implications. In 1937 the experiments at Biggin Hill had succeeded in linking the Chain Home radar technology to a networked command and control system. This development was one of the final steps in the development of the system architecture that enabled the British to build the sophisticated ground-controlled intercept capability that would be decisive during the Battle of Britain. The system enabled the British to use the fighter interceptors with great economy and efficiency, launching them only as needed, vectoring them to targets, and helping them differentiate enemy from friend. The changes and improvements that this technical development had on interceptor tactics cannot be overemphasized. The other British technical development that occurred at this same time would become important in future tactical operations, but had little net impact during the Battle of Britain. In 1939 an airborne radar was integrated into the Bristol Blenheim light bomber, which had been converted to an interceptor role. Too slow to survive in the daylight dogfights with German fighters, the Blenheim was assigned the mission of night bomber interception.

In August 1939, Hitler unleashed his armed forces in an unprovoked attack on Poland. Here all the lessons, tactics, and techniques learned in the Spanish civil war were applied in a massive attack that quickly crippled the weaker nation. The Luftwaffe was central to this campaign. Aerial reconnaissance pinpointed the location of every Polish airfield; massive bombing raids struck deep behind the front lines, destroying airfields and infrastructure, while German fighters established control of the air. Luftwaffe transport aircraft resupplied the army and rapidly airlifted commanders, staffs, and troops to locations as they were needed. Continuing reconnaissance made it impossible for Polish forces to hide their movements, and Ju 87 Stukas with their screaming sirens struck terror into the hearts of military and civilian populations alike. This was the world's introduction to the concept of blitzkrieg (German for "lightning war"), a product of the Luftwaffe's experience in Spain which would soon be seen again in western Europe.

On September 3, 1939, Prime Minister Chamberlain informed the British people that a formal state of war had been declared between Britain and Germany. France followed suit immediately, and World War II was under way. There is reason to believe that Hitler was focused on the defeat and capture of Soviet Russia as his major strategic objective; however, with France and Britain having declared war on Germany, he could not turn east before he had ensured the security of his western flank. He was compelled, therefore, to first attack to the west. Hitler's strategy was fairly straightforward. Rather than attack the French Maginot Line head-on, which would have favored the defensive position that France held, he would attack to the north of the French defenses through the Low Countries, then turn south and roll up the French flank. With France defeated, Hitler believed that Britain, seeing the futility of resistance, would either surrender or, at least, agree to cease hostilities, thus freeing Hitler to attack Russia. The war for the Low Countries and

France began with another German blitzkrieg attack on Holland on May 10, 1940, employing the same tactics that had been successful in Spain and Poland.

The British Expeditionary Force (BEF), sent to assist France in her defense, moved forward into Belgium in anticipation of a German offensive, with French armies on either flank. Four squadrons of Hurricanes were initially sent to support the BEF; these were followed by two additional squadrons. Still later, as the situation deteriorated, four more squadrons were deployed to France. In addition to the units located in France, the RAF was supporting the effort with squadrons based in Britain. An armored thrust succeeded in separating the BEF and the French army on the northern flank from the other French forces to the south, thus isolating them for a killing blow. The Allied armies were effectively trapped by the German army, with the English Channel at their backs. The French requested additional fighter squadrons from Britain, and Churchill, now prime minister, gave the order to comply. At this point, Dowding, commander of Fighter Command, in his "most famous letter," made the case that sending more fighters to France was an exercise in futility that would ultimately result in Britain's inability to defend herself against air attack. He prevailed, and rather than sending more fighters to France, the order was given to the RAF to render all possible assistance to the withdrawal of the French and British soldiers as they made their way toward Dunkirk. British fighters were shortly thereafter returned to Britain to operate out of British airfields while they continued to support the BEF as the withdrawal continued.

On May 26, "Operation Dynamo" was initiated—the extraction of the French and British armies at Dunkirk. Goering had convinced Hitler that the Luftwaffe could administer the death blow to the Allies, and a concerted aerial assault on the beaches began. The RAF initially responded by trying to provide constant cover, but the resources available meant that only small flights could be sent if continuous coverage were to be provided. Air Vice Marshal Park, commander of 11 Group, decided to send larger flights, but this meant that there would be time gaps between flights, sometimes up to an hour. The Luftwaffe quickly took advantage of these gaps, timing their own attacks to coincide with the times when Allied air cover was lightest. It was only when the Germans inexplicably delayed the ground offensive for a period of 72 hours and the flying weather deteriorated that the British gained the margin needed to rescue over 300,000 troops at Dunkirk.

The battle for France continued until June 22. At the end, when France surrendered, the RAF counted among its losses 950 aircraft—over one-half of the operational strength it had when the German offensive in the west began. The Luftwaffe had lost over 1300 aircraft, which was approximately one-third of its strength. At this point both sides needed some time to rebuild and prepare for the coming battle. Just as the Luftwaffe had benefited from the combat experiences in Spain, the RAF gained from its experiences in France. As costly as the operation was, the lessons learned would soon serve the RAF well; the Battle of Britain would begin within a few short weeks.

The possibility of a German invasion of Britain had been raised as early as 1939 by the German navy, but little or no planning was done at that point. After the German armored attack that precipitated Dunkirk had brought German forces to the French side of the English Channel, the issue was once again raised and initial planning was begun. In early July 1940, Feldmarschall Keitel, chief of the German High Command, noted that the führer had decided that an invasion was a possibility—if air superiority could be established. The requirement for air superiority as a necessary condition for invasion would continue to be raised throughout the summer of 1940.

In Britain, the command and control system that integrated the Chain Home radar system with the operational control apparatus and the fighter operations was finally in place. The last elements, IFF together with the operational concepts and doctrine necessary to launch fighters and efficiently vector them to targets, had been added. Now fully operational, this system would prove to be one of the decisive factors in the Battle of Britain. The Luftwaffe, deploying along the northern coast of France and the Low Countries, would not enjoy the element of surprise as it had in its previous battles. The British system would be able to see their formations as they gathered over the Channel. They would be tracked as they approached the English coast, and ground-controlled interceptors would be vectored to meet them. This battle would be far more difficult for the Luftwaffe than had been the earlier assaults where they enjoyed every tactical advantage.

July 10, 1940, is the date chosen to officially mark the beginning of the Battle of Britain. Hitler had, as early as 1939, alluded to a blockade of Britain which would feature attacks on British ports; then in May 1940 he had spoken of the need to retaliate for British attacks on the Ruhr by attacking Britain's aircraft industry. However, no clearly stated set of objectives had been set for the Luftwaffe as they looked toward England in July 1940. The prevailing assumption was that the British would realize the impossibility of their situation and would either surrender or, at least, negotiate a nonaggression agreement. In the absence of a strategy, the Luftwaffe began an effort to simultaneously intercept British shipping in the English Channel and weaken the RAF through a campaign of air engagements over the Channel as a prelude to initiating a concentrated main assault. The idea was to attack British shipping with bombers accompanied by fighters. When the RAF rose to intercept the bombers, the German fighters would attack them and decimate the RAF through a battle of attrition. The pace and intensity of the German attacks grew through the early days of July 1940. On July 10, a large British shipping convoy sailed, setting the stage for a huge aerial dogfight. British Hurricanes and Spitfires, guided and assisted by the Chain Home system, were pitted against German bombers of every stripe, escorted and defended by Bf 109s and Me 110s. This day marked the first, but hardly the last, dogfight in the Battle of Britain in which over 100 aircraft participated. The day marked a significant increase in the intensity of the air war and, as such, was later designated as the official start of the Battle of Britain.

The realities of combat invariably force a reevaluation of the tactics employed by the participants. This was particularly true for the British. The RAF had not had the benefit of multiple battles to work out the details of its tactics as Germany had. The British found that they had to reevaluate their tactics and make numerous adjustments to accommodate the challenges confronting them. For example, the British began to conduct fighter operations from satellite fields near the coast, which reduced the time required to reach the convoys in the Channel and intercept German forces inbound from the continent. This also made it possible to conduct more missions with these forward-deployed aircraft. Other adjustments included the reharmonization of the guns carried by British fighters to converge at 250 yards and the addition of armor to protect pilots and engines. Perhaps the most radical and important tactical adjustment made during this period was the decision by some British squadrons to abandon the formal fighting area tactics and tight three-ship vic formations in favor of tactics resembling those used by the Germans. Initially based on pairs and fours, British tactical formations soon evolved into the loose finger four that the Germans had developed and used so effectively. This change, based on the realities of combat, finally put an end to the fighting area tactics that Britain had clung to since the end of World War I.

This period of conflict also demonstrated to the Luftwaffe that some reevaluation might be required on their part. Even as early as the battle for France, it had become apparent that the Ju 87 Stuka dive-bomber, which was so emblematic of blitzkrieg warfare, was vulnerable to the faster Hurricanes. The fighting over the Channel served only to reinforce that fact. The Stuka was a key weapon in the attacks on British shipping. Its near-vertical dive delivery and the fact that it was relatively slow made it a very accurate platform for bombing the slowly maneuvering ships in the Channel. The Germans attacked British convoys with huge formations of the screaming Ju 87s, but the losses they sustained made it increasingly apparent to the German commanders that the Stuka would not be able to dominate the tactical bombing arena in this battle as it had in Europe. Equally disturbing from a German perspective must have been the fact that the Messerschmitt Me 110 was also showing signs of vulnerability. During the Channel phase, the defensive circles of Me 110s, reminiscent of the Lufbery circles of World War I, became a common sight. A group of Me 110s would form a circle so that each aircraft could provide covering fire for others. The fact that they adopted these maneuvers on a regular basis indicated that the Me 110 too often could not compete with the Hurricanes and Spitfires and was forced into this purely defensive maneuver out of self-preservation. It did not take long for the British fighter pilots to take advantage of their superior capacity for climbing and descending by attacking the defensive circles vertically. Alternately diving into the formations from above, then pulling up to attack them from below, the British fighters exposed the weaknesses of these tactics, forcing a reevaluation by the Germans of the appropriate role for the Me 110 in the air battle.

As previously noted, the Germans had a well-conceived and effective air-sea rescue service. The Heinkel He 59 floatplane was a key German asset in this activity. As this phase of the battle continued, numerous sightings of He 59 air-sea rescue floatplanes were reported in the vicinity of the British convoys targeted by the Germans. Concerned that these aircraft were being used both as rescue vehicles and for tactical reconnaissance, the British Air Ministry concluded that, in spite of the fact that they carried the Red Cross symbol, they would not have immunity from British attack when seen near operational sites. The British began to routinely attack the He 59s in the Channel. This caused considerable consternation, accompanied by charges and countercharges; however, in the end the He 59s continued to operate in the Channel, although they changed to a camouflage paint scheme and were armed to counter the British attacks.

Among German strategists, the possibility of an invasion continued to be a central topic of discussion and debate. Hitler fully expected the British to recognize the hopelessness of their situation after the fall of France and to sue for peace. However, the increasing tempo of combat leading up to the furious aerial engagements on July 10 made it increasingly clear that Britain had no intention of either surrender or entering into any agreement that would permit Germany to turn eastward toward Russia. An internal debate was being waged between the German navy and the army about the best strategy for an invasion, with the army the more optimistic that an invasion was realistic. On July 16 Hitler issued Directive No. 16, stating that in view of the unwillingness of the British to compromise, he intended to prepare for and execute an invasion. The operation would be code-named "Sea Lion." Hitler followed this directive with a public statement three days later in which he urged Britain to accept a last appeal to reason and common sense, threatening invasion as the consequence of a failure to yield. Britain refused. Lord Halifax characterized the statement as an "invitation to capitulate" and Churchill refused to respond at all.

At a commanders' conference in Germany on July 30, Admiral Raeder argued that the earliest date for an invasion would be September 15, although he favored waiting until 1941. Oberkommando der Wehrmacht (OKW), the German High Command, issued a directive on August 5 that alluded to the next phase of the Battle of Britain, referring to an air offensive that would begin on August 5, after which the führer would decide whether an invasion would be necessary. On August 1 Hitler issued Directive No. 17, instructing the Luftwaffe to use all forces available to attack British flying units, their ground support and supply, and the aircraft industry. This was the order that would set in motion the second phase of the Battle of Britain, *Adlerangriff*—"Eagle Attack" in German. Hitler concluded by authorizing the attacks to begin on or after August 5. This was the main thrust that the British had anticipated since the end of the battle in France—the phase we have labeled the "airfield phase."

Two fundamental missions were assigned to German fighters during this phase of the battle. One of these was escorting bombers en route to British targets. In this role they were expected to engage the defending fighter forces as necessary to enable the bombers to proceed to their targets. The measure of success for a fighter escort mission was directly related to the effectiveness of the bombers being escorted. There was another mode of operation, however, referred to as "free chase," in which formations of Bf 109s would be sent specifically to engage and destroy British fighters—in other cases, to provoke the British to launch fighters, thereby preventing those fighters from engaging a bombing raid, which often was to follow. On more than one occasion, a formation of Bf 109s was observed trolling along the British coast, with no apparent mission other than to entice British fighters into the air. In the course of the Battle of Britain, a classic intellectual battle was often waged between German tacticians and British controllers as they attempted to outguess each other and to respond appropriately as they planned their missions, attacked, and countered using the resources at their disposal. The timing of commitments and assignments of mission objectives were a daily concern to planners and controllers on both sides of the Channel.

The search for methods of intercepting enemy bombers during night operations was a matter of some urgency for both sides during this time. The British application of their increasingly sophisticated radar technology to the problem finally paid off. On the night of July 22–23, the first night airborne radar intercept was performed when a Blenheim bomber shot down a Dornier Do 17. Even though neither the Blenheim nor the airborne radar were to be particularly effective during the Battle of Britain, this was a historically important feat in that it confirmed the technical feasibility of airborne radar in the operational environment. Later on in the war, the use of airborne radar in night interception missions would become a key aspect of British tactical capability.

The action-reaction aspect of tactical development responds quickly to the day-to-day changes each side makes either to accomplish revised objectives or to address problems they have encountered during combat. Commanders and their staffs constantly look for ways to gain an advantage, or at least to avoid a giving the other side an advantage. It is not unlike a football or basketball game, in which coaches will try to free their best players for a moment in order to let them score, while the defense tries to match speed with speed and height with height to avoid giving away scoring opportunities. The way that the Hurricanes and Spitfires were employed against the German bombers and Bf 109s illustrates this clearly. The RAF had, from the start, made a general practice of sending the Hurricanes against the German bombers and the Spitfires against the Bf 109s in order to take maximum advantage of the capabilities of each. Later, when the 109s of *Luftflotte 2* were ordered to fly close escort for the bomber formations, Air Vice Marshal Park immediately ordered both the Hurricanes and the Spitfires to concentrate their attacks on the bombers. Sperrle then responded by doubling the escorts, with one group flying close escort for the bombers and the other flying high cover. Park then reverted to the previous policy—Hurricanes on the bomber formations and Spitfires on the 109s providing high cover. Back and forth, give and take, jab and parry—the tactics of war shift from day to day.

Following Hitler's enunciation of Directive No. 17, Goering began to formulate the broad strategy that the Luftwaffe would execute upon the beginning of the *Adlerangriff.* The essence of that plan was a campaign that would start with the destruction of the Royal Air Force in the air and on the airfields, most of which were grouped in the south. The campaign would then progressively move northward until the attacks closed to within 50 kilometers of London. The entire operation was projected to require two weeks. If successful, invasion would either be unnecessary or would not be seriously contested by the soon-to-be-defeated British. August 13 was designated as *Adlertag* (Eagle Day), the beginning of the main assault. This was the campaign that very nearly defeated the Royal Air Force.

Sometimes changes in strategy or tactics may be seen clearly on the battlefield from the outset; in other cases, they are seen clearly only in retrospective analyses. The differences between the buildup to *Adlertag* and then the day itself must have been difficult to see from the perspective of the British fighter pilots. The level of intensity had been growing steadily. The early days of August were marked by increasingly large attacks and massive formations of aircraft. As far as the RAF was concerned, the new campaign was more an increase in degree than in form, in what was already a desperately intense fight for survival.

During this period, several RAF pilots emerged as leading fighter pilots of the day, among them luminaries like "Sailor" Malan, Al Deere, Peter Townsend, and the Czech, Josef František, to name but a few. Malan, particularly, was recognized as an excellent fighter pilot—highly skilled, deadly, and at times lucky. Al Deere, himself recognized as one of the historically great fighter pilots, is said to have stated that he thought Malan to be the best pilot he had ever seen (Hough and Richards 1990). Malan, in the tradition of the Boelcke Dicta years before, codified a set of principles or rules that he recommended pilots use in air-to-air combat. These 10 rules became known as the "Ten Commandments," and they were later adopted as official RAF doctrine to be employed by all. These included the admonition to shoot from close range, use altitude to gain the advantage, always turn to face attacks, never fly straight and level, strike quickly, and take advantage of high cover. Like Boelcke's rules, Malan's Ten Commandments endure to the present as excellent advice for the modern fighter pilot.

In spite of the excellence of the Luftwaffe as a tactical fighting force, the lack of good intelligence about almost every aspect of the British situation was a fundamental weakness that significantly impaired its ability to achieve German military objectives. The reasons for this failure are numerous. The Luftwaffe was poorly organized for collecting intelligence and failed to take advantage of the information available. Senior officers never seemed to fully comprehend the importance of the Chain Home system, even though the evidence was available had they pursued it; and they also routinely and rigidly accepted higher-level staff intelligence estimates without question. The results were that they misinterpreted the importance of key airfields; they consistently failed to correctly estimate Royal Air Force strength; and they were unable to clearly identify key military-economic targets, such as the factories that were building Spitfires and Merlin engines.

A notable example of the consequences of the failure of German intelligence was the attack mounted by *Luftflotte* 5 on the northern coast of Britain. Based on intelligence estimates indicating that the full measure of the Royal Air Force was located in the south to defend against the daily attacks in the 11 Group area, Germany conceived the idea of an attack on what they thought to be the undefended areas north of London. They did not count on Dowding's stubborn refusal to leave the north undefended or his policy of periodically rotating battle-worn fighter crews into the northern group areas to give them rest. These policies ensured that there would be fighter defenses even in these areas that were well removed from the intense fighting in the south. Not realizing this, and believing that they were attacking an undefended target, the Germans sent a force of bombers from Norway escorted only by Me 110s whose rear-facing internal guns had been removed to save weight. To complicate matters, a navigational error resulted in a warning for the British defenders and confusion for the attackers. The result was a total debacle for the Germans. The British single-engine fighters had a field day, inflicting losses of 20 percent on the hapless Germans. It was the last time they attempted a surprise attack on England from their northern bases in Scandinavia; most of the air assets located there were subsequently redeployed to *Luftflotten* 2 and 3.

As this phase of the Battle of Britain intensified, the pace of tactical adjustments and innovations also increased, as both sides sought to improve their chances of mission success. From the German perspective, the primary objective assigned to the bombers was to destroy the airfields and any aircraft or support facilities located on them; the British defenders' primary objective was to render the bomber attacks ineffective. Punch and counter . . . duck and weave . . .

The loss of experienced aircrews was the most serious problem that the RAF faced in the air war. Air Vice Marshal Park, commander of 11 Group, in an effort to conserve his fighter resources and, more important, his fighter pilots, instructed his controllers not to attempt to pursue reconnaissance aircraft over the water and, furthermore, to engage major formations either "over land or within gliding distance of the coast." He also felt it necessary to remind his controllers and fighter pilots that the objective was not to destroy German fighters, but rather to destroy German bombers. In this regard, Park ordered that the minimum number of fighters were to be dispatched to combat German fighters, saving the fighters for bomber engagements wherever possible. Park directed that patrols of fighters be stationed over the aerodromes to defend against attacks when enemy formations were observed inbound. Finally, he further ordered that 12 Group be asked to assist in covering the northernmost aerodromes in the 11 Group area of responsibility when all squadrons of 11 Group were engaged. Meanwhile, the British pilots also developed new air-combat tactics in response to German changes; for example, when the Germans installed armor in their bombers to protect them from attacks from the rear, some British fighter pilots responded by using head-on attacks to disrupt and break the formation discipline of the bombers.

The Germans also developed new and innovative ways to attack British airfields to maximize the probability of successfully destroying the airfields and their aircraft. They began to use carefully timed attacks that would attempt first to lure the fighters into the air, then catch them as they landed and were refueling. They also began to incorporate low-level bombing attacks—sometimes no more than 100 feet above the ground—designed to avoid detection by the Chain Home radar. This, in turn, gave rise to the British use of parachute-and-cable rockets, designed to entangle the low-flying aircraft as they crossed an airfield. German planners also designed bombing attacks to disguise the intended targets. Rather than drive straight in toward an objective, they would form massive formations and then split into smaller groups to attack targets spread over a wide area. On one occasion, 47 different airfields reported being under attack at the same time (Hough and Richards 1990). When that proved less than successful, the Germans began to concentrate their attacks on fewer targets, some of which received extremely intense bombing day after day. This, by the way, was not as successful as it might have been due to the failure of German intelligence to identify those key airfields that were truly important to the British defensive effort. The Germans also initiated a tactic whereby they would overfly a target, then turn and attack it on a heading that would take them toward the European coast, thus attempting to both confuse British controllers about the true objective and reduce the time during which fighters could attack them after they had released their bombs. This was an extremely intense time for both sides, as aircraft pilots and supporting personnel at the airfields were constantly threatened with death or destruction. During those periods when the weather finally deteriorated sufficiently to make air operations impossible, the commanders of both sides would take the time to reassess, make changes, and prepare for the next round.

Manning the fighters—pilots in cockpits—was by far the greatest concern that Dowding faced throughout this period. The problem had multiple dimensions. First, there was the loss of experienced leaders in combat; then there was the problem of simply filling cockpits with trained people. By mid-August, Fighter Command was short of its authorized strength by 400 to 500 pilots—fully one-third of the total needed. Finally, there was the problem of attempting to keep those pilots on whom the burden of combat fell most heavily in reasonable physical and mental shape to enable them to fight on.

To handle these problems, Dowding took several actions. He had already ordered controllers not to send fighters needlessly over water and to try to keep them close enough to land that the pilots could either glide to land or bail out over land. To get more pilots in cockpits, he requested that pilots be transferred to Fighter Command from other commands (e.g., Bomber and Coastal Commands) and he ordered that the duration of the transition training for pilots be curtailed to the shortest period possible. The transition training was eventually reduced to seven days, although, needless to say, the survival rates among pilots with that little preparation was not good. To keep active combat pilots in a reasonable state of physical and mental health, Dowding instituted a policy of withdrawing the units that had seen the heaviest fighting and rotating them into the areas assigned to 12 and 13 Groups where the intensity of the fighting was less. It was, in fact, this policy that had resulted in the RAF success in defending against the Luftwaffe attack on the northern flank of Britain that was launched from Norway on August 15.

In mid-August 1940, Goering held two conferences within a single week with his combat commanders to assess the progress of the war from the German perspective and to make needed adjustments. Goering was extremely unhappy with the lack of apparent progress toward the destruction of the fighters of the Royal Air Force. They continued to mount defenses in numbers that made it clear that they were far from being eliminated as a fighting force. Equally troubling, Germany was losing substantial numbers of aircraft daily; the Luftwaffe lost over 50 airplanes on three consecutive days in mid-August. Clearly, changes were needed.

Goering announced reorganization plans and ordered that the older, more cautious commanders be replaced by the "young turks," the outstanding fighter pilots in the active squadrons. These included names like Galland and Moelders—individuals who would lead the Luftwaffe through the remainder of the war. Further, they decided to withdraw the Ju 87 Stuka dive-bombers from the battle due to the heavy losses they experienced. An order was also given requiring that Me 110s, designed as bomber escorts, themselves be escorted by Bf 109s; perhaps most significantly, they decided to cease the attempt to destroy the Chain Home radar stations, concluding that the effort was too costly and had been largely unproductive. These last two decisions were particularly detrimental to the German cause in the battle that continued in the skies over Britain.

In addition to these decisions, Goering, responding to demands of the bomber crews for more fighters, insisted that the Bf 109s allocate more of their numbers to bomber escort missions, in spite of the fact that free-chase Bf 109s were more productive in terms of British fighter kills. He furthermore announced that most of the Bf 109s assigned to *Luftflotte 3* would be transferred and consolidated in *Luftflotte 2*. This was to take advantage of the narrowness of the English Channel at Pas de Calais and to thereby provide improved escort for the bomber formations bound for the southeast quadrant of Britain.

As Goering and Hitler mapped out the strategy for the bombing campaign against Britain, they were consistent in their demand that the city of London itself would not be targeted without their personal approval. The campaign would march northward to the suburbs of London, but the city itself would remain hostage to the whims of the German commanders. On August 24, the Luftwaffe began bombing the airfields and towns just to the south of London. As day gave way to night, the raids continued uninterrupted in spite of the defenses mounted by Fighter Command. Whether through navigational errors by German crews or due to bombs being jettisoned by bombers under attack by British fighters, bombs fell on the city. Although there appeared to be no well-defined targets and the bombs tended to be randomly dropped in many areas, this was the first time bombs had fallen on the city since the Gothas had last attacked in 1918. It was a harbinger of things to come for London.

Although the August 24 bombing of London was said to be accidental and the German crews were said to have been disciplined, Winston Churchill was determined that England would respond in kind. Within 24 hours, British bombers attacked industrial targets on the outskirts of Berlin. This was followed by sporadic attacks during several successive days. The citizens of Berlin were stunned; they had been assured that this would never happen. Hitler, infuriated, was determined that Britain would pay for this outrage.

As August drew to a close, the intensity of the German effort to destroy the RAF and its facilities increased. The airfields, particularly those to the south and east of London, came under great pressure. While on a visit to the area to inspect the defenses, Churchill registered his displeasure with the speed of airfield repairs. He singled out the field at Manston, commenting that bomb craters remained unfilled days after they had been created, rendering the field almost unserviceable. The unfortunate fact was that ground-support crews, in addition to their responsibilities for maintaining and repairing the aircraft, were also expected to repair the airfields when necessary. As Churchill pointed out, the people were doing their best, but there were insufficient resources available for the job. As was his way, Churchill also suggested solutions. He recommended that larger crater-filling companies be created to take on the work of maintaining the airfields. These should be centrally located, he added, and provided with the means to permit them to travel to damaged sites as needed. He further suggested that civilians be used to relieve the military support crews so that they could concentrate on their primary task of maintaining the aircraft. Churchill even went so far as to suggest that repairs be camouflaged so that bomb craters would appear to still require repair in order to deceive German pilots into passing them by in future raids.

During this period, many pilots, both German and British, had their aircraft damaged to the extent that the pilots were forced to "take to the silk"—bail out. An issue arose regarding the morality and policy of whether a pilot descending helplessly under a parachute canopy was an acceptable target. The two sides generally came down on different sides of this issue. The evidence is that Germans routinely fired at British parachuting pilots. British operating procedure, to the contrary, was not to do so. Needless to say, two men engaged in a duel to the death can be a very emotional event, even if they are in aircraft and using machine guns and cannons, and the issue of shooting pilots after they had bailed out was also highly emotional. Even though the stated British policy was not to shoot, the Poles and Czechs who flew for the RAF, spurred by a visceral hatred of the Germans for the events that occurred when their homelands were invaded, made a routine practice of shooting Germans who had bailed out.

As August ended, the aircrews of the British fighter pilots were exhausted from flying two, three, and sometimes four combat missions every day. During the first week of September, RAF losses were equal to German losses, an exchange rate that could not be sustained. As losses continued, Dowding introduced what he termed a "stabilization scheme" in an attempt to focus his strength at the locations deemed most critical. The most operationally capable squadrons were assigned to 11 Group; the less ready squadrons to other groups. The scheme also included the cross-assignment of pilots from squadron to squadron in order to balance and maintain the readiness of the squadrons located in 11 Group. This had a deleterious effect on morale as flight mates and friends were suddenly posted to different squadrons, but Dowding was desperate—and the RAF was very nearly done for.

In Germany, intelligence estimated that the RAF was down to something on the order of 150 to 300 Hurricanes and Spitfires. Hitler, still smarting from the attacks mounted by Bomber Command on Berlin, announced that he had approved attacks on London in retaliation. Goering, Kesslering, and Sperrle debated whether conditions existed that would make attacks on London beneficial. Sperrle argued that more time was needed to further decimate the fighters, while Kesslering argued that an attack on London would force the fighters into the air and would therefore satisfy both Hitler's desire for retribution and the need to complete the destruction of Fighter Command. Kesselring's arguments carried the day. On September 5, Hitler ordered the Luftwaffe to conduct "attacks by day and night on the inhabitants and defenses of large cities, especially London." On September 7, the Germans redirected their attacks to concentrate on the city of London. The assault on the airfields and Fighter Command was over; the bombing of London would now begin.

Late in the afternoon of September 7, a massive formation of German aircraft was detected. All fighter squadrons within 70 miles of London, having been alerted, were either airborne or ready for takeoff. Those airborne saw a formation of over 300 bombers and over 600 escorting fighters. The formation rose over a mile and a half high and covered over 800 square miles (Mason 1969). It was the first of many raids targeted against London's long nightmare known as the Blitz; they would see 59 consecutive nights of bombing before the first break came. As terrible as this would be, there can be no question about the fact that the turn away from the destruction of the airfields, aircraft, and men of the RAF gave Fighter Command the respite it needed to recover and eventually prevail. It was a critical strategic error on the part of the German leadership.

With the change in German strategic objectives, Fighter Command soon began to recover and attack the daily bombing formations with renewed vigor. The German bomber crews again called for closer escort support from the fighters. Goering again responded by ordering the Bf 109s to provide close escort for the bombers in order to reduce losses. He demanded that the fighters fly in formation with the bombers. This was an extremely unpopular decision in the eyes of the fighter pilots, who viewed this as an order to give up both altitude and airspeed—the two most valuable assets a fighter possesses in air-to-air warfare. With this, Park ordered the Spitfires, which he had normally targeted against the high-flying Bf 109s, to join the Hurricanes in attacking the bomber formations. When the Germans increased the number of fighters in order to provide both close escort and to free chase fighters, Park responded by once again directing his Hurricanes at the bomber formations and Spitfires at the Bf 109 formations. However, the fact that the bomber formations included both bombers and fighters meant that, in many cases, Hurricanes would be matched against both bombers and Bf 109s. During this phase, the German fighters were, for the first time, ordered to disengage if fighter opposition was substantial in order to preserve them for escort missions.

The question of the best tactics for intercepting large bomber formations was a continuing one for the group commanders, especially Park in 11 Group and Leigh-Mallory in 12 Group, the two most closely identified with the defense of London. The tactics employed by each was different and, to a certain extent, reflected the geographical positions they occupied. Park, who had responsibility for the south and east coastal regions, preferred to send squadron formations against the bombers. These smaller units offered the advantages of quicker response and increased mobility, once airborne. Leigh-Mallory, to the north, was a proponent of the Big Wing theory advanced by one of his squadron commanders, Douglas Bader. The Big Wing was a giant formation composed of several squadrons flying together. A Big Wing might comprise up to 60 aircraft. Although it had the advantage of massed firepower, these formations took a long time to organize in the air and they were not as mobile as the smaller formations, once airborne. The issues of responsiveness and relative effectiveness of these two approaches was a source of considerable friction at the command level and, in fact, continued to be an issue of debate until well after the war was over.

September 15, 1940, was officially designated as Battle of Britain Day. Although the bombing of London continued well into the following year, this was the day that the air discipline of the Luftwaffe was finally broken. The respite created by the change in German strategy had renewed the confidence and zeal of the RAF. At the same time, the war was wearing the German crews to exhaustion, and they had begun to doubt the wisdom of the strategies and tactics advocated by their leaders. They had been told again and again that the RAF was down to their last few fighters. When on this day the fighters rose in the hundreds to defend against the bombing raids, the Luftwaffe finally cracked. The formations broke; the discipline that had marked the German effort since the beginning eroded into confusion; and the raids were ineffective. Fighter Command logged over 60 kills that day. Although the full implications would be realized only later, September 15 is the day when it was generally acknowledged that the German daylight bombing offensive was defeated. It was the beginning of the end of the Battle of Britain.

As September ended, the battle took on a somewhat different complexion. The German Bf 109E-7 was introduced during this period. This aircraft carried a single 250-kg bomb, or, alternatively, it could carry a fuel tank under the fuselage, which finally gave the aircraft the long range it had needed since July. The Germans sent what amounted to an almost continuous stream of Bf 109s at altitudes of 30,000 feet or higher. Dropping these single bombs from straight and level flight, these attacks amounted to little more than nuisance raids from a strategic point of view, but they demanded constant and intense defensive efforts from Fighter Command. On some days the Luftwaffe flew over 1000 sorties. Park was finally forced to institute standing patrols of Hurricanes and Spitfires at altitudes of 20,000 and 25,000 feet to counter the high-flying Bf 109s. Toward the end of October, the air war finally surrendered to the British weather and the intensity of the effort ebbed to a trickle as the month ended.

With the end of September, Germany ceased mounting daylight bombing raids. From this point forward, the Blitz would be a nightly event for the citizens of London. The Luftwaffe developed what was referred to as the "*krokodil*" (crocodile) tactic during this period of the battle. The full complement of an entire *Gruppe* (equivalent to an RAF or USAAF wing) or *Geschwader* (several *gruppen,* equivalent to an RAF Group) would be sent along the same path at established intervals. Multiple groups would be launched along different paths, with the net result that a single bombing attack might last up to 10 hours. As airborne intercept radar became better developed and was mounted in better airplanes, the *krokodil* tactic gave way to the concept of the bomber stream, the objective of which was to flood the defenses, concentrating a massive attack on a single point in a short period of time. These intense attacks might last only 20 minutes. Precision in night bombing was accomplished through the *Knickebein* system, which used intersecting radio frequency (RF) beams to mark both the course and the target. When *Knickebein* was compromised, the Germans followed with the *X-Gerät* system, which was even more precise. This clever use of RF technology allowed the Luftwaffe to continue the bombing raids against London in spite of the steadily deteriorating weather.

As the English weather increasingly became a major factor in daily fighter operations, the most significant tactical developments involved the increasingly strident debate between the advocates of the big wings and their opponents. Douglas Bader, the primary proponent of the Big Wing tactic had grown increasingly frustrated over the unwillingness of the Fighter Command to adopt it as a standard tactical procedure. He had managed, through political channels, to raise the debate to the national political level. On October 17, a meeting was convened to discuss this and to attempt to settle it as reasonably as possible. As is often the case in such high-level meetings, much was left unsaid in the minutes, and the debate over the events of this meeting, as well as the argument regarding the effectiveness of the big wings, persisted for years after the war.

The Battle of Britain was officially declared to have ended on October 31, 1940. In a period of three to four months, two well-equipped and motivated air forces had met each other and battled to a decision, rather than a knockout. The fact that Hitler had been unable to either invade Britain or to force her into a nonaggression pact was, by itself, a defeat for Germany. But in fairness to the brave airmen who flew for the Luftwaffe, we must acknowledge that it was a strategic defeat, not a tactical defeat. Well-developed tactics, excellent preparation, and highly experienced aircrews—these were all characteristics of the Luftwaffe on the eve of the Battle of Britain. Germany's failure was in planning, in the choice and timing of objectives, and in leadership. In the air, the German pilots and crews were brave and highly competent, as were the airmen of the RAF. But the leadership of the RAF, made even more effective by the technological advantages they enjoyed, eventually prevailed. There was reason to believe that the Germans might prevail in June 1940 as France fell. But poor decisions at the top, emotionally driven strategies, and leaders who insisted on superimposing their views, however ill-informed, on the tactical employment of the forces at their disposal—these were the strategic and tactical factors that eventually led to the German defeat.

1. Dicta Boelcke

The advent of power projected from the air in World War I soon led to aerial combat in the skies. While the airplane was first used primarily in a reconnaissance role to support ground forces, the tactical fighter (or pursuit) airplane quickly became a weapon in its own right, used to attack the dirigibles and bombers that threatened both cities and populations and also employed to attack other aircraft in the air. This was a new form of combat; one often not well understood even by those involved. The life expectancy of a new fighter pilot was measured in weeks, and it was generally through experience in the crucible of aerial combat and survival that a few tactical pilots began to absorb the lessons that would enable them to prevail.

One concept that World War I pilots soon learned was that their chances of success were much better if they saw their opponent first and could maneuver into an advantageous shooting position before the enemy saw them. If the enemy spotted them, then their chances of survival could be improved if they could cause the enemy to lose sight of them. Keeping the attacking aircraft between the sun and the target, like the pilot of the Fokker Eindekker shown in the illustration, was one of the best ways to make both of those things happen. Those who learned from their experiences and began to see the patterns and techniques that led to success became the aces of their day—von Richthofen, Lufbery, Immelmann, and Rickenbacker to name but a few. The more success a pilot had, the more credibility his theories held. The most forward-thinking combat leaders tried to capture the best ideas and pass those lessons learned on to their squadron mates.

One of those who absorbed the lessons of aerial combat and lived long enough to tell what he had learned was Oswald Boelcke, a German fighter pilot. Boelcke had won the Iron Cross by the time he was 23 years old, and he and Max Immelmann were awarded the *Orden Pour le Mérite,* the famed "Blue Max," in January 1916. He was charged with the responsibility of establishing a new fighter squadron, *Jagdstaffel 2.* Precise and demanding in his approach to flying skills, Boelcke insisted that the members of his new unit take the same disciplined approach to their own flying. His men became known as "Boelcke's Cubs," and they included in their number Manfred von Ricthofen, the famed Red Baron. As part of his rigorous training program, Boelcke developed a set of core lessons that he had learned as a combat fighter pilot. He set these in writing and gave them to the members of his *Jagdstaffel.* They have become known as the *Dicta Boelcke,* or Boelcke's rules, and they represent the first known effort to formalize air combat tactics. Boelcke was killed in October 1916 when he was forced to crash-land after a midair collision during a dogfight.

"Hun in the sun." A victim's view of a Fokker E.III maintaining good sun position.

He had 40 confirmed kills when he died. Although he did not survive the war, his lessons have survived the years. They are listed here:

- Secure the advantage before attacking. Climb to surprise the enemy from above and dive on him from the rear. Keep the sun at your rear as you attack.
- Carry through an attack, once begun.
- Do not fire until the enemy is at close range and well in your sights.
- Keep your eye on the enemy. If he appears damaged, follow until he crashes lest he attempt to deceive you and escape.
- Attack when the enemy is engaged in other duties and when he least expects it. If attacked, turn into the attack; do not turn your back and run.
- When behind enemy lines, do not forget your own line of retreat.
- Foolish acts of bravery bring only death. Attack in groups. When the combat breaks into individual combat, ensure that several aircraft do not follow a single enemy.

These fundamental rules for air combat are as good today as they were in World War I. The technology has changed immensely, but the fundamental principles have not. The advantage in air combat still goes to the aircraft that "sees" the other first, although "seeing" may mean detection using radar or other sensors. Most air combat training that prepares pilots to join operational fighter squadrons is descended from these basic principles. Boelcke got it right; his dicta are excellent examples of tactics that are attributed to a single perceptive innovator and that have endured through the years. Furthermore, Boelcke's effort to capture lessons learned and pass them on to those who follow also broke new ground, and that example continues to be followed today, as well.

2. Three-Aircraft Vic Formations

As World War I wore on and the air war continued, it became increasingly apparent that the "eagles" were dying. Boelcke had made the point in his Dicta Boelcke—warfare in the air was best practiced in groups. The pilots that had fought alone—even the best of them, like von Richthofen—were eventually killed. In the last years of the war, fighters increasingly began to enter combat in groups of two, three, or four aircraft. The formation favored by British pilots was the three-ship *vic*, or vee, a tight group of three aircraft consisting of a lead with a wingman on either side. The formation of SE-5s shown as they execute a right turn illustrates the three-ship vic.

At the time, British tactical squadrons were typically composed of 12 aircraft. The organizational approach taken by the Royal Flying Corps, the air arm of the British army, was to allocate the aircraft into two groups of six that were further divided into two

Vics become standard. S.E.5s turning in Vic formation late in WWI.

groups of three—hence the three-aircraft element became the fundamental tactical flying unit. While hierarchical tradition dictated the size of the fundamental flying unit, it was technology that dictated the in-flight procedures used. In the last year or two of World War I there were no in-flight radios, no voice communications between aircraft in flight. If there were to be any communication at all between aircraft in a formation, it would necessarily be accomplished by means of hand signals. This required that the aircraft be flown close together in tight formations so that pilots could see the signals. With the number of aircraft in the formation fixed at three, and the fact that they would have to fly in close formation necessitated by the lack of communications technology, the only question that remained was the actual configuration of the formation itself—should it be an echelon, with the three aircraft trailing at an angle off one wing or the other of the lead aircraft; a trail formation, with each behind the other; a line-abreast formation, flying wingtip to wingtip; or a vee formation, with one aircraft on either wing of the lead? Clearly, of these choices, only the vee, or vic, permitted both wingmen to see the lead aircraft's signals as the leader gave them; it was the most efficient of the choices available in terms of reaction time and the avoidance of mistakes and confusion that might be caused by relayed hand signals. The three-aircraft vic became the standard formation for the Royal Flying Corps.

The adoption of the three-ship vic was not an unreasonable decision, given the state of the art in communications technology when the decision was made. However, the use of the tight, three-ship formation would eventually lead to serious problems for the Royal Air Force fighters when they met the fighters of the German Luftwaffe. They would find that the tight formation demanded too much of the wingmen's attention, diverting them from scanning the skies for enemy aircraft. It failed to take full advantage of the flexibility that the airplane offered. The failure was not that a poor decision had been made originally, but, rather, that the British were, for too long, unwilling to revisit the original decision and adapt as the technology and tactics of their opponents changed in the 1930s. This issue is developed in more detail in later sections of this book, which discuss the topics of British fighting area tactics, Werner Moelders's development of the "finger four" formation, and Britain's eventual decision to abandon the three-ship vic.

3. Lufbery Circle

Pictured as he instructs newly arrived pilots regarding the techniques and maneuvers needed to survive in the dangerous world of air combat is Raoul Lufbery, a fighter pilot and a born teacher. When America entered the war, he became the combat instructor for the 94th Pursuit Squadron, the famous "Hat in the Ring" squadron. As a natural outgrowth of his position, Lufbery was responsible for development of new tactics and for preparing the young American pilots for combat. Among his students was Eddie Rickenbacker, who was to become America's leading ace of World War I with 26 kills. Rickenbacker was quoted as saying, "Everything I learned, I learned from Lufbery" (National Aviation Hall of Fame 1998). Even though a general movement toward multiaircraft formations replaced the single fighter as the preferred unit in aerial warfare, there remained a tendency for the formations to break up into individual skirmishes once they engaged enemy aircraft. Lufbery is credited with developing the *Lufbery circle,* one of the first tactical maneuvers specifically designed to provide cooperative benefits to the aircraft in a formation, taking advantage of the mutual support that each could bring to the other.

With the end of the lone-wolf era in fighter tactics and the increasing tendency for tactical units to be composed of multiple airplanes, the natural progression was for pilots to experiment with new tactics that might take advantage of the new formations. Lufbery was one of the innovators in this search for new tactics. Never a man bound by convention, Lufbery was born in France, his father an American and his mother French. He left France to see the world at the age of 19, and during his wanderings he came to America to visit his father. He enlisted in the U.S. Army, which gave him American citizenship, and was posted to the Phillipines for two years. He then continued his travels, touring India and the Far East, where he met a French barnstorming pilot, Marc Pourpe. He toured with Pourpe, acting as his mechanic until 1914, when the two returned to France as

Raoul Lufbery explaining the merits of his defensive circle.

World War I began. The only French unit that accepted citizens of other countries was the French Foreign Legion, so Lufbery joined that organization, where he remained until Pourpe pulled strings to get him reassigned to the French air force as his personal mechanic. Pourpe was killed shortly afterward, which led Lufbery to become a pilot in order to avenge the death of his friend. He learned to fly, saw service as a pursuit pilot in 1915, and was then assigned to the Lafayette Escadrille, a French unit made up of American volunteers, in 1916. Lufbery logged five kills within three months, becoming America's first World War I ace. Before he was killed in air combat in 1918, Lufbery was credited with 16 confirmed kills, although it is generally supposed that his actual total exceeded that number.

The Lufbery circle is simply a maneuver in which a group of airplanes sets up a loose circle, each one following the other in a more or less continuous turn about some point at the center of the circle. This tactic is intended to provide mutual coverage against surprise attack, as the guns of each aircraft cover the rear quadrant of the airplane immediately in front. The nature of the maneuver makes it very difficult for an attacker to approach any aircraft in the formation without being exposed to fire from the rear. The formation has some inherent weaknesses—under certain conditions it is vulnerable to attack from both above and below, and it is critical that an attacker not get inside the circle—but it was quite effective in a day when tactical aircraft did not possess the high levels of performance that would become typical by the time of the Battle of Britain. Most World War I vintage fighters were simply not capable of performing the rapid climbs and dives needed to successfully attack and defeat the Lufbery circle.

This formation remained in use into the early stages of World War II; the Germans often used it as a defensive maneuver for its more vulnerable aircraft during the Battle of Britain. Two variations on the Lufbery circle were used in the battle. In one form, the aircraft orbited around a static point on the ground. In the second approach, the aircraft moved from point to point while flying a racetrack pattern that still took advantage of the mutual protection afforded by the nose-to-tail arrangement that characterized the maneuver. The circle was occasionally used by the Me 110s, particularly during the Channel phase, as an offensive loitering maneuver. In this application, they would establish a circle over the Channel as they waited for the British fighters to come out and engage them in defense of the English shipping convoys. The British countered by feeding their fighters into the battle in small groups, forcing the Me 110s to decide when the number of British fighters was sufficient to justify leaving the mutual protection of the circle in order to attack. One could argue that vestiges of the Lufbery circle remain a part of fighter tactics today. When a flight of modern fighter aircraft sets up a loose circle around an intended ground target as they position themselves to attack, it is easy to see that some form of the Lufbery circle has survived, even though it is no longer used as commonly or for the same purposes as in 1918.

4. Training in Russia

Prohibited from conducting any sort of military aviation training after World War I, Germany made secret agreements with Russia allowing German pilots to be trained and aircraft tested at a base near the Russian city of Lipetsk. The picture shows a German student pilot as he receives last-minute advice from his instructor prior to taking off in the Fokker D-XIII. The airmen trained at Lipetsk became the leaders of the new Luftwaffe that would confront the British in 1940.

Even the most casual reading of Part V of the Treaty of Versailles the military, naval and air clauses—leads to the conclusion that these clauses had two fundamental purposes: (1) to reduce Germany's armed forces and military equipment to levels that would make the conduct of war impossible, and (2) to deny Germany access to technology, either through imports or development, that might be used to develop the capability to wage future wars. The various articles included in this part of the treaty required, among other provisions, that the air force personnel of both the navy and army be demobilized (Article 199); the manufacture and importation of engines be forbidden (Article 201); and all aeronautical material including airplanes, seaplanes, dirigibles, engines, munitions, and their components be turned over to the Allies (Article 202). The treaty

Lipetsk, Soviet Union c. 1926: a German student pilot preparing to solo in a Fokker D XIII.

was signed in June 1919, and, with that, the demobilization of the German war machine and its associated technology began in earnest.

Loopholes in the Treaty of Versailles allowed Germany to begin a substantial rebuilding program through sponsorship of civil glider and air sports clubs. The authors of the treaty had failed to understand the role that large aircraft might play in future conflicts, so they had restricted the rebuilding of civil aviation only for a period of six months after the treaty was implemented. This was a major loophole through which Germany's future bomber designers launched the factories and the designs that would enable Germany to develop aircraft capable of conducting bombing campaigns throughout Europe within a decade.

The German military wasted no time in making its own preparations to survive and rebuild in spite of the harshly punitive provisions of the treaty. First and foremost, the core of the German general staff was preserved under the leadership of General Hans von Seeckt, acting as head of the Ministry of the Army. Von Seeckt selected for his core staff such men as Kesselring, Sperrle, Stumpff, and Wever, all of whom would be influential in the rebuilding of the Luftwaffe. He further authored a secret memorandum advocating that the future air arm should be independent of the army and navy. To rebuild the aerial warfighting capability, von Seeckt exploited a relationship that Germany had established with Russia. In December 1923 he negotiated an arrangement with the Russians that allowed Germany to establish a secret training facility at Lipetsk, 220 miles southeast of Moscow.

Out of sight of the Allies and the oversight mechanisms established to ensure compliance, the Germans conducted the air training that was strictly forbidden under the terms of the treaty. The elite of the armed forces—those who would lead the future Luftwaffe—were selected for training at Lipetsk. Among their numbers were Kurt Student, who had organized the German glider clubs during the early 1920s and would command the parachute forces during the invasions of Holland and Crete in World War II, and Hugo Sperrle, who would command *Luftflotte 3* during the Battle of Britain.

More than just a training base, Lipetsk also served as a development and testing center for German equipment that could not be easily disguised as somehow related to civil aviation, such as engines for fighter airplanes. The Germans went to extraordinary lengths to conceal the existence of the operations at Lipetsk. Equipment would often be routed through multiple countries and was marked to deliberately mislead inspectors both as to content and purpose. Engines manufactured in factories outside Germany and destined for Lipetsk for testing might be marked as though for repair and shipped through a third country. Even the bodies of airmen killed during training accidents in Lipetsk were placed in crates marked as spare parts and shipped via sea to Germany to conceal the activities in Russia.

The training and development center at Lipetsk operated from 1924 until 1932. Approximately 240 pilots and observers were trained and returned to Germany to serve as the core of the new Luftwaffe. By this time Germany was on a headlong dash to build the forces that would challenge the Allies once more in World War II. Flying schools were opened in multiple locations throughout Germany, ostensibly to train the crews for the growing commercial fleets that were recognized as among the best in the world.

Aircraft production, often featuring aircraft that had both civil and military potential, was expanding rapidly. Finally, in 1935, Hitler and Goering announced the existence of the new German air force—the Luftwaffe—but it was hardly new; it had been nurtured and rebuilt at Lipetsk under the forward-looking plans devised 15 years earlier by General von Seeckt in response to the Treaty of Versailles.

5. The Bomber Will Always Get Through

A German Gotha bomber releases its bombs over the city of London, bringing the war in Europe home to the doorstep of the citizens of England. In World War I, bombers that could carry their heavy loads across substantial distances first made civilian populations both the targets and the combatants in a new form of war. The ability to project power through the air, bypassing the ground and naval forces that had dominated the battlefield until that time, led to a rethinking of how future wars might be fought. The horrors of World War I, with its trench warfare, gas attacks, and a general inability of generals and nations to achieve the victories promised and expected, was the impetus for military strategists of the day to begin searching for alternative strategies. One of the most influential of these was an Italian general, Giulio Douhet. Italy had suffered 400,000 killed in the Great War, and those losses had profoundly affected Douhet. He theorized that the newly evolving airplane might offer a technological path to new ways of conducting wars in the future. In a seminal work, *Command of the Air,* originally published in 1921, Douhet proposed his ideas about a different kind of war. He theorized that war in the industrialized age could no longer be restricted to the armies of the combatant nations. Douhet postulated that all citizens were combatants in the new warfare, and, to win, the will of the people had to be broken. Armies could not successfully break the will to fight of the population of a nation. Douhet proposed that large bombers be used to conduct *strategic* warfare. Bombers, because of their range, speed, and relative invulnerability, could be used to strike at the industry, transportation, communications, and government of target nations. Using bombs that would include incendiaries and chemical weapons, the bomber would create panic in the population, and the people of a nation attacked in this manner would eventually demand that the government surrender. To those who had experienced World War I, this was a very credible view. Douhet is quoted as having said, "Victory smiles upon those who anticipate the changes in the character of war, not upon those who wait to adapt themselves after the changes occur." In a real sense, he was an advocate of using technology and employing it aggressively in the offense as a tool of modern warfare.

Douhet's theories found support among influential Britons. Sir Hugh Trenchard, chief of the air staff from 1919 until 1929, was developing his own theories and strategies for the employment of air power in the postwar environment. Britain had used bombers against dense population centers in 1918, and Trenchard evolved what he referred to as the "moral effect" of bombing, in which he speculated that the psychological effect of bombing was 20 times greater than the physical damage caused. Trenchard also developed the concept of *substitution,* in which he described his belief that war conducted using aircraft could effect the same levels of damage as that conducted with land or naval forces, but at considerable savings of both money and lives. In fact, Trenchard's success in keeping the young Royal Air Force from being dismantled and reabsorbed back into the army and navy was based largely on his ability to demonstrate the effectiveness of the airplane as a means of policing the far-flung British Empire without the need to send

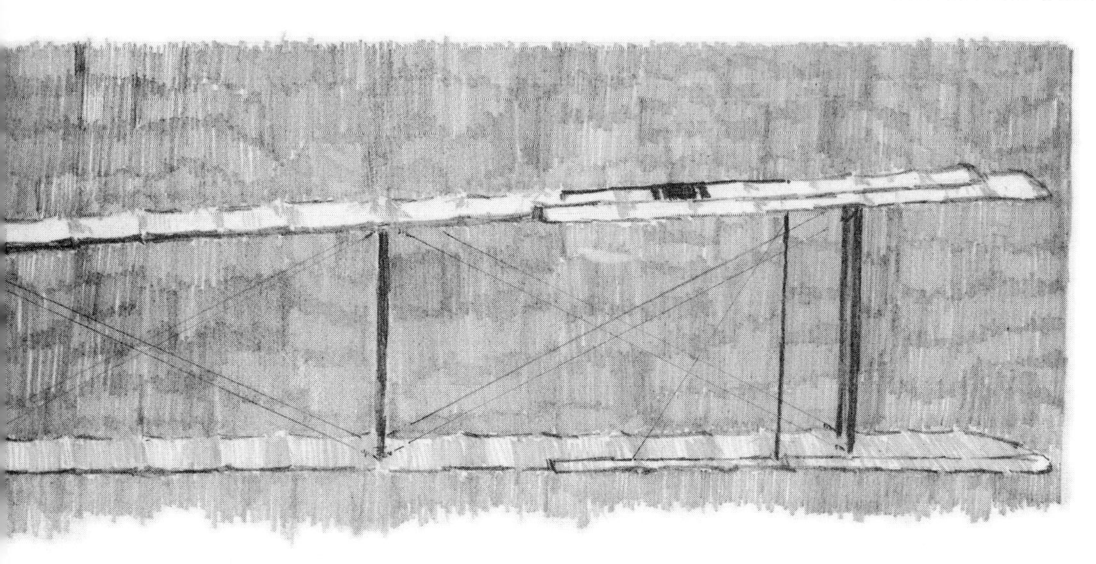

army and navy units. This was a very popular use of air power among politicians in a nation that was trying to avoid military expenditures whenever and however possible. Trenchard's funding policies in the RAF during the 1920s reflected his views about the effectiveness of the bomber: under his leadership, the major portion of the RAF budget was directed toward bomber forces.

As a result of Douhet's theories, supported by military leaders of the stature of Trenchard, Britain's politicians were also firm believers in the primacy of the bomber as the ultimate weapon. Even as Britain began to rebuild her defense forces, force structure plans reflected the prevailing belief in offense as the best defense. The Home Defense Air Force initially authorized in 1923 was to have two bombers for every fighter. The prevailing view was that the bomber was invincible; nothing could stop it. In fact, the bomber was thought to be so dominant that planners assumed offensive attacks would be conducted against

Britain by bomber forces unaccompanied by fighter escorts. In 1932, during a debate on disarmament, Prime Minister Stanley Baldwin made the following comment:

> I think it is well also for the man in the street to realize that there is no power on earth that can protect him from being bombed. Whatever people may tell him, *the bomber will always get through.* The only defense is offense, which means that you have to kill women and children more quickly than the enemy if you want to save yourselves. (Hough and Richards 1990)

This view implied two beliefs that were to significantly influence British air doctrine. The first was that offensive aerial warfare conducted by long-range bombers would be the chosen strategy for any attack on the British homeland; the second was that the bomber was so dominant that attacking forces would need no fighter escorts. On the positive side, the threat represented by the bomber would lead to recognition that England's air defense command and control system would have to be significantly improved, and that the British interceptors would need to perform well beyond the capabilities of anything then on any drawing board. However, the belief in bomber dominance led to British fighter tactics that assumed and even depended on the belief that attacking bombers would need no escorting fighters. This led the RAF to the development of fighting area tactics that were rigid to the point of being suicidal. The unyielding confidence in these ill-founded tactics would cost many fighter pilots their lives before they were finally discarded in favor of more flexible and fluid tactics appropriate to fighter-on-fighter engagements—but these changes were not universally adopted until after the Battle of Britain.

6. Fighting Area Tactics

With the three-aircraft vic having become the standard formation as the RAF emerged from World War I, the next challenge was to determine how best to employ the formation in operational use. Between the wars Britain developed a highly formal and standardized set of formation maneuvers referred to as *fighting area tactics.* The illustration of postwar Hawker Furies shows a three-ship vic executing the standard tactics. The leader is putting his flight through its paces by having the pilots quickly reposition to the various standard formations on his orders.

The concept of the fighting area emerged from the defensive architecture originally devised to protect London during the attacks by the German Zeppelin airships and Gotha bombers of World War I. In that concept, the areas east and south of London were divided into zones in which different kinds of defensive weapons were to be dominant. At the extremes, near the coasts, were the *outer artillery* zones. The task in those zones was to detect inbound aircraft and fire at them from the ground to disrupt and disorganize them as much

On lead's call, "Number 3" in a vic of Hawker Furies repositions to "Echelon right."

as possible. The next interior zone was designated the *fighter area* or *fighting area,* where fighter aircraft were to engage and down as many of the enemy as possible. Finally, the *inner artillery* zone was dominated by ground-to-air artillery as the last line of defense for the city.

After World War I, Britain initially speculated that France would be the most likely source of air attack from the European continent. Based on the assumption that the war and the Treaty of Versailles had ended the German threat and the recognition that France was the most powerful among the European nations, British engagement zones and the fighting area were positioned to anticipate an attack from the south. The area was redefined and reoriented several times as the threat from Germany grew and the fear of France faded, but during the years between the wars, the concept of a defined geographical fighting area remained an integral part of the planned defense of Britain against aerial assault.

As previously discussed, another prevailing assumption that took root during the years following World War I was that the bomber would be the dominant weapon in any aerial offensive. As noted, most British politicians and senior military leaders assumed that bombers would be so dominant that they would attack Britain without any fighter escort and without the need for heavy onboard defensive armament. This combination of factors—the responsibility for engaging and defeating the enemy, an unescorted bomber, in the fighting area—led to the development of a set of formation maneuvers intended to accomplish the anticipated mission. These maneuvers—standardized, numbered, and rigid—were the fighting area tactics.

There were six basic fighting area tactics:

1. *One three-ship vic versus one bomber.* Fighters attack in line astern from directly behind bomber, with each aircraft firing in turn.
2. *One three-ship vic versus one bomber.* Fighters attack from below and to side of bomber, with each fighter firing in turn, initially using 40 degrees of deflection (angle off).
3. *Two three-ship vics versus three bombers in vic formation.* Fighter vics line up one behind the other. Each fighter in lead vic fires on corresponding bomber. Once their ammunition is expended, second vic moves into position to repeat attack. If to one side, vics form into line astern and attack as in Tactic 2, reforming into vics after attack.
4. *Two three-ship vics versus three bombers in vic.* Combines Tactics 2 and 3. If line astern, then reform into three-ship vics after attack and repeat as necessary until all bombers are downed.
5. *Two three-ship vics versus bombers in large vic.* Last three bombers in formation are attacked first, after which fighters reform into echelon formation to continue attack under direction of lead.
6. *Squadron of 12 fighters in vics of three versus squadron of bombers.* Form into sections of three-ship vics in line astern to approach target, then vics swing out to line up behind bomber vics. As bombers are downed, fighters weave across leader's line to attack survivors. Requires substantial turning and forming and reforming as attack continues.

These maneuvers were adopted during the early 1920s and continued to be the standard tactics for the Royal Air Force through the period of the Battle of Britain.

Fighting area tactics had several distinct weaknesses. The first, and possibly most glaring, was that they were based on an assumption about the enemy that turned out to be absolutely wrong. The fundamental assumption was that Germany would attack using lightly armed bombers with no fighter escorts. In this scenario, a fighter could approach a bomber, choose a target, aim carefully, and shoot. No requirement for countering enemy fighters existed. However, when the German forces attacked, they did so with massive formations of heavily armed bombers accompanied by an excellent and aggressive fighter escort. Fighting area tactics had not prepared the RAF for the reality of the war it was fighting. Furthermore, there was the absolute rigidity with which the maneuvers were to be executed. Aircraft in the fighter vic were expected to fly in tight formation so that signals and commands from the lead aircraft could be immediately acted upon and the three aircraft could essentially act as a single unit. This continued even after radio communications between aircraft became common. All direction came from the leader; the job of the wingman was to maintain position in the formation. This lack of flexibility required that the wingmen devote the majority of their attention to watching the leader and maintaining their own positions relative to the lead aircraft. These formations allowed for none of the fluidity that engagements with opposing fighters demand, nor did the tight formations allow the wingmen to help the leader scan the skies for enemy aircraft. Another weakness was exposed when the three-ship vic turned. The aircraft on the inside of the turn would have to reduce power and hang on the leader's wing to maintain position. There are two things a fighter pilot wants to hold onto in flight—altitude and airspeed. Today's fighter pilot, confronted with the need to slow the plane's forward speed, will trade airspeed for altitude and climb, thus exchanging speed (kinetic energy) for altitude (potential energy). Then when forward speed is needed again, the pilot will dive, trading altitude for airspeed. Energy management is a significant issue in fighter maneuvers. The losses of energy that the tight three-ship vic imposed on the aircraft in the formation were unnecessary and dangerous to the pilots.

British fighter pilots initially went into World War II with the three-ship vic as the standard formation and fighting area tactics as the standard tactical maneuvers. Early engagements with German fighters during the last stages of the battle for France exposed the weaknesses of these tactics. As a result, some individual fighter pilots began to modify and, in some cases, abandon these tactics as they observed the German tactics and gained experience in combat. The changes consisted of looser formations, weaving maneuvers, and the gradual development of dogfighting (fighter-on-fighter) tactics. These individual initiatives notwithstanding, official RAF policy was to advocate the use of fighting area tactics until well after the Battle of Britain had begun.

7. Condor Legion

When civil war broke out in Spain in 1936, Germany was quick to offer support to General Francisco Franco in his opposition to the Republican forces. The clandestine force sent to Spain was known as the Condor Legion (*Legion Kondor*). Justified as necessary to prevent a communist takeover of a western European power, the German intervention in the Spanish civil war provided a perfect laboratory for the development of both air and ground doctrine and tactics, and it also offered an opportunity to test new aircraft recently developed in Germany. A formation of the new Ju 87 Stuka dive-bombers is pictured here enroute to a target in northern Spain. The knowledge and experience gained in

Ju-87's with Condor Legion markings heading for a target in Spain, c 1936.

the Spanish civil war would soon be applied in German blitzkrieg assaults on Czechoslovakia and other continental European countries in the late 1930s.

Despite the fact that the nations of Europe formed a broadly based committee specifically to discourage foreign intervention in the civil uprising underway in Spain, the military leadership in Germany saw an opportunity to extend the training program begun in Russia and to expose their growing military forces to combat conditions. Generalmajor Hugo Sperrle, a graduate of the training conducted at Lipetsk, in Russia, was given command of the Condor Legion. Volunteers were recruited from the Luftwaffe; those who agreed to serve in Spain were advanced one rank. To avoid direct antagonism of the European nations committed to nonintervention, Germany sent this newly organized force to Spain wearing civilian clothes and sailing as "cruises" that carried names like Strength Through Joy and Union Travel Society.

By May 1937, the Condor Legion was flying approximately 200 aircraft, including about 50 Ju 52/3 bomber-transports, 50 fighters, and various ground attack and reconnaissance aircraft. These were organized as one bomber wing, one fighter wing, and a reconnaissance squadron, each with an associated ground support organization. These ground support functions were an important part of the total Condor Legion experience. The intricacies of ground support operations were learned, organizational needs were identified, and structures were developed to handle them. For example, airfield staffs, medical units, and supply units were organized and manned, and the procedures necessary to their function were developed.

Operationally, the Spanish civil war was the proving ground for future Luftwaffe leaders, for the aircraft that would be employed in the Battle of Britain, and for the tactics that were used both during the Battle of Britain and in earlier battles as Germany waged blitzkrieg warfare throughout Europe. In addition to Hugo Sperrle, the commander of the legion, many of the individuals who would emerge as leaders and fighter aces in the Luftwaffe received their introduction to aerial combat in Spain. Among these was Adolf Galland, who at age 29 became a general officer in the Luftwaffe and who would later become one of the most admired fighter generals on either side. In addition to Galland, there were men like Hannes Trautloft, Walther Balthasar, Günther Lützow, and Hajo Herrmann—all of whom were later selected to lead fighter wings during the Battle of Britain and many of whom became aces, with five or more kills to their credit. These were the "young Turks" that replaced the older, more conservative leaders when Goering became convinced that his fighter leadership was not equal to the task at hand. Also included in this illustrious group was Werner Moelders, who is credited with developing the *Schwarm*, or "finger four," formation, which is still the basic four-ship formation used today. By 1940, the experience these young officers gained during the Spanish civil war, added to the experience gained in other European campaigns, made them among the best operational leaders in any air force in the world.

Spain also offered the prospect of operational testing for the many new weapon systems being produced in Germany during the mid-1930s. True, the bulk of the aircraft originally assigned to the war effort were Ju 52s and He 51s, but Germany later introduced the Ju 87 Stuka dive-bomber, the He 111 bomber, and even the Bf 109. The dive-bombing techniques that would come to characterize the blitzkrieg style of warfare were developed and perfected in Spain. The Republican forces were flying the Russian-made Polikarpov I 16, which was a good fighter for its time. It featured a low-wing monoplane design with a single radial engine that could power the little aircraft along at speeds near 250 mph. It was an excellent opponent for the Luftwaffe to cut its combat teeth on—good enough to require skill and cunning, but not so good that it was dominant. In addition to the testing of aircraft systems, the civil war gave Germany a chance to test other innovative concepts. The He 51 was the first aircraft to be equipped and fitted with drop tanks. Had the Bf 109 been similarly equipped during the Battle of Britain, the story might have ended differently. The Germans also experimented with tanks filled with gasoline and oil, which were dropped with effects much like those of the napalm bombs still used today during close air support missions.

No discussion of the Condor Legion and the German experience in Spain would be complete without mentioning the development of tactics. Spain provided the perfect venue to explore variations on the conduct of aerial warfare, and the tactical elements of blitzkrieg warfare employed in World War II were developed there. The use of air forces to support ground operations, with the heavy demands for precise bombing and well-coordinated attacks, was refined. Dive-bombing tactics that take full advantage of the capabilities of the He 51 and the Ju 87 Stuka were developed, and such perverse additions as the sirens attached to the extended landing gear that gave the Stuka its screaming effect during dives were added as psychological weapons. Moelders developed the highly effective finger four formation, which was far superior to the fighting area tactics then being practiced by the British.

It was in Spain that Germany developed the concept of mobility as the key to modern warfare. The world had seen and was familiar with the trench warfare of World War I. Germany broke the mold and based its operations in Spain on a highly mobile force that could move rapidly and strike anywhere. Galland describes a concept in which a fighter squadron might be quartered on a train, ready to move at a moment's notice. The aircraft would be flown to the base nearest the operational area and the pilots and support organizations would arrive by train. In this way new units and full operations could spring up overnight from almost anywhere. Similarly, aircraft were used to transport ground troops from one point to another, so that entire units could move rapidly across vast areas in hours, a truly revolutionary concept to a world that had recently fought a static war in the trenches of northern Europe.

The Germans also used their time in Spain to develop the support activities and operations that make a combat thrust successful and sustainable. They refined supply and maintenance procedures to handle repairs, fuel, and food for the fighting forces. They used reconnaissance aircraft to serve as the eyes for the ground commanders, providing them the intelligence necessary to find and assess the opposing forces on the ground. They also used the insights about their own operations to target or create weaknesses in enemy operations; for example, they employed rear-area attacks to confuse and disrupt the enemy's ability to supply and maintain its forces.

All said, the Germans, through the operations of the Condor Legion, used the Spanish civil war as a graduate school for the conduct of war. By 1938, as Germany prepared for the initial assaults on Czechoslovakia, Poland, and other European countries, the Luftwaffe had developed the leadership (at least at the operational level), the organization, the equipment, and the skills to conduct war as the world had never seen it. Once unveiled and launched, the effect was paralyzing, and the German Wehrmacht marched through Europe without significant opposition—until Germany attacked Britain in 1940.

8. The Moelders "Finger Four" Formation

The four Bf 109s shown flying in formation represent one of the most significant tactical innovations developed between the world wars. Werner Moelders, one of the greatest of all German fighter pilots, developed the formation while fighting in the Spanish civil war. The formation, called *Schwarm* by the Germans, was referred to as the "finger four" by the English pilots who first encountered it while flying over France. It was called the finger four because the positions of the four aircraft, when viewed from above the formation, appear to be in positions that resemble those of the four fingertips of a human hand when the fingers are extended and the hand is viewed from above. This formation revolutionized fighter tactics, enabling the pilots of the four aircraft to take on defined roles and provide mutual support while still taking full advantage of the flexibility and agility that the fighter offered.

Moelders was one of the young fighter pilots who rose to prominence fighting with the Condor Legion in Spain. He arrived in Spain in 1938, replacing Adolf Galland, who is quoted as having stated that Moelders was "the best man the Luftwaffe had" (Heaton 1997). It was during the last stages of the Spanish civil war that Germany began to bring its newest aircraft to the Spanish theater to test them under battle conditions. As a Bf 109 pilot, Moelders scored 14 kills in Spain, and by July 1941 he became the first Luftwaffe fighter pilot to score 100 kills in World War II.

Conditions in the Spanish civil war, where highly trained and experienced German pilots flew advanced aircraft against a relatively weak and inexperienced foe, permitted the

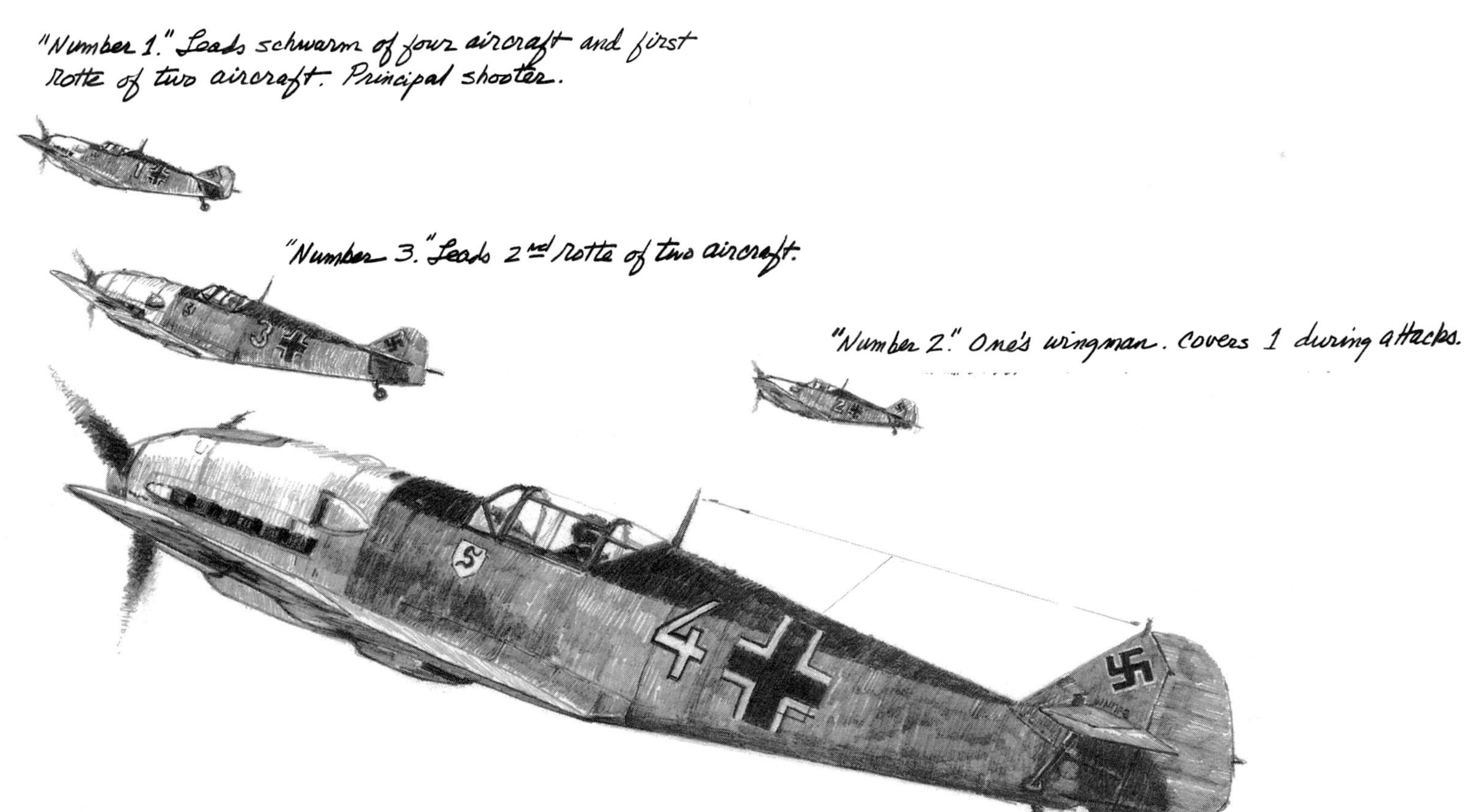

MOELDER'S "FINGER FOUR" FORMATION

Germans to experiment with different tactics in fighter-on-fighter engagements and evaluate which worked best. Moelders had become convinced that the optimum unit for fighter aircraft was the pair (or *Rotte*), rather than the three-ship vee, or *vic*, formation that was then in popular use among most air forces. The two-ship pair was under the command of a senior pilot as lead. The other aircraft, the wingman, flew on either side, positioning himself to take advantage of the sun and other conditions. The wingman flew with sufficient separation from the lead aircraft that he could search the skies for enemy aircraft and yet close enough that he could react as necessary to protect his leader. This was a complete departure from the tight formations that the English were flying, which demanded the complete attention of the wingman to maintain position in the formation without colliding with the flight leader.

To add additional firepower to the tactical fighter formation, the practice of flying in four-ship formations was developed. The finger four was made up of two pairs of fighters. When flying in this formation, each member of the flight had a specific responsibility:

- The leader of one pair was the flight lead; he made all decisions—to attack or not, maneuvers, and when to return to base for refueling. It was understood that all other members of the formation would follow the flight leader in whatever decision he made.
- The leader's wingman (the other aircraft in the lead pair) flew on the sun side of the leader. His responsibility was to protect the leader. He flew at an altitude slightly below that of the leader so that other pilots in the formation would not have to stare into the sun to see him.
- The second pair flew on the other side of the lead aircraft. The leader of the second pair was responsible for ensuring that no surprise attacks came out of the sun, so his focus was the sky around the sun.
- Finally, the wingman of this second flight had the responsibility for protecting the lead of that pair.

The four-ship formation, called the *finger four* or *fluid four*, is flown loosely—like the pair. If the leader makes a turn, the other aircraft maneuver as necessary to maintain their positions relative to the sun and to maintain their speed. The two pairs are very stable in comparison to the three-ships vics that the British were using during the same period. When a vic engaged the enemy, the tendency was for each member of the formation to separate and engage the enemy one-on-one. The nature of the pairs is such that, when a four-ship formation engages, the leader of each pair will engage with his wingman following; thus, two aircraft engage, with one protecting the other. The British would learn how effective this concept was when they met the Luftwaffe over France.

Ju 52 dropping paratroopers ... a key principle of Blitzkrieg warfare wa

Moelders once remarked that Galland was the Richthofen of World War II, while he was the Boelcke. Clearly, like Boelcke, he changed fighter tactics forever. Moelders went on to fly on every front—Poland, Britain, North Africa, and Russia. He commanded *Jagdgeschwader* (fighter squadron) 51 and later became a general officer in the fighter arm of the Luftwaffe. He went on to amass 115 victories before he was killed. Ironically, Moelders, this greatest of fighter pilots, died as a passenger on board a passenger aircraft while returning from the funeral of Ernst Udet.

9. Blitzkrieg Tactics

Blitzkrieg!—the name says it all. Literally "lightning war," this was a new and revolutionary approach to combat developed by the Germans in Spain and subsequently used in one attack after another as Germany swept from Czechoslovakia to France at the outset of World War II. As the name implies, blitzkrieg is characterized by surprise, rapid maneuver, and massive firepower. The symbols of this style of warfare include the Ju 87 Stuka, appearing as if from nowhere and diving with sirens screaming to bomb selected targets. The Stukas would be followed by fast-moving armored forces and very rapid deployments

of troops behind enemy lines to take defenders off guard. The Ju 52 shown as it drops airborne troops was one of several innovations that the Germans brought to this new form of warfare, enabling them to attack the enemy suddenly and in undefended areas.

Dating back to World War I, Germany had used airplanes as an adjunct to the Wehrmacht, the German army. Initially employed primarily as reconnaissance vehicles, aircraft gradually shifted to an offensive role, first attacking ground troops of the opposing armies, and then later attacking the aircraft and cities of the opposition—but always as an extension of the ground effort. After World War I, as the Germans emerged from the restrictions of the Versailles treaty, they began to develop the new fleet of aircraft that would carry the war to both the west and the east in World War II. However, the thinking among the leadership in Germany remained rooted in the tradition that air forces existed to assist the ground army, and this had a decided impact both on the types of aircraft Germany developed and the capabilities of those aircraft. For example, the slow but precise Stuka grew out of a desire on the part of the German leadership to execute precise bombing in support of army operations where ground troops were nearby. The Bf 109 was developed with severely limited range because the prevailing assumption was that it would be used to support army operations, primarily on the European continent, where long range was not a requirement. This "ground as the primary form of war" thinking also shaped the German bomber fleet. Few German leaders saw a need for long-range four-engine bombers, which implied independent air operations; instead, they favored producing more twin-engine bombers that could be allocated to the ground commanders, even though such aircraft carried smaller bomb loads and had less range. These decisions that shaped the Luftwaffe as a fundamentally tactical air force would all come to appear shortsighted when examined in hindsight after the Battle of Britain.

The tactics used by this tactical fleet—the elements that came to characterize blitzkrieg warfare—were developed and refined in Spain. The Spanish civil war was Germany's laboratory for development of the operational tactics that would be used in the world war. Blitzkrieg must be understood as one of the earliest—and best—examples of combined arms tactics, in which complementary ground and air movements are closely coordinated in order to take maximum advantage of the unique contributions that each brings to the battlefield. An assault might typically begin with the massing of infantry, artillery, and armor on the border of the country targeted for attack. These were the days before satellites could monitor such movements from space, so it was difficult for a targeted country to gauge the precise size and timing of the planned attack. An artillery barrage intended to soften up targets within a few miles of the front would announce the beginning of the attack. The artillery barrage would be accompanied by air strikes against deeper targets, disrupting the ability of the opposition to mobilize and reinforce front-line troops. Resupply and support facilities were also targeted. Stukas, with sirens screaming, hit key targets to simultaneously destroy them and to terrorize people on the ground. The Germans then attacked using armor supported by infantry, moving rapidly through the front lines of the enemy. The tactics used in blitzkrieg attacks, such as deep penetration bombing to disrupt reinforcement and resupply of the front lines and the use of air transport to move troops large distances within hours, were woven together to create an atmosphere of chaos among the defenders and invincibility on the part of the German attackers. Airborne troops were often dropped behind the front lines to further disrupt and confuse the situation. So rapid and so massive were these blitzkrieg attacks that most countries capitulated within days.

These tactics were successful not only in winning battles against the immediate target, but also in establishing an aura of invincibility around the German war machine. The German machine was viewed throughout Europe as a superior fighting force, well armed and well organized—a force that many believed futile to resist. Using blitzkrieg tactics, the Germans rolled through Czechoslovakia, Poland, the Netherlands, Belgium, and finally France. Only when they arrived on the northern coast of France at the shores of the English Channel and contemplated the next target—Great Britain—did it become obvious that the tactics of the next battle would have to be different. The Channel would force German planners to consider a battle in which the army would no longer be in a position to contribute to the total campaign. The blitzkrieg tactics developed in Spain and perfected on the European continent would not be suitable for the battle against Britain. This next battle would have to be fought primarily by the air force until an invasion force could be landed. In this battle, the short-range air force designed to augment the Wehrmacht would have to fight a long-range campaign, carrying the battle to England. The decisions that tailored the Luftwaffe to be an integral participant in land-war campaigns using blitzkrieg warfare—and which failed to consider the possibility of any other type of aerial campaign—would come to haunt Germany as the air battle over Britain was engaged.

10. British Experience in France

From 1936 into 1938, the Spanish civil war afforded Germany the opportunity to refine the doctrine and tactics that would be employed throughout Europe in World War II. The combat skills gained in Spain were further honed when Germany used overwhelming force to attack neighboring European countries, beginning in late 1939. Within a few months the German Luftwaffe would encounter the Royal Air Force in the skies over France. While this was not a battle that Dowding, as the head of Fighter Command, would have chosen, it offered the single benefit of providing his untested fighter pilots an opportunity to gain some measure of combat experience against the foe they would soon face in the Battle of Britain. The drawing depicts two British pilots, one using the universal language of the pilot—talking with his hands—to describe the close call he had on the mission from which he has recently returned. The bullet holes in the tail of his Hurricane tell the tale; he narrowly escaped with his life. Combat experience is invaluable—if one lives to talk about it.

As late as 1937, the Royal Air Force was woefully unprepared for modern aerial warfare in comparison to the German Luftwaffe. While the Germans were flying Bf 109s and were gaining combat experience in Spain, the RAF was still flying biplanes and performing fighting area tactics against imaginary fleets of unescorted and unarmed bombers. Two years later, the picture was improving substantially. Deliveries of Spitfires had begun in October 1938, and production had accelerated as airfields were prepared and squadrons were equipped with the new aircraft. Deliveries of the Hurricane likewise had begun, and by July 1939 there were 12 combat-ready squadrons. The RAF had made the transition from a World War I force to what was approaching a modern air force, and its strength was building to levels that would enable it to mount a credible defense against air attack. The objective force of 52 squadrons had not been achieved, but progress was definitely being made, in spite of continuing demands that Dowding commit resources for various activities that he considered superfluous to his primary mission.

As discussed previously, the German attack on Poland had brought both France and Britain into the war; they both declared war on Germany on September 3, 1939, three

Gaining experience... a Hurricane pilot shares his first taste of combat in France.

days after the attack on Poland began. Though his ultimate objective was Russia, Hitler initially turned to the west to secure that flank before committing to an eastern campaign. Before the declaration of war, France and Britain had reached an agreement that included the formation of a British Expeditionary Force (BEF) that would be deployed to France to assist in her defense in the event of war. The agreement further required that four squadrons of fighter aircraft be deployed to support the BEF. This requirement to send Britain's precious fighters to France was painful to Dowding as he continued the process of building the RAF to the levels of strength necessary to defend the homeland, but he had little choice in the matter. Four squadrons of Hurricanes were dispatched to France in September 1939, and, to Dowding's consternation, these were followed by two additional squadrons in November.

Hitler's westward turn brought the Low Countries, France, and ultimately England into his sights. In preparing for the war, France had, as countries too often do, assumed that any new war would resemble the last war and had, accordingly, built a line of static defenses—the Maginot Line—along and parallel to the German border. The Germans, however, had built their army and air force around the concept of the highly mobile blitzkrieg type of warfare, relying on movement and surprise. Hitler's generals had no intention of becoming bogged down in the static trench warfare that had characterized World War I. Their strategy was to attack to the north of France's defensive line—through neutral Holland and Belgium—then to turn the French flank and proceed south to Paris. It was brilliant, and it caught the Allies, like the Czechs and Poles before them, completely by surprise.

As the Germans swept into Holland and then across Belgium, the Luftwaffe, in typical blitzkrieg fashion, sent bombers, fighters, and reconnaissance aircraft to complement the ground thrust led by heavy armored forces. Initially, the British fighters had little contact with the advancing German air forces, but in December the air war began to get serious. These early engagements gave the British fighter pilots their first look at the front-line German aircraft. The Hurricanes met the German Bf 109s, the Me 110s, and the Do 17s; the learning curve was very steep. The early engagements with the Bf 109s underlined the need for identification markings to help identify friend from foe in the melee of the air battle and gave urgency to the need to install the variable-pitch propellers on the Hurricanes to improve their performance relative to the Bf 109. Here the British also learned the wisdom of coloring the undersides of their airplanes the light blue used by the Luftwaffe to make it more difficult to detect them from below. Air-to-air engagements with the Me 110s taught the British pilots that the 110 was faster than the Hurricane and had a higher service ceiling, but it also showed them that the Hurricane could turn faster than the 110, an advantage they could exploit when attacking. They also learned that the Do 17 was a sitting duck to a good fighter like the Hurricane. It was in these dogfights over France that the British pilots began to realize that their fighting area tactics were far inferior to the German finger four, and some began to experiment with alternative tactical formations similar to those used by the enemy. The battles in France also led to the first recommendations that guns should be harmonized at 250 yards rather than the 400 yards standard to British fighters at the time and to the consideration of the use of armor plating in the cockpits to protect pilots. In fact, many of the upgrades and improvements that continued through the period of the Battle of Britain had their genesis in the lessons learned in these fighter engagements in France.

The experience on the continent also changed the tactics associated with daylight precision bombing. Britain, like other nations whose leaders subscribed to the views advanced by Giulio Douhet, had assumed that the bomber would be dominant on the battlefield and that it would prevail without the need for fighter escorts. Events contradicted these assumptions, and, during two memorable missions in May 1940, Britain lost 16 of 17 bombers dispatched. Following these disastrous events, Bomber Command concluded that daylight bombing missions could be conducted only with dedicated fighter escorts. The immediate impact of this decision was to increase the workload imposed on Fighter Command to provide the needed escorts. Eventually these early experiences led Bomber Command to a decided preference for night bombing, which influenced British bombing tactics for much of the duration of the war.

The lessons learned in this campaign came at a considerable cost to Fighter Command. When the invasion of the Low Countries began, an additional four squadrons were sent to augment the six already there, and the RAF was, in addition, required to provide support from bases in England. By May the situation on the ground was desperate, and the political pressure to send more aircraft and additional squadrons to France was enormous, with requests for yet more reinforcements coming from the French premier, the king of Belgium, the queen of the Netherlands, and the commander of the BEF. To prevent more aircraft from being destroyed in what was apparently a hopeless cause, Dowding sent his "most famous letter" to the Air Ministry, expressing his conviction that the very survival of Britain would be imperiled if Britain did not cease the deployment of air assets to France and begin to conserve her fighters for the coming battle for the homeland.

In the final analysis, Fighter Command lost over 950 aircraft based both in France and England during the battle of France—over half of the front-line fighters it had when the German invasion of the Low Countries initially began. The losses included 386 Hurricanes and 67 Spitfires—the very aircraft that Dowding so desperately needed for defense against the attacks that everyone knew could not be long in coming. But as appalling as the loss of aircraft was, by far more serious was the loss of the men who had flown them. Experienced fighter pilots became the scarcest resource that Fighter Command had before the Battle of Britain ended.

On the positive side, for those who survived, the experience gained over France was invaluable. The lessons learned—the tactics, the appreciation gained for maintaining and supplying fighters under combat conditions, and the confidence gained by the fighter pilots that they could hold their own against the vaunted Luftwaffe—all would be vital in the battle that loomed just a few weeks into the future. In the assessments that followed Vietnam, one of the conclusions reached by the Tactical Air Command of the United States was that the likelihood of a fighter pilot's survival increased very rapidly after the pilot had completed an initial 12 missions. The genesis of the Red Flag tactical exercises that fighter pilots in the U.S. Air Force are routinely exposed to today is rooted in that conclusion. These exercises are intended to simulate combat conditions to improve the probability of the pilots' survival in actual combat—in essence, to give them an experience base. The experience in France provided the RAF with the experience and lessons learned that would carry them through to eventual victory in the Battle of Britain. In that sense, and despite the strains that the losses imposed, the experience was positive.

11. Dunkirk

In May 1940 as the Germans moved into Belgium, bypassing the French Maginot Line, the British moved forward to meet them with the French Seventh Army on their northern flank and the French Ninth Army to the south. The plan was to leave a gap between the British Expeditionary Force (BEF) and the French Ninth Army so that the Belgian army could retreat into this gap to complete the Allied front. The German commander, von Rundstedt, attacked with such ferocity that the Belgian army was routed. He then drove forward into the gap between the BEF and the forces to the south using classic blitzkrieg tactics and raced to the coast, isolating the BEF and French Seventh Army from the rest of the Allied forces. The BEF found itself trapped with the French on the northern shore of France, facing the German Wehrmacht to the south and the English Channel at its back. The Germans turned north and began a systematic advance that promised to crush and destroy an entire French army and the bulk of the British army. The BEF began a withdrawal to the only port still available in the area—Dunkirk.

At Goering's request the Luftwaffe had been awarded the privilege of destroying the English forces on the beach, and each day brought new attacks by German fighters and dive-bombers against the hastily prepared British positions. Just a week or so before, Dowding had won his point regarding the demands to send additional fighter squadrons to France; Churchill had agreed that no more fighter units would be deployed to the continent. With the sudden reversal in Belgium, however, the order was sent that Fighter Command was to render all possible support to the armies trapped at Dunkirk. Air Vice Marshal Park, commander of 11 Group, which had responsibility for operations in the south and east of Britain, was faced with the puzzle of how to stretch his limited resources to provide the coverage required. Park initially attempted to provide round-the-clock coverage, but that necessitated sending single-squadron patrols—easy pickings for the Luftwaffe now massed for the kill. On May 27, 11 Group flew over 200 sorties and claimed 37 kills, but this was still inadequate to halt the continuous Luftwaffe attacks from the skies around Dunkirk. Park lost 14 planes and decided to strengthen his patrols—first to two squadrons, then to four—but this required time gaps between patrols, sometimes up to an hour. The picture showing a number of Spitfires returning to base as a group illustrates both Park's solution and the problem it caused. With the Spitfires deployed in larger groups there would necessarily be time required to refuel and repair the airplanes before they could once again return to the Dunkirk area. The Luftwaffe seized on the opportunities created by the gaps in coverage to attack the British troops virtually unopposed. The German forces advanced steadily, and by May 24 they were poised to administer the coup de grâce that would crush the British army. Then, for reasons that are still not entirely comprehended today, the German advance was halted.

Since May 20, when the drastic plight of the BEF had become obvious, the British Admiralty had been rounding up every naval vessel, fishing craft, and pleasure boat capable of sailing across the Channel in anticipation of an emergency extraction of the troops on the beach at Dunkirk. On May 26 this armada of small boats set sail. With the Luftwaffe having been awarded the privilege of administering the killing blow to the Allies trapped on the beach, aircraft from bases in Holland, Belgium, and France set to work

Downside of large patrols at Dunkirk. Everyone returns to refit at the same time ... leaves gaps in the coverage.

scouring the beach—a classic shooting-fish-in-a-barrel situation. On the evening of May 26 General Guderian, commander of the German Panzer divisions in the area, reinitiated the armored assault on the perimeter around the beachhead.

For the next week "Operation Dynamo," the extraction of the BEF and French from the beaches at Dunkirk, proceeded in spite of the combined German air and armored attacks. The French Seventh Army established and held the perimeter while the small-boat armada loaded and sailed time after time with its human cargo. By June 4 only 4000 British soldiers remained inside the perimeter with 100,000 French soldiers. In the final analysis 338,226 British and French troops were loaded at the beach at Dunkirk and carried back to England. Virtually all of their equipment was lost, but total disaster had been avoided. The RAF flew over 650 bomber sorties and over 2700 fighter sorties in support of the extraction. In spite of these impressive numbers, the average British soldier saw little evidence of the air force while the Germans were prowling the beaches. The RAF, limited in resources available to provide coverage and often required to engage German fighters over the waters of the English Channel and above the clouds that had descended over Dunkirk, found itself the object of severe criticism from the soldiers being extracted. The hard feelings that arose out of the misperceptions surrounding this event continued for some time after the debacle at Dunkirk had ended. In spite of the unfortunate fact that many of the exhausted British troops felt that the "flyboys" had abandoned them, there is little question that this operation would never have achieved what can only be considered a miraculous outcome had the RAF not carried the fight to the Luftwaffe as it did.

12. Order of Battle

France surrendered to Germany on June 22, 1940. There could be little question that Britain would be the next country to be targeted. The map showing primary airfields and operations centers provides a notional bird's-eye view of England as viewed from the air above northern France. The defeat of France, the withdrawal of British Expeditionary Force at Dunkirk, and the return of the surviving aircraft of the Royal Air Force to England left the airfields in the Low Countries and in France open for occupation by the Luftwaffe. Goering lost no time in positioning his forces and organizing the Luftwaffe for the coming campaign. Operating out of the bases in Belgium, Holland, and northern France would place the Luftwaffe on the doorstep of Britain, substantially reducing the problems of range that would have been introduced had the aircraft operated from Germany. Bombing missions launched from Germany would have precluded fighter escorts, justifying some of the earlier British assumptions regarding the unescorted and lightly armed bomber as the primary threat. The situation in June 1940 made it clear that they would face well-armed bombers escorted by the high-performance Bf 109, one of the finest fighters in the world at the time.

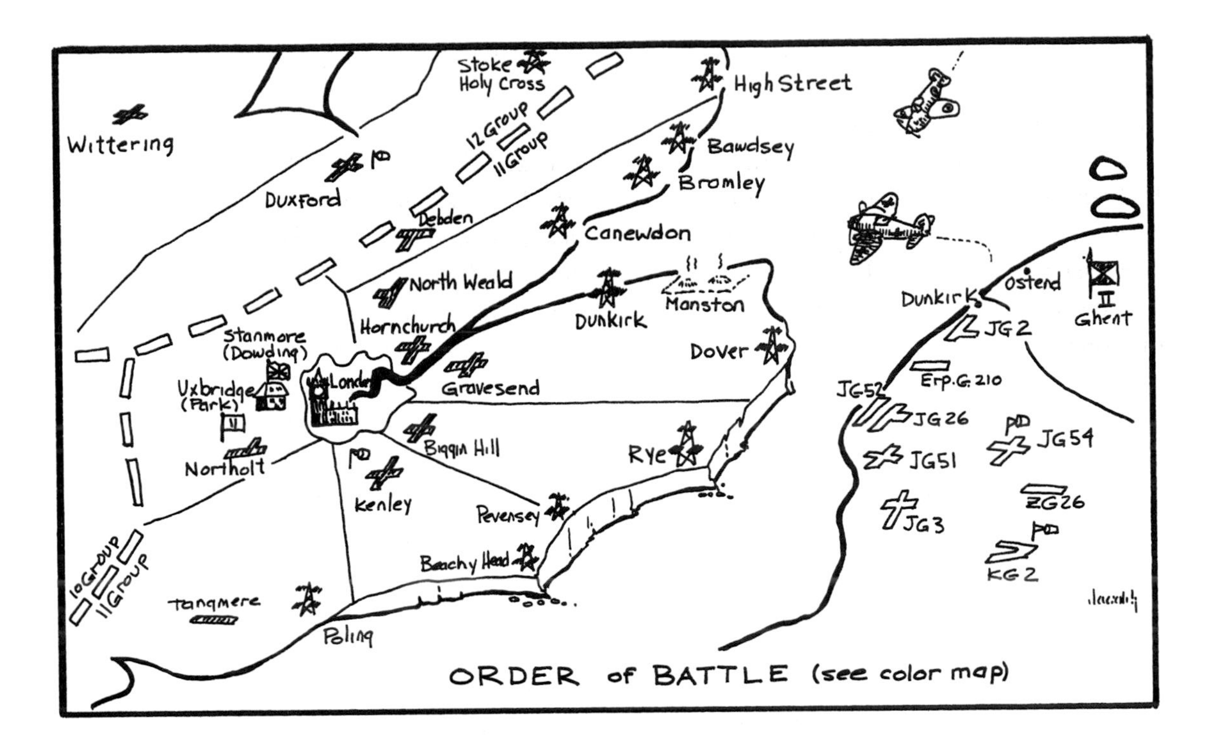

The Luftwaffe was organized into air fleets (*Luftflotten*), each of which was assigned a full range of aircraft—fighters, bombers, reconnaissance, and transport—enabling it to conduct independent operations. In preparation for the coming battle with Britain, *Luftflotten 2* was deployed to bases in Holland, Belgium and northeastern France under the command of Generalfeldmarschall Albert Kesselring, an extremely able leader who later held various high command positions in the Wehrmacht. Generalfeldmarschall Hugo Sperrle, a graduate of Lipetsk and a veteran of the Condor Legion, was given command of *Luftflotte 3,* located in northwestern France. A new fleet, *Luftflotte 5,* was created for the campaign against Britain and was positioned in Scandinavia under Generaloberst Hans-Jürgen Stumpff. The two remaining *Luftflotten* in the German air force, 1 and 4, continued to occupy positions in Poland and Germany.

Estimates of the number of aircraft deployed in *Luftflotten 1, 2,* and *5* vary because there was understandable confusion associated with the rush to occupy the recently abandoned bases in the Low Countries and France and preparations for the coming assault on Britain. Some estimates have placed the total number of aircraft deployed for the attack against Britain at approximately 2800, of which about one-half were bombers and about 800 were Bf 109s. Estimates of combat-ready aircraft range considerably below these levels; some estimates have placed combat-ready aircraft as low as 50 percent of the total aircraft assigned.

On the British side of the Channel, three operational commands opposed the German forces—Fighter Command, Bomber Command, and Coastal Command. Both Bomber and Coastal Commands played important roles in the war effort, but it was Fighter Command that bore the majority of the burden during the Battle of Britain. Fighter Command, under the direction of Air Chief Marshal Sir Hugh Dowding, was organized into groups, each of which had a defined geographic responsibility. Fighter Command headquarters, located at Bentley Priory near London, exercised central control. Initially, there were three groups: 11 Group had responsibility for the areas south and east of London, 12 Group had the areas north of London and stretching from coast to coast, and 13 Group was to defend the northern areas of Britain and Ireland. A fourth group had already been authorized, and 10 Group, responsible for the areas in the south and west of Britain, became operational in July.

The British commanders were, like their German counterparts, able and outstanding leaders. Air Vice Marshal Keith Park, a brilliant New Zealander, commanded 11 Group. By virtue of his assignment to defend the southern and eastern coasts, he, with Dowding, is generally recognized to have been one of the central figures in the successful defense of Great Britain. Air Vice Marshal Trafford Leigh-Mallory led 12 Group. He was an extremely able commander although often a critic and competitor of Park's during the Battle of Britain. Air Vice Marshals Richard Saul and Quintin Brand commanded 13 and 10 Groups, respectively.

On the first day of July 1940, Fighter Command had a complement of some 700 Hurricanes and Spitfires, of which somewhat fewer than 600 were combat ready, and approximately 1200 pilots. The losses in the campaign in France had cost both countries dearly in terms of both aircraft and experienced pilots; furthermore, the Germans had yet to clearly define their strategy and objectives for the next battle. Both countries needed time to complete deployments and make other preparations for the coming battle, but the certainty that war was coming to the shores of Britain was clear. The only question left on the first day of July 1940 was how and when it would begin.

13. Ground-Controlled Intercept

The development of Britain's integrated air defense system had begun in World War I. The basic architecture of the system actually changed very little between the wars; the fundamental composition of the system consisted of the detection, tracking, communications, command and control, and interception elements. However, the way those functions were accomplished underwent vast changes in the late 1930s as radar, IFF, and new aircraft like the Spitfire and Hurricane were introduced. The cartoon identifies the primary components of the system that was central to British success in the defense against German attacks. If the Germans had fully appreciated the effectiveness of the

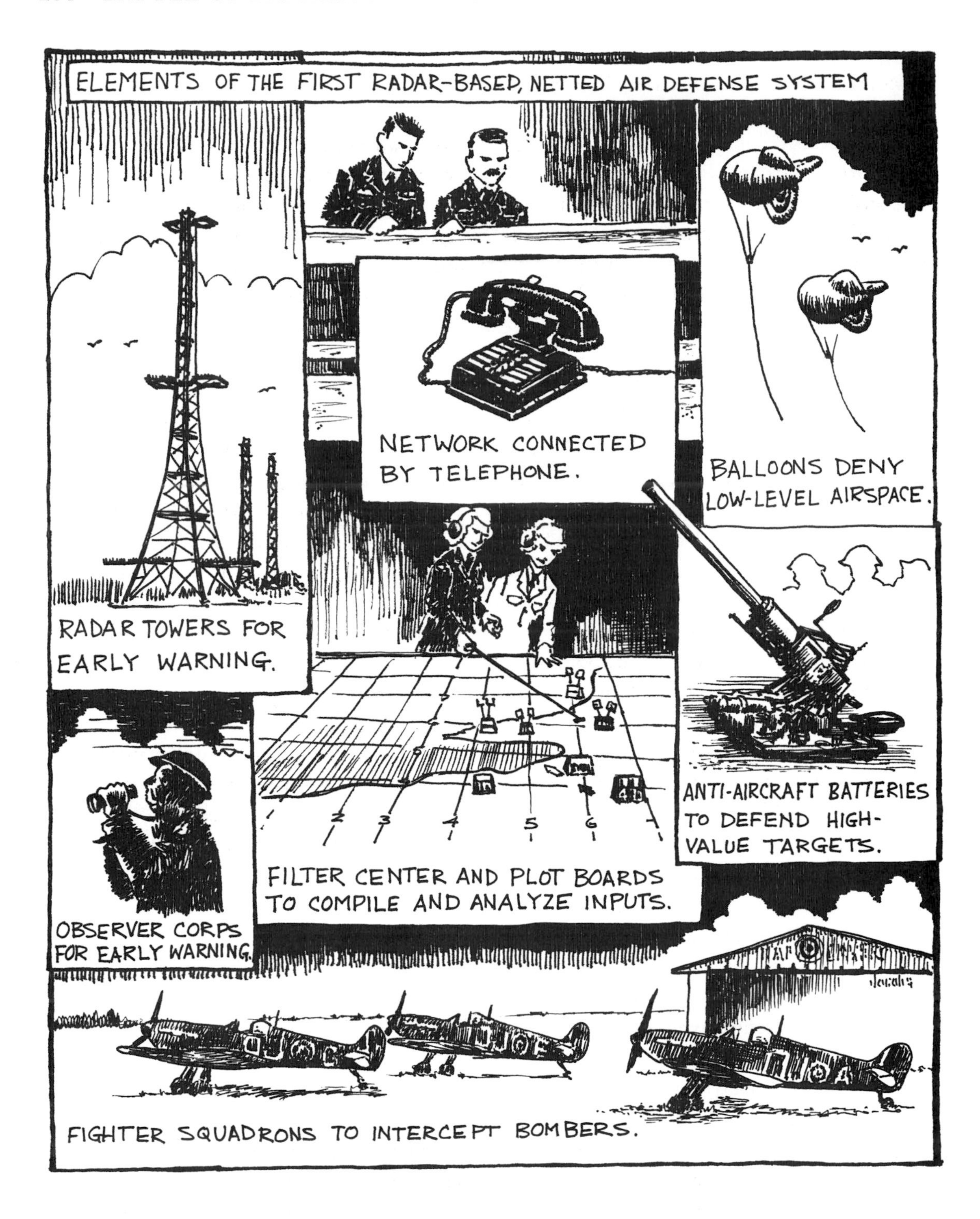
ELEMENTS OF THE FIRST RADAR-BASED, NETTED AIR DEFENSE SYSTEM
NETWORK CONNECTED BY TELEPHONE.
BALLOONS DENY LOW-LEVEL AIRSPACE.
RADAR TOWERS FOR EARLY WARNING.
ANTI-AIRCRAFT BATTERIES TO DEFEND HIGH-VALUE TARGETS.
FILTER CENTER AND PLOT BOARDS TO COMPILE AND ANALYZE INPUTS.
OBSERVER CORPS FOR EARLY WARNING.
FIGHTER SQUADRONS TO INTERCEPT BOMBERS.

integrated air defense system and the central role that it played in the air defense of Great Britain, they would surely have placed it among the highest-priority targets. Without it, the British could not have prevailed.

As the RAF prepared for the attack that everyone knew was soon to come, the old outer artillery zone of the World War I defense architecture was abolished and the responsibility for engaging the enemy at and beyond the coast fell to the fighter aircraft of the RAF. The geographical area assigned to each group was divided into sectors. There were 15 sectors in all—each with three fighter squadrons. The Chain Home radar stations, augmented by Chain Home Low, were in place and operational by July 1940. When a radar return from an inbound formation was received at a sector radar station, the officer or noncommissioned officer (NCO) on duty would immediately report the azimuth, altitude, position, and estimated number of aircraft to the filter room at Fighter Command headquarters by landline telephone. It was the job of the filter room to identify duplicative reports from adjacent stations and to generally ensure the accuracy and completeness of the information to the extent possible. Based on the information available, decisions were made regarding which group, or groups, would be alerted to react to the incoming attack. The information and orders would then be passed to group operations and to sector operations. At each of these locations the same information was plotted with color codes to indicate the currency of the information. Group headquarters subsequently made decisions about which sectors to alert for the interception. Sectors, in turn, controlled the fighters to the point of interception.

Augmenting the Chain Home detection system, a substantial observer corps was organized whose task it was to assist in the tracking of inbound bomber formations once they had crossed the coastline. Observer posts were linked to group operations control centers so that reports could be immediately relayed to Fighter Command and these position reports could be integrated with the radar reports. Searchlights, as well as both heavy antiaircraft guns for high-flying aircraft and light guns for defense against low-flying aircraft, were deployed along the expected attack routes. Barrage balloons were also provided at major ports and population centers.

With the establishment of the detection, tracking, and communications elements of the defense system, the challenge remained to equip the necessary numbers of fighter squadrons to complete the interception element of the system. The defense plans of 1938 and 1939 reflected an agreement between Dowding and the Air Ministry that 46 squadrons would be authorized to defend against what they estimated would be an attacking force of 1750 German bombers. The demands to support the British Expeditionary Force in France, compounded by requirements to deploy squadrons in the north of England and in Ireland, reduced Dowding's effective forces. Dowding responded with a number of rancorous exchanges with the Air Ministry, but the evidence shows that the disagreements between the commander and his headquarters were the normal tensions between people and organizations committed to the same objectives, but with different responsibilities and different priorities. By April 1940, Fighter Command had a total of 47 squadrons, of which 22 were equipped with Hurricanes and Spitfires.

The last elements of the integrated ground-controlled intercept systems were added when "Pipsqueak" and the identification-friend-or-foe (IFF) capabilities were added. These enabled ground controllers to discriminate between friendly and enemy aircraft in flight and to control friendly aircraft up to the point at which visual contact could be established. With the elements of the system—detection, tracking, communication, and interception—in place and, equally important, with the operational procedures established, Britain stood ready for the battle that began in July and gradually increased to the firestorm that it became before the summer of 1940 was over.

14. July 10, 1940: The Battle Begins

Wednesday July 10, 1940, is the date chosen to officially mark the beginning of the Battle of Britain, although the choice of that date is really a convention rather than recognition of some particular event or attack. German attacks on English shipping in the Channel steadily increased in intensity beginning immediately after the defeat of the Allies in France. On July 10, a large convoy of ships—code name Bread—departed the Thames estuary and attracted the attention of German reconnaissance aircraft in the area. Once the Germans had spotted the convoy, information was passed to their command and control elements in the Pas de Calais, whereupon a huge formation consisting of bombers and fighter escorts was massed for an attack. The large German formation was detected on the Chain Home radar, and British fighters were launched to defend against the attack. As shown in the illustration, a massive dogfight ensued, where for the first time—but hardly the last—over 100 German and British fighters engaged in air-to-air combat over the Channel. July 10 was later chosen as the day to mark the official start of the Battle of Britain, primarily in recognition of the increase in the intensity of the air war that the events of that day heralded.

As July began, nothing in the experience of the Germans would have led them to consider the possibility that Britain would not eventually yield to the might of the German war machine. Clearly, the simultaneous air and ground assault that characterized the attacks on other countries could not be mounted against Britain; the English Channel made that impossible. On the other hand, the German High Command had every confidence that German bombers, attacking key targets in mass, could achieve the desired results. The theories advanced by Douhet seemed to have been validated by the German experience to that point. They had sent the bombers into Spain, Poland, the Low Countries, and France, and in every case the targeted country had surrendered within a matter of months or, in some cases, weeks. Hitler fully expected Britain either to surrender out-

July 10, official start of the Battle of Britain; First 100-plane dogfight.

right or, at a minimum, to agree to a nonaggression pact that would give him the freedom to turn eastward toward Russia. To complete the plan, once the Luftwaffe had established control of the air, an invasion force would follow to ensure the defeat of Britain and end the threat of interference from the Allies.

The air activity leading up to July 10 had begun almost immediately after the fall of France. The records from both sides show the British consistently destroyed more German aircraft than they lost in the daily air engagements. However, it is also true that the British losses were almost entirely Hurricanes and Spitfires, whereas the German losses were spread across both bombers and fighters, with the Dornier Do 17s being the German aircraft most often destroyed in combat. The records further indicate that most of the Bf 109s destroyed and damaged sustained the damage in landing accidents rather than in combat—a sobering thought for those who would face these fighters in the air war to come and testimony to the difficulty of handling the narrow wheelbase that was characteristic of the 109 design.

As the intensity of the air campaign increased, marking the official beginning of the Battle of Britain, the leadership of each side watched for the indicators that would signal the directions of the coming battle. Dowding must have realized that the German attacks on British shipping in the Channel had dual objectives. The longer-term objective was to impair Britain's vaunted maritime capability; but the more immediate—and, for Dowding, more worrisome—objective was clearly the attrition of the air assets of Fighter Command. These early engagements over the Channel gave clear indications that the Germans would aggressively target British fighters, both in the air and on the ground. Preserving the assets of Fighter Command would soon become the primary concern for the RAF leadership and for Dowding, in particular. In fighter-on-fighter engagements, the Germans appear to have more than held their own, although the total aircraft exchange ratio significantly favored the British. Given that they had over 1300 bombers as the campaign started and could afford losses, the Germans must have taken substantial comfort in the performance of their Bf 109s relative to the British Spitfires and Hurricanes. As things stood on July 10, 1940, the Germans no doubt felt a justifiable level of confidence that they would prevail in this last battle against the western nations. The Battle of Britain had begun.

15. Germans Attack Convoys

A Stuka dive-bomber rolls in to begin an attack on a ship in the English Channel. To increase the chances of getting a direct hit, the pilot would align the attack along a line drawn from bow to stern. Ideally, the pilot would prefer to attack from the rear to maximize the probability of success; in any event, the pilot would prefer the longitudinal attack to an attack across the beam, which offered a smaller target. A slow-moving ship may have seemed, at first glimpse, to be a relatively easy target for German aircraft, but sinking a ship

Ju87 dive bombing shipping in the English channel.

from the air was, and still is, extremely difficult. The mere fact that the target was moving complicated the issue, particularly at the time of the Battle of Britain, since aircraft-mounted radar was not yet developed sufficiently to be operational, and the bombsights of the day were basically designed to handle static targets. In their own defense and to complicate the problem, alert captains would maneuver along their routes of travel, often performing a series of S turns. The fuses on most bombs were anything but sophisticated, so a near miss usually meant that the bomb did no damage. In many cases, torpedo bombers were more effective than were the conventional bombers, which were normally used to attack land targets. Even if the aircraft scored a hit, a ship might still escape if the crew was able to contain the damage to a relatively few sealed compartments. All in all, it was a very difficult mission for an air force in 1940. One might well ask why the Luftwaffe started the Battle of Britain by attacking shipping. The answer was almost certainly that the ships were only incidental targets; the real targets were the British fighters that came from the bases in southern England to defend against the German attacks.

Today it is common to refer to the different phases of the Battle of Britain, the first of which is generally recognized to be the Channel phase or *Kanalkampf.* This probably gives a more strategic flavor to the initial stage of the battle than is justified. In fact, Hitler and Goering had not, by the early days of July, established a well-thought-out strategy for the battle to come. As early as May 1940, Hitler had identified Britain's aircraft industry as the primary target for German aerial attacks in response to earlier raids by Bomber Command in the Ruhr Valley; however, little other than sporadic attacks took place until France surrendered. With the fall of France, the German Luftwaffe began to occupy the airfields of northern France and Belgium, and the serious planning got under way for the coming attack on Britain. The somewhat ad hoc nature of the German offensive at this stage is no doubt due at least in part to the absence of a coherent strategy and plan of attack.

While the Luftwaffe waited for a plan of attack to be developed and communicated to the operational levels, Kesselring and Sperrle, the commanders of *Luftflotten* 2 and 3, had already begun to think about the objectives that would need to be defeated if the campaign against Britain were to succeed. Kesselring was, without doubt, a far better military thinker than either Hitler or Goering; it is not surprising that he and Sperrle would have identified Fighter Command as the primary obstacle to be overcome in the coming campaign. They no doubt concluded that they could use the time available and the numerical advantage they enjoyed in every type of aircraft to deplete the British defenses while the leaders and their staffs pondered this last major effort in the west. It is important to recall that the fighters of the Luftwaffe had been uniformly successful in each battle they had undertaken to that point in the war. They were the most experienced fighter force in the world; they had in the Bf 109, one of the finest airplanes in the world; and they had already defeated the Allies when they met them in France. There was no reason for the Luftwaffe not to believe that they could defeat the RAF, so the idea of luring them into engagements over the Channel while waiting for the battle plans to be finalized must have seemed inviting.

This is not to imply that the attacks against the ships in the Channel were not pursued with all the intensity that the Luftwaffe could muster. As July wore on, the British persisted in sending convoys through the Channel, in spite of the German attacks—perhaps more as a political statement of will than a necessity. As the German deployments and preparations for the campaign to come were completed, they in turn increased the size and frequency of their attacks. The German attacks, while focused primarily on shipping, also included key airfields and ports, like Portsmouth and Dover, located on the south and east coasts of England.

To carry out the attacks, Germany used the complete array of aircraft types that they intended to use when the main assault on the English homeland began. The record shows that Do 17s, He 111s, Ju 88s, Ju 87s, and Me 110s were all regularly used during this phase of the battle. One of the by-products of the action over the Channel for both sides was knowledge gained about the capabilities of their own aircraft relative to those of the opponents. For their part, the British began to understand that the Defiant, with its rear-facing turret guns, was no match for the Bf 109; on July 19, an entire squadron of the aircraft was essentially annihilated. The Germans learned that two of the designs most favored by the leadership of the Luftwaffe, the Ju 87 Stuka and the Me 110, were far more vulnerable than they had thought. The Ju 87 had not previously been engaged in an environment where the preponderance of the opposition flew aircraft of the quality of the Spitfire and Hurricane, and it was becoming ominously clear that the slow-moving Stuka was particularly vulnerable as it entered and recovered from its dive while delivering bombs. The Me 110, often referred to by Goering as the "Destroyer," proved unable to handle the nimble British fighters, and this was a major concern, because the Germans were also becoming aware that they could not consider sending unescorted bombers into action; the British fighters were far too lethal for that, and the Me 110 was to have been the long-range escort fighter. By the end of July the RAF had lost 69 fighters, and the Luftwaffe had lost approximately 155 aircraft of all types.

16. Satellite Basing

Lying in the open waiting for the signal that will place them on alert or send them speeding over the Channel to meet enemy bombers, these pilots and their aircraft have been deployed to forward bases. These remote satellite fields were sparsely provisioned, often consisting of little more than a grass field. Aircrews and maintenance crews were rotated into these fields, located as close to the coast as possible, in order to reduce the time required for them to intercept incoming German aircraft and deny them easy attacks on British shipping and airfields.

Most forward satellite airfields had aircraft, air crew, and little else.

The defense, whether one is speaking of sports or war, always suffers the disadvantage of having to be prepared to react and take action at the time and place that the offense chooses. In the case of air defense, this becomes a time-distance problem. If the enemy can be detected at a sufficient distance so that the defensive interceptors can take off, climb to altitude, and intercept at some point before the incoming aircraft can launch its missiles or drop its bombs, then the probability of a successful defense is relatively high. If, on the other hand, the attacking force can be positioned to launch its weapons before being intercepted, then the defense fails. Solutions to this fundamental problem can be approached both technologically and tactically. For example, in modern warfare the development of stealth technology is intended to delay the point at which an inbound aircraft can be detected, thereby lessening the time that the defense has to react. On the other hand, a defensive tactic might consist of mounting round-the-clock patrols by interceptors so that the time to intercept is reduced once the target is detected.

The Channel phase of the Battle of Britain offered several unique challenges to the RAF that dealt explicitly with this time-distance issue. From their position in the Pas de Calais area of northern France, the Germans could, depending upon the weather, actually see British ship convoys as they proceeded through the Channel. Detection was further aided by a continuous flow of reconnaissance aircraft that kept watch over the Channel and reported to the German command and control elements in the Pas de Calais. The Germans could then assemble bomber squadrons from the various bases in northern France, mass them into formation, and attack at the time and place of their choosing. The total distance from the French coast to convoys in the Channel varied from 30 to about 100 miles, so even if the British Chain Home radar system detected the bombers as they massed for the attack, the time before the Germans could be in position to attack was relatively short.

To provide protection and to lessen the time needed to react to German attacks, the RAF tasked Fighter Command to provide fighter escorts for important convoys. The geographic division of responsibility that characterized the Fighter Command group and sector organizational structure meant that 11 Group, which was responsible for the southeast quadrant of England, and 10 Group, which defended the southwest quadrant, had primary responsibility for most convoy escorts as well as defensive interceptions in the Channel area. Squadrons in the sectors that bordered on the Channel were naturally the ones generally tasked to escort convoys and intercept inbound German formations; however, the bases within these key sectors were often a considerable distance from the coast. In those cases, even though the alert from the ground-controlled intercept system may have been timely, the time and distance that the fighters had to fly in order to reach the Channel rendered their defensive efforts ineffective. To reduce the time required for fighters on the ground to react to German attacks in the Channel, Dowding ordered Fighter Command to institute a practice of redeploying fighter squadrons to forward satellite bases. In addition to reducing the reaction time, basing fighters at forward locations also offered the benefit of increasing the number of sorties that an aircraft could fly during any given day. Life at the satellite fields was often difficult, but forward basing was an excellent tactical response to a difficult situation.

These bases were sometimes as primitive as a cleared field with tents, or they could be true airfields that were still under construction. They were, under any circumstances, located close to the Channel so that aircraft taking off could arrive over convoys under attack with minimum time delays. Depending upon the rotation, pilots might stay for a few days, or, in other cases, they might rotate in and out of the bases daily. Conditions at the fields invariably required that pilots bring any clothes they might need for the duration of their stay, and the food available generally consisted of sandwiches or other fare that required little or no preparation. Satellite bases were natural targets of German attacks, and the dangers to both the flying crews and the servicing crews were substantial. Servicing crews, in fact, were generally rotated only periodically, and, in that respect, their lot was even worse than that of the pilots.

17. RAF Abandons Three-Ship Vic Formation

While the German Luftwaffe adopted Moelders' finger four formation and made it an integral part of their standard fighter tactics by the late 1930s, the RAF had clung to the three-aircraft vic, or vee, formation and the highly standardized fighting area tactics described previously. When the RAF was required to deploy fighters to France in support of the British Expeditionary Force, the British fighter pilots quickly learned that the tight three-ship vic formation was inferior to the German formation in every respect. It was not as flexible as the finger four; it did not afford the mutual protection that fighter engagements demanded; and it focused the wing-position pilots' attention on the leader to the exclusion of watching for enemy attacks. It was not long before some British flight commanders began to experiment with alternatives in the search for tactics that would not place them at an automatic disadvantage. The formation of Spitfires flying line astern shows one of the tactical variations employed as the RAF transitioned from the tight three-ship vic to formations that increasingly emulated the German finger four.

Evolving formations...some RAF commanders used "line astern" between switching from "Vics" to "finger four" formations.

Some of the squadrons in France began to borrow from the French air force tactic that featured a five-ship formation consisting of a spread-out three-ship vee formation with the two other aircraft positioned behind them. The two behind the formation were expected to weave back and forth to protect against attack from the rear. Another variation on the weaving tactic involved joining two three-ship vics to make a flight, then having one or two aircraft (the number-three aircraft from one or both vics) weave behind the flight. These "tail-end Charlies" were vulnerable if they dropped too far behind the flight, and the Germans often managed to isolate them and pick them off in combat. This led to the use of complete sections—three-ship vics—to act as the weavers, protecting the aircraft in a full squadron formation. Only after Dunkirk, where casualties were high among the weaving tail-end Charlies, did the RAF abandon the use of weavers behind their fighter formations.

As the Battle of Britain began, the three-ship vic remained the officially approved tactical fighter formation within the RAF. This fact notwithstanding, pilots—especially those with experience in France—continued to experiment and search for improved tactical formations. In the early phase of the battle, 111 Squadron was known to form into very shallow vics or line-abreast formations and attack the Luftwaffe bomber formations head-on. The idea was to cause the bomber formation to break, after which the fighters would maneuver to pick them off—but to say that this tactic required nerves of steel is to understate the situation, at the very least, and it did not address the shortcomings of the tactical formation inherent in the three-ship vic.

By the end of the Battle of Britain there were actually only a few squadrons—for example, 501 and 605—that had actually, on their own initiative, adopted the true finger four developed and perfected by the Germans. That fact notwithstanding, many were looking for alternatives. "Sailor" Malan of 74 Squadron was apparently instrumental in leading Fighter Command away from the three-ship vic. He began to explore variations such as placing his sections in line astern, or trail, formations, and he is credited with having been one of the first to begin to experiment with breaking the standard squadron of 12 aircraft into three flights of four, rather than the standard four flights of three. Other records show that 19 and 54 Squadrons were among the first to routinely begin using formations other than the three-ship vics during the Battle of Britain. 54 Squadron adopted the use of pairs, rather than threes, when combat losses made flying in threes impossible. Somewhat later, 152 Squadron, under the leadership of Squadron Leader P. K. Devitt, began to make a standard practice of using formations made up of "pairs" and "fours" (Mason 1969). The standard unit in this construct was the pair, where the lead had responsibility for decisions and the wingman for defending against attacks from the rear.

In spite of the fact that some of the squadrons were openly abandoning the three-ship vic, the official policy of the RAF remained unchanged, and most RAF squadrons, as a result, continued to practice and use the three-ship formation in combat. Finally, in the spring of 1941, well after the Battle of Britain had ended, the official policy of the RAF was changed to make the four-ship section—and the finger four formation—the standard, but far too many pilots lost their lives using the three-ship vic formation during the Battle of Britain. Some pilots later charged that the failure to move more rapidly to the finger four was attributable to the fact that those in positions to set tactical policies and practices were not veterans of encounters with the Luftwaffe and the Bf 109s and, as such, were inclined to place more emphasis on rigid control than on flexibility in the air.

18. Me 110s Use Lufbery Circles

So taken was the Nazi leadership with the Me 110 when it was designed and developed by Messerschmitt that it was chosen by Goebbles, the minister of propaganda, as one of the aircraft regularly featured in movies and broadcasts touting the speed and armament of the new long-range fighter. At 350 mph, with a range of over 500 miles and carrying two cannons and four machine guns, the airplane certainly seemed superior to the fighters of the day. Goering particularly liked the look and performance of the new airplane, and it was he who named it the *Zerstoerer*—the Destroyer.

The first sign of weakness in the design showed up during the air war in the battle for France, when British pilots shot one down and reported that even though the Me 110 was faster than the Hurricanes, it was less maneuverable. They indicated that a Hurricane could turn more rapidly than the Me 110. In a dogfight, the aircraft that can turn faster will often prevail. If that airplane is in front, the faster turn will make it impossible for the pursuing aircraft to bring his gunsight to bear on the target; if the pursuing aircraft has the faster turn, that aircraft can easily maneuver into a firing position. The report of the British Hurricane pilots indicated that the design of the Me 110 might make it vulnerable to the more maneuverable Hurricanes and Spitfires.

By the time the Battle of Britain began, the earlier thinking that "the bomber would always get through" was no longer operative. It was generally recognized that any bomber attack would require fighter escort. In a nutshell, the success or failure of the German attack would depend largely on the ability of the German fighters to defeat the British fighters sent to intercept and down the bombers. The German fighter fleet as the battle began consisted of about 1000 aircraft, of which about 200 were Me 110s. As soon as the larger engagements started, the Me 110s were observed to periodically form themselves into Lufbery circles, the nose-to-tail circling defensive maneuver conceived during World War I and named for the inventor of the maneuver, Raoul Lufbery. The Me 110s shown in the illustration have formed a Lufbery circle, and the defensive nature of the maneuver is evident. It was a harbinger of things to come, and it did not auger well for the vaunted *Zerstoerer.*

The Me 110's vulnerability shows when they begin to revert to Lufbery circle in July.

The circles flown by the Me 110s generally took two forms. In the classic Lufbery circle, performed strictly as a defensive tactic, the aircraft in the circle would fly around a static point on the ground. In the second form, the circling aircraft actually moved along a flight path while circling. This form of the maneuver afforded the same defensive cover as the first, but it enabled the formation to progress toward safety—often cloud cover or the French coast. In either case, it was a sign of weakness relative to the British fighters, particularly the Hurricane, which had a shorter turning radius than the Spitfire.

When the Lufbery circle was originally conceived in World War I, aircraft of the day did not possess the speeds and agilities that were common to the aircraft of the Battle of Britain. The Lufbery circle assumed that attacks against the aircraft in the circle would primarily come from a horizontal aspect by airplanes operating at essentially the same altitudes as those in the formation. British pilots soon learned that they could position themselves above the circles and then dive on the aircraft flying the horizontal path that defined the formation, passing through the formation at speeds too fast to allow those in the formation to react. As soon as they had passed to the outside of the circle, the attackers would reverse their vertical direction with a high-*g* pull and attack again from below. This was an extremely effective tactic that left the Me 110s in the circle in the position of lumbering targets. Many Me 110s were shot down as they flew in this formation that had once been thought the perfect defensive tactic. They were simply unable to defend themselves from what was essentially a vertical attack, and the losses of the 110s soon mounted to levels that could only have been considered disastrous. It starkly illustrated the idea that tactics and technology must be matched if they are to be successful in battle.

By mid-August, the weaknesses of the Me 110 were regularly exploited by the Hurricanes, and in one agonizing week the Luftwaffe lost 79 of them plus many damaged. Goering, at a conference held with his commanders, ordered that they be used only when the range requirements exceeded the capabilities of the Bf 109; otherwise, they were to be escorted by the 109s in combat. This was patent nonsense—an escort for an aircraft that itself had been designed as an escort—and was symptomatic of the meddling by the German High Command in what should have been purely tactical decisions. The 110 continued in use, performing as both a long-range fighter and a fighter-bomber throughout the war, and although it performed relatively well later on during the air war over continental Europe, it never lived up to the expectations that the German High Command held for it as the Battle of Britain began.

19. Fighting over Own Territory

To the British pilot whose aircraft was so damaged that he had to bail out, the descent by parachute may have been the most hazardous part of the entire ordeal. Having survived the gun or cannon shells that downed him and the fire that often forced him to leave the aircraft, the pilot would struggle to open his canopy, unbuckle his seat belt, and pull himself out of what might well be an aircraft completely out of control. There were no ejection seats in 1940. One might think that the pilot could heave a sigh of relief once he had cleared the airplane and his parachute had deployed and could then concentrate on landing without further harm. The problem, from the German perspective, was that a pilot who parachuted to British soil represented a pilot who would likely return to the battle in another airplane before the day was over. The pilot in the picture here, twisting in his harness as he tries to see and somehow avoid the attacking Bf 109, illustrates the situation that was a fact of life (and death) during the Battle of Britain—British pilots were fair game for German fighters even when they were dangling under their parachutes.

War is often characterized by seemingly odd rules observed by either side, or both, depending on the tactical circumstances. The Germans routinely targeted British pilots who had "hit the silk," much to the dismay and anger of British fighter pilots, but Fighter

RAF pilots bailing out over England... potential future combatants to the Luftwaffe...

Command policy was unequivocally opposed to allowing the British pilots to reciprocate. The policy was rigidly enforced from the command level down. The exceptions were the Polish and Czech pilots, who hated the Germans with a ferocity that would not be denied. They often targeted German aircrews in parachutes as a token response to the violence perpetrated on their countries when Germany invaded. In a document published after the war, Dowding addressed this issue, stating that he fully understood the German position on shooting British pilots during parachute descents and thought that their actions were defensible, given that British pilots fighting over British soil were clearly still potential combatants, whereas German pilots descending over Britain were not (Dowding 1946). Of course, the British position was that attacks on air-sea rescue aircraft and boats in the Channel were appropriate, both because it was suspected that the aircraft were acting in reconnaissance roles and because a crewmember rescued in the Channel could be returned to the battle just as certainly as could a crewmember bailing out over home soil. The rule was consistent; any crewmember likely to return to combat was a valid target. It was a deadly serious game, but even then Dowding was able to say that in some cases the German pilots refrained from shooting descending British airmen and instead "greeted a helpless adversary with a cheerful wave of the hand." In spite of the fact that warfare had become increasingly technical and complex, some of the spirit of the "brotherhood of eagles" that was born during World War I lived on in the skies over England.

The fact that the majority of the air combat took place over British territory also had a tremendous effect on pilot morale and incentive. British pilots understood well that this was a battle for the survival of their homeland, and that alone accounted for a willingness to accept personal sacrifice that contributed immeasurably to the effectiveness and aggressiveness of their attacks. British pilots routinely encountered forces that far outnumbered them, yet they plunged into the battle day after day. To the German pilots, on the other hand, this was yet another battle in a series that had started two years earlier—a series that would not end even if they won the battle. Most of them had been fighting intermittently since the Spanish civil war, and they had fought steadily since the occupation of Austria. A particularly difficult issue for German crews was the danger represented by the English Channel. The Bf 109s had severely limited range; rarely could they linger over British soil for more than 15 or 20 minutes before they would see the "fuel low" lights begin to blink, warning them that they must either return to the French side of the Channel or risk having to bail out or ditch in the cold waters of the Channel. Repeated exposure to this possibility often left German crews with a condition referred to as *Kanalkrank,* a psychological condition brought on by concentrating obsessively on the possibility that one might have to bail out into the Channel. Over time, the combination of morale and incentive that the British derived from defending England and the corresponding lack of incentive that German crews felt as the war wore on became a significant contributor to the British victory in the battle.

20. Hitler Directives 16 and 17

The cartoon showing Hitler as he contemplates Germany's next adversary in the campaign in western Europe captures the essence of two of the most important directives he issued during the Battle of Britain. During the battle, Hitler often announced top-level policy decisions or strategies by issuing formal, numbered directives to his field commanders. These directives were relatively few, and they were treated as orders that the commanders were to support, formulating their tactics as necessary to accomplish the tasks described in the directives. These are analogous to the National Security Decision Directives used in the United States today to communicate presidential decisions on defense-related issues to the president's staff and commanders. Hitler's Directives 16 and 17, issued during July and early August 1940, were of particular interest and importance to the conduct of the Battle of Britain.

Directive 16 dealt specifically with German plans for an invasion of England. As mentioned earlier, Hitler had expected—and clearly would have preferred—to strike an arrangement with the English that would have allowed him to turn his attention to the east and Russia without having first to fight another major battle in the west. Even today, questions remain about the extent to which he was actually committed to an invasion. Some have suggested that the issue did not receive the priority and planning necessary because the German general staff viewed the Channel crossing as a "river crossing on a broad front." The use of this phrase perhaps gives the impression that the Germans thought that crossing 30 or more miles of open water was equivalent to crossing a river. The more likely explanation is that Germany had never faced the tactical problems involved in a full amphibious assault, nor did Germany at this point in time have the benefit of having observed the techniques used by the United States and its Allies later in the war. They had little to use as a reference other than the tactical river crossings that any army must be prepared to execute. What is evident is that Germany never prepared adequately to execute the kind of operation that would have been required to cross the Channel, invade, and hold territory on the English homeland.

Initial invasion planning was done in the autumn of 1939 by the German naval staff under the direction of Admiral Raeder. Raeder apparently undertook this planning as a contingency in the event that Hitler should decide on an invasion as a part of some future strategy. There is no solid evidence that invasion was part of a coherent plan at this point, and as the planning for Norway, the Low Countries, and France intensified, this effort was put on hold. Then, in May 1940, when the German army arrived at the coast of northern France, Raeder reopened the issue directly with Hitler, not because he was enthusiastic about the idea, but because he did not want to be caught flat-footed in the event that Hitler should decide that the navy would lead such an operation. As the Channel phase of the Battle of Britain got under way, Hitler and his staff were preoccupied

Hitler's Directives

DURING THE BATTLE, HITLER ISSUED TWO DIRECTIVES WHICH INFLUENCED THE CONDUCT OF THE CAMPAIGN...

with developing a strategy to guide the effort. In meetings with the navy, Raeder argued that an invasion should be undertaken only as a last resort, and then only with total air supremacy. Raeder strongly urged the senior staff to consider a naval blockade. In early July, in similar meetings with the army, Hitler was told that, if the navy could transport the army across the Channel and the Luftwaffe could provide cover, the army could successfully establish and hold a beachhead in Britain. The prevailing assumption was that the British army had been so weakened by the experience at Dunkirk that it would not be able to resist effectively once the Wehrmacht was on British soil.

On July 16, Hitler issued Directive 16, in which he stated that as a result of England's unwillingness to recognize her hopeless situation and negotiate an agreement to Germany's satisfaction, he had decided to prepare for and, if necessary, execute an invasion. The operation was given the code name *Seelöwe*—"Sea Lion." On July 19, at an awards ceremony for those who had led the campaign in France, Hitler made what he termed his "final appeal to reason"—another appeal to the British to capitulate, which was refused outright in a response from Lord Halifax, the British foreign secretary, on July 22. Hitler then directed that the Luftwaffe target the British navy and coastal facilities during the *Kanalkampf* to weaken the British navy while planning went forward for the main aerial attack, yet to begin.

The internal debate regarding the timing and conditions for invasion did not stop with the issuance of Directive 16, however. In late July, at a meeting with Hitler, Raeder stated that the earliest possible date that the navy could be ready for an invasion would be September 15, and he further argued that he personally preferred May 1941 to give adequate time for planning and preparation. Hitler responded that preparations would have to be complete by September 15. But nothing was ever finalized on this matter, and on August 1 the German High Command (OKW) issued a statement that "eight or fourteen days after the beginning of the air offensive against Britain, scheduled to begin on 5 August, the Fuehrer will decide whether the invasion will take place this year or not" (Hough and Richards 1990). Given that this statement placed a decision point around mid-August and that navy preparations would likely not be completed to the satisfaction of Raeder, one has to question the level of commitment to an invasion at the highest levels in Germany.

On August 1, Hitler also issued Directive 17, in which he stated that in order to establish the conditions necessary for the final defeat of England, the Luftwaffe should direct its attacks against the RAF, focusing on the flying units, the airfields, the ground installations and supplies, and the English aircraft manufacturing capability. Hitler reserved for himself the right to initiate terror attacks in reprisal for British attacks on Germany. The main attack would begin on or after August 5, depending upon the state of preparations and the weather. Directive 17 set the stage for the second phase of the Battle of Britain and the enunciation of Goering's *Adlerangriff* (Eagle Attack) plan, describing the broad tactics that would govern the conduct of this second phase of the battle, which would very nearly exhaust and defeat Fighter Command and the Royal Air Force. The expectation on the part of the German High Command was that *Adlerangriff* would make invasion unnecessary.

However, to be prepared to support the invasion in the event it were required, the German navy began requisitioning barges, tugs, steamers—whatever type of boat appeared to have the capability of carrying either men or matériel across the Channel—from ports in northern Europe. That the Germans decided to depend upon such ill-suited and inadequate craft speaks volumes about either the seriousness of their plans or their lack of understanding of the problem. (Compare these preparations, for example, with the preparations of the Allies to invade France in 1944. Allied preparation, planning, and execution were far more complete and thorough.) By July 19, they began moving these craft toward the planned departure ports, which naturally convinced the British that invasion was imminent. It is important to recall that Britain had moved over 300,000 men from Dunkirk to Britain, albeit without their equipment, so it is not at all unreasonable that they viewed these preparations very seriously, no matter how inadequate they may appear with the benefit of hindsight. On September 5, Bomber Command began attacks on the invasion ports in an effort to reduce the German capability to invade, and on September 7 the British chiefs of staff issued the code word "Cromwell," meaning invasion was deemed imminent and all troops were to report to battle stations.

What the British could not have known was that the earliest invasion date had been delayed from September 15 to September 21. Then, on September 11, Hitler delayed the decision for three days, and on September 14, he delayed it an additional three days. On September 15, later commemorated as Battle of Britain Day, it became apparent to all concerned—German and British alike—that Germany would not prevail in the air war. German air supremacy, a condition that had been established early on as a necessary condition for a successful invasion, would never be achieved. Within 48 hours, Hitler postponed the invasion indefinitely. Directives 16 and 17 set in motion events that defined and characterized the Battle of Britain, but in the final analysis, the objectives described in them were never achieved. As shall be seen, the failure to eliminate Fighter Command as a factor in the battle and the consequent total reliance on offensive bombing in the absence of supporting ground forces ensured the eventual failure of the German offensive.

21. Free-Chase Fighter Sweeps

The broad tasks assigned to Luftwaffe fighters could generally be separated into two distinct types: bomber escort and free-chase missions. These two missions were differentiated by the ultimate objective of the sorties themselves. Bomber escort missions had as their objective ensuring that the bombers reached their targets and dropped their bombs

Low level sweep (with high cover) to draw RAF fighters into a decisive engagement, 1940.

successfully. The free-chase mission essentially uncoupled the fighter from the bomber. It was a mission devoted strictly to the destruction of British fighters as the ultimate objective. German fighters assigned these missions were free to maneuver as necessary to engage and destroy the opposing fighters. The attrition of RAF fighters was an important objective of the Channel phase of the battle, and the complete destruction of the RAF and its airfields was established as the central objective of the second phase. German free-chase tactics were an important aspect of achieving these objectives from the outset.

The picture showing a flight of Bf 109s flying high cover while a second flight attempts to lure British fighters into the sky illustrates German free-chase tactics. If the British rise to the bait, the aircraft flying high cover will dive on them, taking advantage of their altitude as the British fighters climb to engage the lower flight. From the earliest days of the Battle of Britain, German free-chase fighters attempted to get British fighters to take off and engage them in air-to-air engagements, often as a prelude to a bombing attack. The plan was to have the fighters exhaust their fuel, forcing them to return to the airfields for refueling just as the bombers appeared. These tactics would not only decrease the number of fighters available to intercept the bombers through combat losses, but would also leave the fighters exposed on the ground where the Germans might destroy them during refueling and maintenance operations.

When Britain first mounted standing fighter patrols near the Channel to protect the shipping convoys that were the targets of German attacks, the Luftwaffe began to respond by sending out medium-altitude fighter sweeps to attack these standing patrols. By the end of July, it was apparent—from the British perspective—that the German free-chase fighters represented the single greatest threat to Fighter Command's continued existence as a viable force. Somewhat later, the Germans established what amounted to a continuous stream of free-chase fighters flying some 20 miles off the English coast and parallel to it. From this stream they would randomly turn toward the coast to simulate an attack on an airfield or radar station in an effort to tempt RAF fighters into the air. Some have speculated that, had Germany continued to emphasize and mount free-chase tactics throughout the airfield phase of the battle, this may have eventually destroyed Fighter Command. Certainly, had the Bf 109s been equipped with drop tanks that would have extended their range and endurance, the use of free-chase tactics would have been even more effective. As it was, these fighter sweeps could generally not remain over British soil for longer than 15 or 20 minutes before low-fuel conditions demanded that they depart for the French coast.

By mid-August the Germans had defined their main attack (*Adlerangriff*) strategy, the initial objective of which was the destruction of the RAF and its facilities. The bombing tactics employed had also evolved to combine both the bomber escort and the free-chase missions. Bf 109s escorted each bomber formation, with free-chase fighters supporting them, operating at some distance from the bomber formations. As British fighters rose to meet the attacking bombers, the free-chase fighters led attacks to deflect and destroy them whenever possible. The majority of British losses were to the free-chase fighters. As British losses mounted, Air Vice Marshal Park, commander of 11 Group, was finally moved to publish Instruction No. 4, which unequivocally identified German bombers as the primary objective of the defense and directed controllers not to send fighters to engage reconnaissance aircraft and other fighters and to attempt to time defensive intercepts so that British aircraft could reach British soil if they were damaged or pilots had to bail out. On August 19, a formation of approximately 100 Bf 109s cruised along the southeast coast in an attempt to goad the British into launching their fighters. True to their directions, no British fighters responded to this overt provocation. When British fighters then responded less frequently to the provocations of the free-chase fighters, Goering and Kesselring, with the advice of Galland, interpreted the lack of response to mean that they had essentially achieved the air superiority that had earlier been established as a condition for invasion.

There is little question that Germany had found effective tactics that would probably have resulted in the eventual defeat of the RAF had events been allowed to run their course to the end. It is further evident that the free-chase fighters were at the heart of the success the Germans were experiencing. But before these tactics had the full impact that they might have had, Hitler changed the direction of the war. In a decision based primarily on emotion—the desire for vengeance in response to British bombing attacks on Germany—but also at the urging of his Luftwaffe, Hitler turned away from the RAF as the primary objective just as victory was within his grasp. Of the several serious errors that Hitler made during the battle, this was the strategic error that ensured the survival of Fighter Command—and Britain.

22. German Bombing Tactics

Bombers, by their nature, have several problems to overcome. They are usually slower and less maneuverable than the fighters charged to intercept them, so they cannot normally choose the time and place to fight; they must be ready to engage enemy interceptors at any point in the mission. If they are successful in penetrating enemy airspace to the target area, they must then locate the specific target that they trying to destroy—often in conditions made difficult by night and smoke that obscures the target, not to mention antiaircraft fire from the ground. Finally, having located the target, the challenge becomes to actually drop the bombs near enough to destroy it. It is not an easy task; it requires all of the bravery and commitment of the fighter missions but enjoys little of the glory.

During the Battle of Britain, German bomber crews developed various tactics to increase their odds of survival and improve the likelihood of successfully destroying targets. Like their counterpart flying fighters, they recognized the advantages of flying in

Ju88 bomber formation ... provides interlocking defensive fire and tight bomb impact area on the ground.

formations. German bombers normally took off from bases in northern Europe and gathered into massive strike forces as they climbed to altitude before heading north and west toward England. This allowed them to establish their formation positions and avoid British fire while they were slowly climbing to altitude. Once the formation was in place and they were operating at cruise speeds, they would take up the attack heading. All German bombers were armed, typically with machine guns operated by gunners located forward, rear, top, and bottom. By grouping multiple aircraft into formations, the bombers were able to accomplish the same sort of mutual coverage that the fighters sought. As fighters approached the bombers, their positions would be called out by radio so that all aircraft in the formation could direct fire at them. The interlocking fire from multiple aircraft meant that any approaching fighter would receive fire from more than a single bomber. Formations also offered the advantage of delivering the bombs dropped in more tightly grouped patterns. The three Ju 88s shown in formation illustrate both the mutual protection afforded by interlocking defensive fire and the potential for tightly grouped bomb patterns as they all drop their loads simultaneously.

The size of the German formations varied, sometimes relatively few aircraft, but often very large by today's standards. During the airfield phase of the battle, the formations were smaller, as might have been expected, since they were attacking many targets, often simultaneously. However, in the later phases, as the strategic objective turned to the destruction of London, there are descriptions of bomber formations that were massive almost to the point of disbelief. On September 7, 1940, the first day of the campaign against London, a formation of approximately 400 bombers escorted by over 600 fighters departed Europe en route to London. The formation was over 1½ miles high and covered 800 square miles (Mason 1969). Imagine yourself as a British fighter pilot when you first caught sight of that armada! It would be a sight that you would never forget.

During the London phase of the battle, the Germans experimented with various types of formations. One, sometimes referred to as the *Krokodil* (crocodile), consisted of very long streams of bombers flying along the same path, with elements of the formation spaced at 4-minute intervals (Chant 1979). At times, multiple units might fly attacks along multiple paths chosen to avoid having them interfere with each other. These *Krokodil* tactics produced attacks that lasted for hours with bombs falling every 4 or 5 minutes. As the British improved their ability to perform night intercepts using airborne radar, the Germans responded by using more tightly grouped formations intended to flood and overwhelm the defenses. Attacks from these formations lasted less than an hour, but they were much more intense than the attacks from the more spread-out formations.

Of course, it is seldom satisfactory to just drop bombs on a city without regard to precisely where they fall. In every attack, there are defined targets—factories, harbors, airfields, for example. The Germans used several techniques to improve the accuracy and therefore the effectiveness of their bombing raids. The use of beam-flying technologies, *Knickebein* and *X-Gerät,* were used to great advantage to improve bombing accuracy during both poor weather and night operations. The Germans also used fire to create aim points for bomber formations. The most accomplished and deliberate practitioners were the crews of *Kampfgruppe 100,* a pathfinder unit whose job it was to locate targets using *X-Gerät,* then to drop incendiary bombs, creating fires that could be seen even through smoke and clouds. The bomber formations that followed would then aim their bombs at the fires, although they might never actually see the target. During cloudy conditions, the British countered this technique by setting fires in uninhabited areas outside the city in hopes of drawing the bombers away from the population centers.

For the German bomber crews, the trip back to their home bases could be as dangerous as was the approach to the target. Again, they had to run the gauntlet of British fighters determined to take down as many of them as possible while they headed for the Channel and safety. In many cases this meant re-forming on the run, since searchlights, antiaircraft fire, barrage balloons, and the relentless attacks of the defending fighters often worked to scatter the formation in the target area. The missions assigned to bomber crews were difficult in the extreme, and, as noted, bomber crews seldom reaped the glory that the fighter pilots enjoyed. It is useful to remember that in the final analysis the success of most aerial offensives is measured by how well the bombers do the job they are assigned. "Bombs on target" is a common measure of success in tactical and strategic operations.

Coordinating a massive bomber attack, or the defense against such an attack, is a daunting task. In the first place, it demands an intelligence capability that correctly assesses the condition of targets so that attacks are not wasted on targets that are either already destroyed or have little value. An information system is necessary that enables the commander to rapidly assess the combat readiness of resources, and a planning capability must be in place to decide which units will participate, which targets are to be allocated to the units involved, and what weapons are to be carried. Routes must be selected, radio procedures verified with code names, and call signs established. Rescue must be coordinated for downed crewmen and, although not a consideration for the Germans during the Battle of Britain, refueling planned. Then there are the tactics within tactics—how the attacking or defending forces are to be fed into the battle to maximize or minimize the damage done. For example, a pattern observed during the Battle of Britain involved a three-phased attack in which small incendiary bombs were initially dropped to mark the target, after which larger incendiaries were dropped. A delay would follow to encourage damage-control organizations to reach the fires, after which high-explosive bombs would be dropped, obviously intended to destroy not only the targets themselves but also the crews charged with controlling fires. In addition to this tactical chess game, diversions were also employed to draw the defenders away from the main attack. The commander in charge of such operations must truly possess the broadest knowledge about the many resources available, and that knowledge must be accompanied by excellent organizational and personal skills in order to send forces into battle when all know that some will not return.

Modern bombers are still faced with the same fundamental problems that the Germans faced in the Battle of Britain. They must be able to penetrate, find the target, and hit it. Many state-of-the-art technologies involved in air warfare today are focused on solving these problems. Defenders today use both radar and infrared (IR) sensors to detect inbound aircraft. Stealth technology is intended to make penetrating aircraft more difficult to detect; flares are dropped to defeat IR-guided weapons. Targets may be located using airborne radar, or they may be illuminated using laser beams. The weapons dropped today are often guided to the target using data links, laser guidance, and satellite positioning. The march of technology aimed at improving the ability to accomplish the fundamental tasks will continue, as will the countermeasures to defeat them, but at the end the challenges remain essentially the same as they were in 1940, whether one refers to those that confront the individual bomber or the challenges involved in planning a campaign.

23. Matching Capability with Responsibility

A Spitfire attacks a BF 109, while a Hurricane engages a cell of German bombers. That this would occur was no accident. Faced with the need to target both German fighters and bombers, Air Vice Marshal Park explicitly directed this division of responsibilities to ensure that a mismatch in performance did not give the Germans an advantage in the air that could be avoided by thoughtfully matching the capability of his two primary fighters with the targets they were responsible for defeating.

One of the most difficult tasks facing any combat commander is figuring out how best to deploy forces to maximum advantage. As mentioned in the overview to this chapter, the considerations that go into these decisions are not unlike those that face the coach of an athletic team. The job assigned to any individual, unit, or, for that matter, aircraft, must be consistent with the ability of that element of the force to handle the job assigned. The history of the Battle of Britain is rife with examples of aircraft that eventually had to be withdrawn from combat or given less critical assignments because they lacked performance relative to the adversary they faced. This inability to find an appropriate assignment for a given aircraft is almost invariably accompanied by wasted resources, both in terms of the time and effort spent to develop the airplane and in lives lost unnecessarily. Specific examples include the Stuka, which could not survive against the RAF Hurricanes and Spitfires and, on the British side, the Defiant, which could not survive its encounters with the Bf 109. The experience of the Battle of Britain further demonstrates that the capability imbalance is not resolved by placing larger quantities of a relatively poor design into combat. The high-performance airplane will force the adversary into conditions where the lack of performance becomes a lethal liability.

As Air Vice Marshal Park considered his fighter assets in view of the task facing 11 Group, he faced just this sort of decision. Arrayed against him was a German force substantially superior in terms of numbers and that could roughly be categorized as a mix of bombers and high-performance fighters. Park's solution was to assign his two frontline fighters, the Hurricane and the Spitfire, different targets—but targets closely matched to the performance capabilities that each possessed. The Hurricane was a very stable shooting platform but was slower and lacked the agility and altitude capability of both the Spitfire and the Bf 109. Park therefore assigned to the Hurricanes the primary task of intercepting and defeating the German bomber forces. The Spitfire, on the other hand, was extremely maneuverable, capable of rapid turns, quick climbs, and operational ceilings above 30,000 feet. To the Spitfires, Park assigned the responsibility for defeating the German fighters, specifically the Bf 109. The illustration on these pages demonstrates this division of responsibilities. This is not to say that Hurricanes would not engage 109s, given the opportunity; they did on many occasions, just as the Spitfires would gladly take on bombers. Park's guidance was really intended more for those who directed

Faster, tighter turning, the Spitfire was better suited to engage fighter escorts.

A rugged, stable gun platform, the Hurricane was better suited suited to engage bombers.

and controlled the battle, giving them direction and rules of thumb that would guide the selection of targets for specific aircraft to maximize the overall effectiveness of 11 Group's scarce fighter resources.

24. *Adlerangriff*

Adlerangriff—Eagle Attack! The very name conjures up the image of a strike from the sky—fierce, rapid, deadly, and overwhelming. Such was the vision shared by Goering and Hitler as they contemplated the main attack by the Luftwaffe on the Royal Air Force. The primary elements of the strategy that Goering intended to follow are included in the cartoon. *Adlerangriff* was Goering's implementation of Hitler's Directive 17, which had called for a comprehensive attack on the flying units of the RAF, including their ground installations and supply organizations, as well as the British aircraft industrial capacity. The attack was to begin on or after August 5, and the German High Command estimated that the primary objectives would be achieved within a matter of about two weeks. With the completion of the operation, the expectation was that either Britain would negotiate a nonaggression treaty on Hitler's terms, or an invasion would be launched. Should invasion be necessary, they assumed that Britain would be so weakened that it would be essentially unopposed. However, before indicting German planners and commanders for their failure to appreciate the difficulty of what they had undertaken, it is important to recall that Germany had, at this point in World War II, rolled over Czechoslovakia, Poland, Belgium, and France—each within a matter of weeks. The failure was that they did not understand the depth and quality of their primary foe—the RAF—nor did they fully appreciate the differences between this campaign and the combined-arms blitzkrieg battles they had fought earlier.

That the German leadership deluded itself that the defeat of Britain would be both swift and simple is testimony to the lack of adequate intelligence available to planning staffs and decision makers. German intelligence had indicated that the British had 400 to 500 fighters available to confront the Luftwaffe in this main phase of the battle. But when one considers the total assets available to Dowding, including not only his southern groups, but also his northern-based aircraft, the number was actually substantially higher—about 750. Furthermore, German intelligence failed completely to account for the repair and replacement production rates that the British were now beginning to generate. This capability, under the direction of Lord Beaverbrook, the minister of aircraft production, resulted in Dowding actually having more fighters at the end of the battle than he had at the beginning. In fact, Dowding, in a document written after the war, describes Beaverbrook's contribution as nothing short of "magical" (Dowding 1946). This inability to correctly estimate the size of the opposing fighter force led to sarcastic grumbling among German pilots—both fighter and bomber pilots—that every day they were required to face the "last 50 British fighters." This kind of sarcasm eats away at aircrew morale and lessens respect for command authority as well. But as serious as these failures were, perhaps the greatest was Germany's lack of appreciation for the central role that the Chain Home system played in the defense of Britain. While the *Adlerangriff* phase started off with attacks on the Chain Home radar towers, had the Germans fully comprehended that these stations enabled the RAF to concentrate their forces against the Luftwaffe with maximum efficiency, those attacks would have been much more focused. The destruction of the Chain Home system should have been a priority target on a par with the destruction of the RAF itself.

In broad terms, the strategy was to destroy the aircraft and facilities of the RAF through a series of attacks that would start with targets in the areas between 100 and 150

Adlerangriff Strategy

FIRST. "WIPE THE BRITISH AIR FORCE FROM THE SKY."

LUFTFLOTTE 5 RESPONSIBLE FOR NORTH AND MIDLANDS, LUFTFLOTTES 2&3 FOR SOUTH.

SECOND. BOMB AIRCRAFT FACTORIES AND PORTS.

THIRD. DESTROY THE ROYAL NAVY.

kilometers south of London. After five days, the attacks would be moved inward to cover targets within the southern areas 100 to 50 kilometers from the city. The final five days would cover areas in all quadrants within a 50-kilometer radius around London. Goering issued several clarifying orders to *Luftflotten 2, 3,* and *5* during the early days of August, which included the requirement to achieve air superiority and to include naval targets in order to ensure that the British navy did not interfere, should an invasion be necessary. The attack date, called *Adlertag*—Eagle Day—was initially set for August 10. The campaign got off to a slow start because of poor weather, but by August 13 the weather had cleared and the Luftwaffe began a phase of the battle that would almost defeat the RAF. *Adlertag* is officially recognized as August 13, 1940.

The attack started with low-level bombing attacks on the Chain Home towers. The Germans apparently realized that the ability of the British fighters to show up in mass at precisely the right time and place was somehow related to the strange twin towers along the coast. There is no evidence that they comprehended the full capability of the integrated air defense system with its command and control links, but they obviously sensed that destruction of the Chain Home system would contribute to their objectives. To attack the towers, the Germans employed some of their most highly trained and experienced specialized bombing units, and though they struck the targets with precision, little significant damage was done because of the steel lattice construction of the towers. The Germans could report that the Chain Home system had been struck with devastating accuracy, but the British were back in operation almost without interruption.

The importance of integrated air defense systems and command and control facilities as strategic targets has not been lost on history. Modern armed forces are highly dependent on information—detection, surveillance, and command and control—to manage and direct their combat forces. For this reason, these have become the primary targets during the first stage of almost every war fought recently. In a typical scenario in modern combat, stealthy aircraft are sent into enemy territory specifically to destroy the eyes and ears of the adversary. Only when the enemy has been rendered incapable of detecting attacks and communicating direction to its own forces does the main attack aimed at destroying its combat forces begin. Had the Germans adhered to that strategy, the outcome of the Battle of Britain might well have been different.

Adlerangriff would continue day and night until the second week in September. Before it was over, Fighter Command would be on the brink of exhaustion, with a shortage of experienced combat pilots so severe that German victory was almost certain. The choice of the RAF, its bases, and British aircraft production capabilities as the primary objective was a well-founded strategy that, pursued to its conclusion, might well have produced victory for the Germans. The key failures were, again, not those of the combat unit commanders and the crews who executed the plan. The failure of the *Adlerangriff* was fundamentally one of intelligence, compounded by high-level interference that finally culminated in abandonment of the strategy at the precise moment when victory was within the grasp of the German eagle.

25. "Sailor" Malan's Ten Commandments

Every war since World War I has produced fighter pilots whose natural abilities have exceeded those of their contemporaries, earning for themselves lasting recognition as being unique and gifted in the deadly business of air-to-air combat. "Sailor" Malan was one of the small group of British fighter pilots who is remembered even today as a pilot, leader, and tactician. Born Adolf Gysbert Malan in Wellington, South Africa, in 1910, he was assigned to the training ship *General Botha* as a cadet in 1924. After completing his training, he joined the Union Castle Steamship Line in 1927, where he performed as a junior deck officer. In 1935 he applied for a short service commission in the Royal Air Force; he was accepted and began flight training in England in 1936.

It is a long tradition among pilots in many military organizations to give newly assigned officers nicknames based on their background or, sometimes, on events in which they play some central role. In Malan's case, he became "Sailor" based on his earlier experience as a seaman. He completed his flying training and was assigned to 74 Squadron in December 1936. The squadron was flying the Gloster Gauntlet aircraft, which had entered service in 1935, and was recognized for its excellent handling qualities and was considered by many to represent the zenith of biplane design. Malan soon made a name for himself as an outstanding aerobatic pilot and sharpshooter. He was promoted to flight lieutenant in March 1939. Eventually, 74 Squadron transitioned to the Spitfire, and by August 1940 Malan had risen to the rank of squadron leader.

Like Boelcke before him, Malan was a student of his craft, and as squadron leader he took the responsibilities of command seriously. Those responsibilities included the preparation of his pilots for combat. To give them the benefit of his experience and abilities, Malan, like Boelcke, developed a list of rules for air combat that was circulated under the heading "Ten Rules for Air Fighting." The panels shown in the cartoon summarize Malan's rules. A somewhat more complete description follows:

1. *Wait until you see the whites of his eyes. Fire short bursts of 1 or 2 seconds, and only when your sights are definitely "on."* This was Malan's exhortation that his pilots not fire until they were very close to the opposing aircraft, that they not exhaust their ammunition in a single burst, and that they ensure they have a firing solution before firing.
2. *Whilst shooting think of nothing else, brace the whole of your body, have both hands on the stick, concentrate on your ring sight.* In other words, concentration is the key to good shooting. Every nerve and every fiber must be trained on only one thing—the target.

MALAN'S 10 RULES
GET CLOSE BEFORE SHOOTING
CONCENTRATE WHEN SHOOTING
ALWAYS KEEP A SHARP LOOKOUT
HEIGHT GIVES YOU THE ADVANTAGE
ALWAYS TURN AND FACE THE ATTACKER
ACT QUICKLY
NEVER FLY STRAIGHT AND LEVEL FOR MORE THAN 30 SECONDS IN COMBAT
WHEN DIVING, LEAVE TOP COVER
INITIATIVE, AGGRESSION, TEAMWORK, DISCIPLINE IMPORTANT
GO IN—PUNCH HARD—GET OUT

3. *Always keep a sharp lookout. Keep your finger out.* The gun trigger is located on the control column. The reference to keeping the finger out is an admonishment to always be ready to squeeze the trigger—but never to fire rounds inadvertently. Things happen very fast in the air; pilots must be ready to take advantage of any opportunity while avoiding surprise attacks on themselves.
4. *Height gives you the initiative.* Back to Boelcke. As has been observed, the pilot that has the energy advantage also has the tactical advantage in air combat.
5. *Always turn and face the attack.* This is another dictum straight from Boelcke. Turning away from an attacker automatically places the enemy in position for a potential kill, whereas turning into the attack tends to give equal opportunity to both.
6. *Make your decision promptly. It is better to act quickly even though your tactics are not the best.* Split-second decisions and reactions are much more difficult to anticipate and defeat than are slow, smooth maneuvers.
7. *Never fly straight and level for more than 30 seconds in the combat area.* Called *jinking*, this is similar to rule 6. Constant turns and changes in altitude make it very difficult for an enemy to creep up behind and shoot one down by surprise. It also encourages the pilot to keep looking up, down, and side to side.
8. *When diving to attack, always leave a portion of your formation above to act as top guard.* This also harks back to Boelcke's dictum that, although attacks should be accomplished in groups, several aircraft should not follow a single aircraft. Malan's rule demonstrates his grasp of the way tactical formations can be employed to provide mutual protection and the way pairs can reinforce each other.
9. *Initiative, aggression, air discipline, and teamwork are words that mean something in air fighting.* Here Malan was making the point that these phrases are used often and routinely in organization-speak, but in air combat, crewmembers do not survive for long unless they practice the concepts.
10. *Go in quickly—punch hard—get out!* The most effective attack is one where the attacker roars into the engagement, shoots to kill, and disappears before the defender can react. Air combat is not a timid contest; give away no advantage.

These rules were widely adopted throughout the RAF and eventually became standard operating procedures. Often referred to as the "Ten Commandments," these rules, like Boelcke's dicta, are as good today as they were when Malan conceived them in 1940.

Malan went on to a position as wing leader at Biggin Hill and led large groups of fighters in sweeps over the Channel and northern Europe. After being relieved of front-line combat flying operations, he toured the United States, giving lectures on the war effort. He returned to command the RAF Gunnery School and the Advanced Gunnery School. He retired from the RAF in 1946 as a group captain, after which he returned to South Africa with 27 enemy aircraft destroyed and 7 shared kills. In 1959, Malan returned to visit 74 Squadron for the last time. By this time he was suffering from Parkinson's disease and his health was failing. He died from the disease in September 1963. Britain benefited tremendously from the contributions of foreigners during the battle of Britain. These included outstanding pilots from Poland, Czechoslovakia, Australia, Canada, and New Zealand. Few, if any, contributed more than did Sailor Malan.

26. Northern Attack from Norway

The He 111 slowly drops behind its formation as another jettisons its bombs in preparation for a dash to safety, as British fighters prepare to attack. The German force sent from *Luftflotte 5* in Norway suffered disastrous losses when they attempted a surprise attack on the northern areas of England. So decisive was this engagement that Germany never again mounted attacks from *Luftflotte 5* during the Battle of Britain.

The defense mounted by the RAF during the Channel phase, which lasted through the first couple of weeks in August, was substantially more effective than the Germans had expected. Intelligence estimates had placed the total British fighter strength at between 400 and 500 fighters. The stubborn defensive effort displayed in 10 and 11 Groups in southern England led Goering's chief of intelligence to conclude that the RAF must have moved the fighters stationed in the northern group areas to the south in order to reinforce 10 and 11 Groups. In his view, this was the only way they could have resisted the German assault as they had.

The original deployment of the Luftwaffe that placed *Luftflotte 5* in bases in Denmark and Norway was intended to put all areas of Britain within the reach of German bombers. The forces in *Luftflotte 5* included approximately 130 He 111 and Ju 88 bombers, plus about 35 Me 110 long-range fighters, intended as escorts for raids on England. Now convinced that there were few, if any, fighters left in the northern regions of England, the German High Command (OKW) set about planning an attack that would force the British to redeploy their fighters to the north. This would reduce the number of fighters in the south and open up the airfields and factories to the attacks described in the *Adlerangriff* strategy that would take the Luftwaffe to the outskirts of London.

The fact is, however, that the German conclusions were based on guesses rather than on solid intelligence estimates. The British defenses were organized around four groups, as previously explained, with 10 and 11 Groups to the south; 12 Group was to the immediate north of the southern groups and had six squadrons of Hurricanes and five Spitfire squadrons, plus one each of Blenheims and Defiants; 13 Group had the northernmost areas, with six squadrons each of Hurricanes and Spitfires. Dowding had resisted any suggestion that he deplete the northern-based fighters to reinforce the south, so these fighter units remained in place in the north in spite of the heavy demands made on the southern groups.

15 August: Escort, Range, and Navigation problems made Raids from Norway impractical.

Most of the pilots in these squadrons had fought in France and northern Europe. They were experienced in combat, aware of the intensity of the fighting to the south, and they longed for a piece of the action. In an effort to give his forces in the south some respite, Dowding had initiated a policy of periodically rotating the northern squadrons into 10 and 11 Groups, pulling the squadrons from the areas where combat had been most intense and sending them north for a period of rest and recuperation. The result was that the northern groups were manned by experienced combat pilots, some of whom wanted an opportunity to demonstrate that they were as ready and able as those assigned to the southern groups, and others who had recently been in combat were now rested and ready for action. The Germans had seriously misjudged what would wait for them in their planned attack on the northern regions of Britain.

The German attack, planned for August 15, was to consist of two primary thrusts. A force of some 60 Ju 88s, flying from bases in Denmark, would attack Driffield, a base in the 12 Group area, while some 70 He 111s, escorted by 20 Me 110s flying from Norway, would attack targets in the northernmost 13 Group area. To draw any British defenses away from the northern thrust, a feint would be mounted consisting of a force of seaplanes that would fly toward the English coast north of Edinburgh. These seaplanes would turn and retreat to Norway before reaching the coast, their purpose having been only to draw British fighters away from the main attack.

As the southern attack force approached the coast they were met by 616 Squadron, flying Spitfires, and 73 Squadron in Hurricanes. Seven Ju 88s were shot down as they approached, and three more were downed by antiaircraft fire in the vicinity of the target. Three others crash-landed on return from damage sustained. In spite of these losses, the German attack was basically successful as they struck four hangars and destroyed 10 British bombers at Driffield.

The attack on the more northerly targets was not as fortunate—if the loss of 10 aircraft out of 66 could be termed fortunate. As the seaplane force approached the coast, the fighters in 13 Group were launched to intercept what appeared to be an attacking formation. Twelve Spitfires from 72 Squadron and six Hurricanes from 605 Squadron hurried to meet what appeared to be a formation of about 30 attacking aircraft. Meanwhile, the real attack force had become somewhat disoriented while inbound and had drifted well north of their designated inbound path. This northerly drift had taken their route very near that of the seaplane formation, and it resulted in the British fighters being placed relatively near the main attacking force. When the British fighters arrived at altitude, they saw—not 30 inbound aircraft, but nearly 100 bombers and escorting long-range fighters.

So confident were the Germans that they would meet only token fighter defenses that they had made unusual preparations for the relatively long bombing mission from Norway. They had reduced the bomb loads to 3000 pounds each for the He 111s, and they had slung long auxiliary fuel tanks, called "dachshund bellies," under the Me 110s. In addition, RAF pilots reported that they received no rearward fire when approaching the Me 110s, which further suggests that the gunners had been eliminated to save additional weight.

Two additional RAF squadrons, 41 and 79, were launched to supplement those already in the air, and the battle was engaged. Once again, the Me 110s proved unable to deal with the British single-engine fighters with their superior rate of climb and agility. Before the bombers had managed to complete their bombing runs and start the long trip back to Norway and Denmark, they had lost approximately 20 percent of their number. In spite of the 10 RAF bombers destroyed at Driffield, this raid proved immensely expensive to the Luftwaffe in terms of both men and machines. It demonstrated the futility of attempting to conduct long-range bombing missions without high-performance fighter escorts, but, more important, it demonstrated to the Germans that they could make no assumptions that parts of Britain might be left undefended. Shortly after this raid, the Luftwaffe redeployed some of the bombers in Scandinavia to reinforce *Luftflotten* 2 and 3 in northern Europe. No further raids on England were attempted from the bases in Norway and Denmark, and, with that, the role of *Luftflotte* 5 was essentially minimized for the duration of the Battle of Britain.

27. Head-on Attacks

The picture showing Hurricanes diving head-on into a formation of He 111s illustrates a tactic that was occasionally used by the pilots of Fighter Command as they attempted to perform the function that would determine the success or failure of the defensive effort—stopping the bombers. There is a tendency when considering air-to-air tactics to think in terms of fighter on fighter tactics how one fighter maneuvers to defeat an opposing fighter. However, the effectiveness of an offensive air campaign is often measured by the extent to which bombers are successful in delivering their loads. The task for a defender, conversely, must be to stop the bomber. Dowding and Park recognized this fact early during the defense of Britain and ordered their controllers to focus on stopping the bombers rather than engaging the German Bf 109s that trolled the English coastline, hoping to tempt the fighters into the air.

Bomber formations are conceived partially to concentrate the bombs dropped, but primarily to provide mutual support among and between the bombers so that any fighter attacking the formation will come under fire from several of the bombers. A bomber formation typically consists of cells—often three bombers—arrayed in vee formations, with different units flying at different altitudes, all within the same larger formation. With air craft deployed both horizontally and vertically, an attacking fighter could depend on being fired at no matter how the pilot approached the formation. German bomber for-

Head-on attacks were an effective way to break up bomber formations.

mations routinely included 100 or more bombers, and the combination of formation discipline and fighter escort provided by Bf 109 fighters made any attack on the formation a dangerous undertaking.

Fighters, looking for potential targets, seek to separate aircraft from the formation or to break the discipline of the formation, causing the aircraft to scatter. If mechanical problems or combat damage cause a bomber to drop out of the formation, the mutually supporting defensive fires are no longer effective and the bomber can then be picked off by the fighters. It is not unlike the tactics used by wolves as they track a herd, attempting to separate a weak member and isolate it for the kill.

As the pilots of Fighter Command contemplated the problem of breaking up the Luftwaffe formations so that individual aircraft could be isolated and attacked, some turned to head-on attacks. The bomber formations in general were unprepared for head-

on attacks. They often did not have forward-firing guns, and, furthermore, head-on attacks created a difficult situation for the escorting Bf 109s, since they would find themselves firing toward their own bombers if they tried to kill the attacker. Literally from the beginning of the Battle of Britain, certain units and individuals were recognized for employing this death-defying tactic; 111 Squadron was known to have used line-abreast or shallow vic formations in head-on attacks, but the squadron was eventually forced to abandon the practice as its combat losses mounted. Head-on attacks were employed even more frequently once the Germans installed armor plating in the bombers to protect the crews and engines—but only from fire that came from behind the aircraft. Dowding initially disapproved of the head-on tactics, but by late August he relented and even recommended their use to overcome the effects of the German armor plating.

To imagine the mechanics of a head-on attack, think in terms of two aircraft closing at speeds approaching 600 miles per hour, which equates to closures of around 900 feet per second. This implies that an attack begun 900 yards from the bomber would have only 2 seconds to aim and shoot plus 1 second to pull away before colliding with the bomber. Since many fighters had their guns harmonized at 250 yards, it would be important to press the attack to the point at which collision was 1 second (or less) away to improve the chance of a hit. Of course, the head-on attack was successful if the bomber formation broke, whether or not a hit was registered on the initial pass. The fighters could then return and pick the bombers off one by one if they were alone. The head-on attack was so dangerous to execute that it was never considered or recommended as an accepted tactic; at best, it was a desperate measure adopted for desperate times.

28. Low-Level Bombing Tactics

A single Do 17 approaches the English coast, flying just above the waves to avoid detection, either by radar or by other means. By the time the Battle of Britain had started, the Chain Home (CH) radar system had been modified to enable detection of aircraft at altitudes well below 5000 feet, but some gaps in coverage remained at very low altitudes.

The Chain Home system was in place before the Battle of Britain began. By February 1940 some 29 stations were in place, with coverage on the seaward side that provided detection up to ranges of 150 miles. A weakness that had been exposed—the fact that the system could not detect aircraft below 3000 feet—had been remedied through the addition of the Chain Home Low (CHL) system. Whereas the CH transmission towers stood 350 feet high and the receiving towers 240 feet, Chain Home Low featured transmission and receiving towers mounted on 20-foot-high structures. CHL operated on a wavelength of 1.5 meters (CH used wavelengths of 13.6 m). The lower-wavelength transmissions enabled the use of an antenna that was small enough to be rotated (CH could not). Normal procedure was to have CH pass azimuth information on targets that were flying low to the CHL operators, who would then track these targets, rotating the antenna as necessary to keep them in contact. CHL could track low-flying aircraft to altitudes as low as 500 feet at ranges to 30 miles and at greater ranges if the targets were higher.

Germany had never completely grasped the full significance of Britain's radar and integrated air defense system. German intelligence estimates continued to predict that the skies would soon be cleared of British fighters, yet the fighters always seemed to show up to challenge the bombers as they approached the English coast. The use of radar by the British illustrates perfectly the concept of *force multiplier,* a term used often in defense circles. A force multiplier is something that makes it possible for a combat unit of some defined size and capability able to have an impact in excess of what would normally be

A DO-17 approaching the coast at low level to avoid Radar detection.

expected of such a force. The combination of CH and CHL was a huge force multiplier for the British.

From the beginning of the battle, the Luftwaffe had employed highly specialized units to conduct low-level bombing of selected targets. Probably the most famous of these was *Erprobungsgruppe 210,* an elite unit that flew a mix of Me 110s and Bf 109s to attack designated targets. *Erpro 210* almost always operated at altitudes just a few hundred feet above the surface to penetrate and bomb specifically identified high-value targets. *Adlerangriff* had established the fighters of Fighter Command and their airfields as primary targets, and by mid-August 1940 the Luftwaffe had adopted tactics that promised significant success for bombing British airfields that combined conventional bombing with very low level attacks. In adopting attacks from very low levels, the Germans found the last remaining gap in the British defenses and began to exploit it—very low level bombing raids. The attack directed against the RAF field at Kenley on August 18 provides an example. The attack was to include three waves of bombers, each attacking from a different altitude. The first wave was to be carried out by dive-bombers; the second was to be a high-level conventional attack by Dornier Do 17s; and the third consisted of nine Dorniers that would approach at altitudes of 100 feet or lower, well below the levels at which CHL could detect them. The illustration that shows the Do 17s approaching the British coastline at altitudes just above the waves shows the aircraft as they might have appeared in executing the low-level attack. This was the first recorded attack in which a formation of attacking bombers flew the entire mission at very low altitudes.

Flying at extremely low levels offered challenges for both the attacker and the defender. Probably the most immediate problem for the pilots of the low-level bombers was the fact that any error in flying or lapse of attention that resulted in losing altitude could be fatal. It is no great issue to lose 100 feet of altitude when one is flying at 18,000 feet; that is hardly the case when flying at 100 feet. Another problem for the attacker is that maneuverability in the event a fighter intercepts is limited to the horizontal plane; one cannot dive to pick up airspeed when flying at 100 feet. Navigation is also very difficult at extremely low levels since the pilot can typically see only a very limited geographical area. The greatest challenge for the defender is seeing or detecting the attacker. Whether searching for the attacker visually or by radar, it is often difficult to see an aircraft against the ocean background. Visually, the aircraft tends to blend into the gray-green colors of the water, and radar returns can become obscured by what is referred to as *ground clutter*—caused by energy reflected back from the ocean waves. During the Battle of Britain, the Luftwaffe's use of very low level bombing attacks on the airfields led to the development of the parachute-and-cable rocket defenses described elsewhere in this book. After the Battle of Britain, the British actually improved the Chain Home system to eliminate the weakness in very low altitude coverage by the addition of radars optimized to detect targets in this region. These were known as Chain Home Extra Low (CHEL).

29. Other Bombing Tactics

As Germany pressed the attack on Britain, starting first with the Channel phase and then continuing into the main attack that targeted the fighters and the facilities associated with them, the Luftwaffe often varied its bombing tactics. This was necessary both to find the most effective tactics and to prevent the British from anticipating German actions on a day-to-day basis. One of the criticisms of the U.S. bombing attacks on North Vietnam in the 1960s was that they too often repeated the same attack routes, altitudes, and times, leaving the pilots vulnerable to defenses planned on the assumption that the attacks would not vary from day to day. The Germans sought to change their tactics often, which kept the British off balance and avoided having Fighter Command predict when and where they might show up. These variations, while preserving an element of surprise, could also introduce weaknesses. For example, simultaneous attacks on multiple targets might spread the defenses, but they also reduced the concentration of firepower that masses of bombs on a few targets could achieve, as illustrated.

A favorite ploy used by the Luftwaffe was to vary the timing of bombing attacks to take maximum advantage of the British need to refuel, rearm, and sometimes recover from the damage done by earlier raids. For example, the Germans would send bombers in large enough numbers to force the British to mount a significant fighter response. A typical fighter mission might require 15 minutes to take off and climb to altitude, leaving 30 minutes or less at near maximum power for fighting before leaving the scene to refuel and rearm. The Germans would often send in a second attack force specifically timed to catch the fighters and their support crews on the ground during these turnaround operations. By late August, the Germans were flying what has been described as an almost continuous stream of aircraft off the English coast, some of which would suddenly turn toward the coast to force the British to launch fighters. Subsequently, raids would be timed to catch them as they refueled and rearmed. A variation on this theme was the mix of attacks by various aircraft flying at different altitudes, described in the previous section of this book. Such attacks made it very difficult to set altitudes for anti-aircraft guns; furthermore, the smoke and dust from one attack could serve as a target identification marker for the formations to follow. It is also true that bombing attacks were often timed by the type of bomb load carried to achieve specific purposes. For example, a raid might carry incendiaries intended to start fires, then later another attack would follow carrying blast and shrapnel bombs intended primarily to destroy and damage the fire-fighting units that would, by that time, be in the streets trying to control the fires set by the first raid. On August 30, Kesselring varied the timing yet again, launching a series of sequential attacks that started at 1300 hours (1:00 P.M.) and continued straight through for five hours, with a new attack every 20 minutes. This seemingly unending flow of bombers as the primary tactic continued for several days, but then on

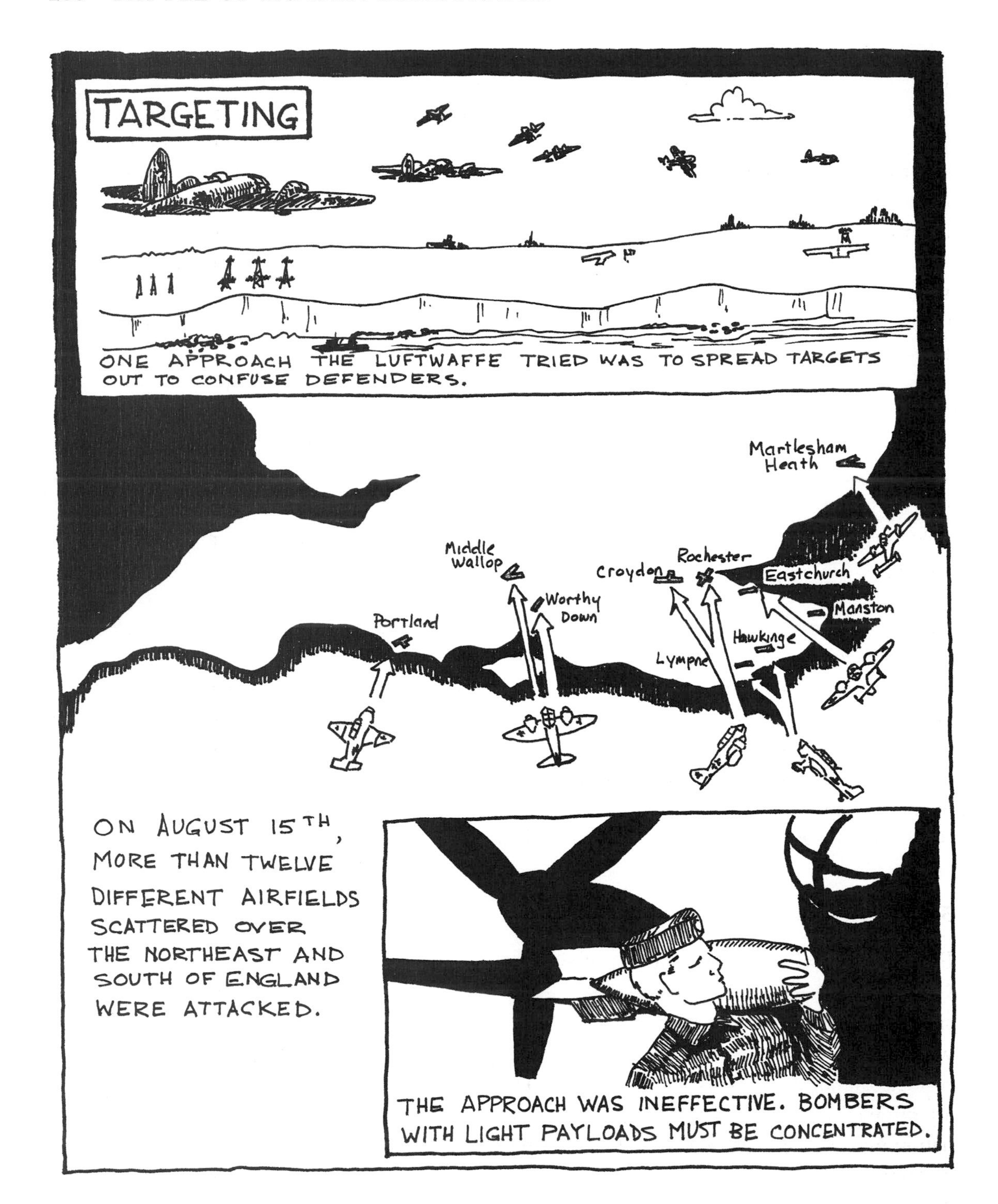
TARGETING
ONE APPROACH THE LUFTWAFFE TRIED WAS TO SPREAD TARGETS OUT TO CONFUSE DEFENDERS.
Martlesham Heath
Middle Wallop
Croydon
Rochester
Eastchurch
Worthy Down
Manston
Portland
Hawkinge
Lympne
ON AUGUST 15TH, MORE THAN TWELVE DIFFERENT AIRFIELDS SCATTERED OVER THE NORTHEAST AND SOUTH OF ENGLAND WERE ATTACKED.
THE APPROACH WAS INEFFECTIVE. BOMBERS WITH LIGHT PAYLOADS MUST BE CONCENTRATED.

September 5 Kesselring switched yet again, sending what amounted to 22 separately identifiable raids against multiple targets.

As the airfield phase of the battle got under way, the Germans initially attempted to overwhelm the defenses by attacking many targets simultaneously. The idea behind this was to fly toward Britain as though headed for some key target, then at the last minute split the formation into several smaller units that would attack a number of airfields almost simultaneously. This tactic offered the attacker the benefit of disguising the actual destination, making it difficult for the controllers to decide which units to commit and when. On August 30 there were 48 different stations reporting bombers in their areas (Hough and Richards 1990). The disadvantage from the German perspective was that this dispersion lessened the concentration of bombs on any one target and lessened the mutual-support aspects that the larger formations of bombers afforded. Later, this tactic was alternated with heavier attacks on a limited number of targets.

In addition to the unit tactics already described, one also should appreciate the individual tactics employed by bomber pilots to increase their own effectiveness and probability of survival. The variations are far too numerous to cover in any comprehensive way. A few examples include a tactic often used by the pilots of the Ju 88. As they approached the target area, most Ju 88 pilots would put their aircraft into a long, shallow dive. This had the twin benefits of enabling them to make aiming corrections with ease during the run into the target and increasing their airspeeds substantially, which made it much more difficult for British fighters to hit them. Another tactic used by some German pilots was to fly past a designated target, giving the initial impression that they were bound for targets further inland, then turning around and attacking. This was intended not only to confuse the defenses, but also to place the bomber on a heading toward the Channel and the European coast, thus lessening the time that the aircraft was exposed to British fighters during the return leg—a period of high vulnerability because the formations were seldom as tight and disciplined after dropping their bombs.

30. Dowding's Rest Policy

Physical exhaustion is as much a part of combat as is danger. The deployment of Fighter Command resources into four geographically defined groups guaranteed that the units stationed in 10 and 11 Groups, which occupied the southern and eastern quadrants, would bear the brunt of Luftwaffe attacks launched from Europe. Flying two, three, and sometimes four combat missions in a single day with nerves on edge, muscles knotted, and attention totally focused left pilots exhausted. Repeated day after day, it could create a situation in which pilots were a danger to themselves and others. The story is told that one Spitfire pilot of 111 Squadron rolled to a stop after landing at his assigned airfield, whereupon the ground crew noticed that the aircraft simply stayed in position with the

September '40. Exhausted 11 Group pilot arrives at 13 Group base to begin rest and refit.

engine running rather than taxiing in to the servicing bay. When the crew approached the aircraft, expecting to find the pilot with mechanical problems—or, worse, wounded and unable to proceed—they found instead that the pilot had fallen asleep in his cockpit. Every combat operation in which prolonged and continuous demands are made on crews has shown that at some point pilots are no longer capable of reacting and handling even the most ordinary flying tasks. The challenge for the commander is to gauge when exhaustion has progressed to the point that it has begun to seriously detract from the ability of the fighting unit to perform the missions assigned.

By mid-August Dowding had concluded that some of the squadrons in the 10 and 11 Group areas needed rest if they were to survive to the end of the battle. In spite of the pressures brought on those groups, he insisted that selected squadrons be rotated northward into 12 and 13 Group areas where the combat demands were generally less intense. The picture shows an exhausted pilot who has just landed at a 13 Group base after his unit was rotated from 11 Group, where combat was most intense. It was precisely this policy that had resulted in there being ample and experienced forces in the northern areas when the Luftwaffe attacked from Denmark and Norway expecting to find those areas undefended. Individual pilots were also at times singled out and rotated for combat rest when they began to show evidence of exhaustion—heavy drinking, inability to sleep, or other neurotic behavior.

The decision to rotate experienced units was particularly difficult, because the new units posted southward to replace those rotated for rest almost invariably suffered greater initial losses than those replaced. The evidence suggests that experience is the best predictor of the probability that a pilot will survive in combat. Until the replacement crews gained experience, they were often vulnerable. Nevertheless, Dowding adhered doggedly to his rotation policy, intuitively understanding that the overall situation would only have deteriorated had he allowed experienced crews to become victims of total exhaustion rather than accept the initial spike in losses that usually accompanied a unit rotation. The rotation policy resulted in a broader experience base throughout Fighter Command and simultaneously enabled the most experienced combat pilots to rest and regain their edge.

A unit rotation was a cumbersome and complex operation, involving the movement of personnel, machines, and other equipment as well. To speed up the process, Dowding eventually began to require rotating units to leave their support staff and equipment in place, and he would rotate only pilots, aircraft, and key staff members. This had the effect of lessening unit cohesion, however, and allegiance to units was one of the strengths of the RAF. Park, in 11 Group, was even less disposed to preserving whole units; he recommended rotating only pilots from unit to unit. Dowding, however, initially rejected this suggestion as too deleterious to unit morale.

By September Dowding was forced to introduce what he referred to as his "stabilization scheme," which was brought on by the combination of combat exhaustion and losses, which are highly intertwined. Under this plan, all operationally ready units were designated as Category A and were placed in 11 Group and the Duxford and Middle Wallop sectors of 12 Group—the groups that bore the brunt of the daily demands of combat. Category B units, those that were partially ready, were positioned in 10 and 12 Groups. They were to be brought up to strength and prepared to replace Category A units as needed. Category C units, those with few experienced pilots and not yet ready for combat, were generally placed to the north in 13 Group. Also, the assignment of pilots from one unit to another in order to maintain a balanced mix of experience soon became the norm, as losses mounted and the very survival of Fighter Command appeared to be in doubt. Had the Germans not suddenly abandoned the RAF and its airfields as the primary objective in early September, even the desperate measures taken by Dowding to rotate crews and units would not have prevented defeat.

31. Goering Tactics Conference

On August 15, the Luftwaffe had tried and failed to surprise the British with attacks in the northern regions of England. The Germans had flown over 1750 sorties on that day. Combined with their losses in the southern areas, the total Luftwaffe aircraft lost on that day was well over 50. On August 16, they followed that effort with another intense day during which they flew another 1700 sorties and again lost over 50 aircraft. August 18 was a repeat, with heavy losses on both sides, and although the British became increasingly concerned that the losses of combat-experienced crew were becoming critical, Goering was dismayed that his vaunted Luftwaffe was not achieving the kind of overwhelming results Germany, and Hitler, had come to expect. Goering had convened a commander's conference on August 15 to discuss strategy with his unit leaders. After four days of combat during which he had lost almost 200 aircraft and crews, Goering called another conference on August 19 to revise the aerial tactics. The cartoons indicate several of the critical tactical decisions that were made at this conference—the Ju 87 was withdrawn, only one officer was to fly on any aircraft, Me 110s were to be escorted, and the attacks on the Chain Home radar stations were halted.

Goering had predicted that two or three weeks would be sufficient to defeat Fighter Command, yet the British fighters were still able to mount defenses that numbered in the hundreds of aircraft. At the conference of his commanders, he began by criticizing his own fighter commanders, blaming their lack of aggressiveness for the Luftwaffe's failure to achieve the results expected within the time period predicted. He blamed the fighters for failing to provide adequate coverage to the slow, vulnerable Ju 87 Stuka dive-bomber. Goering followed his tirade with an announcement that the Stuka would be withdrawn from the air battle, although he left open the possibility that it would be used again once air superiority was gained and the invasion got under way.

GOERING'S TACTICS CONFERENCES ... 15,19 August
AFTER ADLERTAG, GOERING CALLED TWO CONFERENCES TO DISCUSS STRATEGY. FOUR CHANGES WERE DECREED.
FIRST, JU87s WERE WITHDRAWN FROM THE BATTLE.
SECOND, ONLY ONE OFFICER WOULD FLY IN EACH BOMBER CREW.
THIRD, Me110s WOULD BE ESCORTED BY Me109s... ESCORT THE ESCORTS!
FOURTH, STOP ATTACKING THE BRITISH RADAR SITES.

The losses of Me 110 *Zerstoerer* during the devastating days that immediately preceded the conference were particularly galling to Goering. The aircraft had been a favorite of his from the beginning, and he refused to accept the fact that the design was ill suited to the combat environment in which it was employed. Instead, he blamed both his fighter pilots and his fighter unit commanders for failing to protect it. He proceeded to dismiss several of his fighter commanders, replacing them with younger and—Goering hoped—more aggressive and able men. Among those he selected to replace the older commanders were Galland and Moelders, who in retrospect were quite successful as unit leaders. Goering then went on to insist that henceforth the Bf 109s were to escort the Me 110s. Not only was this humiliating to the Me 110 crews, it was an extremely poor use of fighter assets—in effect using an aircraft whose primary mission was bomber escort (the Bf 109) to act as an escort for an aircraft that had originally been designed as a long-range escort (the Me 110). Goering's commanders argued that the problem lay in the fact that the Bf 109s were already too restricted. They argued that free chase flights would prove a more effective tactic, but Goering insisted that they needed to be tied more closely to the aircraft they were escorting. Goering further ordered that the majority of the fighter units located in Sperrle's *Luftflotte 3* be transferred to Kesselring's command in *Luftflotte 2*. The Bf 109s were to be located in the Pas de Calais in order to place them closer to the British coast, thereby reducing some of the range disadvantages that the design suffered, lessening the time required to reach the combat areas, and making it easier for the fighters to escort bombers during attacks on the southern and eastern regions of the country.

The difficulty associated with destroying the British radar towers was acknowledged at this conference. The Germans still did not fully appreciate the vital role that the Chain Home radar played in enabling the relatively few fighters of Fighter Command to be employed with maximum efficiency. Goering ordered that, henceforth, no further effort would be directed at attempting to destroy the towers. He then issued orders that the attacks should focus on fewer targets, concentrating their primary efforts on destroying the fighter forces, with the British aircraft industry as the second most important target. Finally, he delegated to the Luftwaffe the authority to select specific targets, but he reserved for himself the authority to order attacks on London and Liverpool.

The tactics conference that culminated with the meeting on August 19 in effect solved no problems and for the most part served only to make things worse for the Germans. The worst decisions were those that tied the Bf 109 to close escort duties and abandoned the effort to destroy the Chain Home system; the best were those that clarified the target priorities and decentralized target selection. Every participant in combat—every crewmember—sees the battle through the filter of his or her own experiences. It should come as no surprise that the Luftwaffe bomber crews sought closer fighter escort. They resented the fact that the fighters were often out of visual contact and felt abandoned when the British fighters attacked. The fighter pilots, on the other hand, understood that they could not fly at the low altitudes and airspeeds that the bombers flew and adequately defend them. The fighters needed the advantages that altitude and airspeed provided to contest the British Hurricanes and Spitfires. The role of combat commanders is to take the inputs of those they command and make rational decisions that reflect the greatest good for the overall combat effort, not allowing themselves to be swayed by either the emotions of their subordinates or their own prejudices. In this Goering failed. Influenced by his own unrealistic views of the Me 110 and attempting to appease the demands of his own bomber crews, his decisions made the German cause even more difficult than it already was.

32. Goering Orders Close Support for Bombers

One of the outcomes from the mid-August tactics conference was the decision to reposition most of the Bf 109s then under Sperrle's *Luftflotte 3* to the Pas de Calais, where they would operate under Kesselring (*Luftflotte 2*). The primary purpose of this move was to improve the ability of the fighters to perform the bomber escort mission. Bowing to the increasingly vocal calls by the bomber crews for closer support from the fighters, Goering, to the dismay of his fighter commanders, ordered that a substantial portion of the fighter force be committed to the close escort mission. The Bf 109s shown weaving above a formation of German bombers represent the result of this directive. Whereas the fighters might have previously positioned themselves several thousand feet above the bombers, ready to dive down upon any attacking British formations, this new order placed them essentially at the same airspeeds and altitudes that the bombers flew. This had the effects of lessening their effectiveness as escorts and causing them to burn fuel at needlessly high rates as they wove back and forth near the bomber formations.

As has been observed previously, airspeed and altitude represent different forms of energy, and one may be converted, or traded, for the other. What is critical to fighter pilots is that, if they cannot have both, they have one or the other. Altitude and airspeed are equivalent to hard currency in the world of air-to-air combat. When the German pilots were forced to fly at lower altitudes than their normal levels—and, furthermore, were forced to slow to speeds corresponding to those of the slower bombers—the effect was that the fighters started at a disadvantage in any engagement with the British Spitfires and Hurricanes.

When 11 Group fighters encountered the formations composed of both bombers and Bf 109s, Park was forced to reconsider the policy he had established earlier, which had directed that Spitfires be targeted against German fighters and Hurricanes against bombers. Park's response was to direct his controllers to begin targeting both Hurricanes and Spitfires against the bomber-fighter formations.

As more Bf 109s were repositioned to bases in the Pas de Calais area, Kesselring was gradually able to mount formations consisting of bombers and close escorts and, in addi-

Close support drawback. Fighters must weave over formation to keep speed up... but consume fuel.

tion, to once again position additional Bf 109s as high cover for the formations. These high-cover formations generally operated at altitudes well above 20,000 feet. Park, ever flexible, then once again ordered that the previous policy be reinstituted—Spitfires against the high fighters and Hurricanes against the lower formations. Of course, since the lower formations were composed of both bombers and Bf 109s, it goes without saying that the Hurricanes would face both routinely. Under these circumstances, with substantially increased numbers of German fighters positioned to defend the bombers whether from high altitude or from the formation itself, the pressures on the British fighters became significant. This is, once more, the kind of flexibility normal to tactical combat operations—with move and countermove—as each side in a conflict attempts to deal with the situation encountered. As shall be shown, and as might be expected, these tactical shifts continued throughout the battle as both sides probed and sought any advantage, no matter how short lived it might be.

33. Defiant Withdrawn

The Boulton-Paul Defiant proved completely unsuited for the dynamic environment that characterized day fighter operations. Shown here limping back to home base with combat damage, the airplane was finally withdrawn after having been decimated by the highly maneuverable German Bf 109s. The shortcomings of the Boulton-Paul Defiant are discussed in some detail in Chap. 2. The separation of the gunnery function from the pilot made maneuvering and aiming very difficult to coordinate; the gun turret mounted just aft of the cockpit restricted the gun from firing directly forward; the turret, with its difficult entry, made quick-response takeoffs almost impossible; and to further compound matters, the performance of the airplane was inadequate in comparison to the Bf 109 fighters it most often encountered. As the intensity of the battle for Britain increased, the dangers for the crews assigned to the Defiants were heightened, and the Defiant's days as a day fighter—a role for which it was never intended—were numbered.

As early as July 1940 the signs pointed to eventual disaster for both the Defiant and its crews. On July 28, just weeks after the Battle of Britain had officially begun and shortly after the aircraft was introduced into combat, 141 Squadron lost six of nine Defiants assigned when it encountered a squadron of Bf 109s. In spite of this disastrous initiation and the clear indications that there were serious design deficiencies associated with the aircraft, a second squadron, the 264th, was deployed in the 11 Group area on August 22. The question as to why an obviously deficient aircraft was deployed forward into the most intense combat area of all four Fighter Command groups remains unanswered. 264 Squadron was based at Hornchurch, a Spitfire base. With no trained ground support at its assigned base, the squadron was required to be self-sufficient, much in the same sense that today's expeditionary forces are expected to carry their own support and be totally

28Aug40; 264 Sqn aircraft limps home after partisipating in last daylight attack by Defiants.

self-sufficient. Given that it was able to operate independently, 264 Squadron was then ordered to operate out of Manston on a daily basis. Manston was a base that was located on the forwardmost land areas along the routes that led directly from the Pas de Calais to London. It was attacked more often than any other base in 11 Group. Not only was the Defiant a marginal design, but it was deployed to the most active combat area in all of 11 Group—a location that demanded quick-response takeoffs, rapid climbs to altitude, and the ability to engage both fighters and bombers. The Defiant was lacking on every count; it was a doomed airplane and a death sentence for the crews assigned to fly it.

On August 24, 264 Squadron was being refueled when an inbound raid was detected. Unable to climb to altitude and prepare for a coordinated attack before the Germans were upon them, three aircraft and crew were lost, including the squadron commander. Then, on August 26, three more aircraft were lost when two squadrons of Bf 109s attacked as the squadron was lifting off in response to an alert. This was followed by another disastrous day on August 28, when the 264th suffered seven Defiants either destroyed or damaged in combat. This finally signaled the end for the Defiant. Having lost approximately 12 airplanes and 14 crewmembers, with still others wounded, the Boulton-Paul Defiant was finally withdrawn from daytime operations and assigned duties as a night fighter.

34. London Attacked

The last week in August and the first week of September marked the most difficult period of the battle for Fighter Command and the period when defeat loomed as a distinct possibility. The German objectives—the fighters and the airfields—had not changed, but the intensity with which they were assaulted increased substantially. As bombing resumed on August 24 after several days of poor weather, it was clear that the Germans were determined to move the locus of their attacks northward. On that day, the attacks started in the morning and continued through the day. Numerous airfields in the areas surrounding London were targeted, a distinct change from earlier attacks which had focused almost exclusively on the airfields well south of the city.

11 Group defenses were committed, and Park called upon 12 Group to provide cover for airfields north of the Thames. The defensive pressures brought against the bomber formations caused various of them to break; in several instances, crews jettisoned their bombs before they were able to deliver them against the targets planned.

In spite of Goering's insistence that the decision to bomb London was his alone, bombs fell onto the city, and fires broke out all over London. Portsmouth was particularly hard hit, with over 200 bombs dropped on that area and over 300 killed. This was the first time that bombs had fallen on the city proper since 1918, when the Gothas had last attacked. German contrails over London, as depicted in the accompanying drawing, were

September 7th. The Germans change their objective to London... and seal their fate.

not a sight the inhabitants of the city had prepared for. They would become distressingly common before the Luftwaffe finally abandoned its campaign—but that was not to happen until the spring of 1941.

While the Germans insisted that bombs were dropped on the city only as the result of navigational errors and accidents, Churchill's view was that the selection of targets located on the outskirts of London guaranteed that any error would result in the bombing of heavily populated urban areas. On the evening of August 25, one day after the German bombs dropped on London, Churchill ordered Bomber Command to carry the attack to the German homeland. Targeting industrial sites on the outskirts of Berlin, Bomber Command delivered its loads on three of the following four nights. Although the damage done was largely negligible, the emotional impact was huge. Goering had assured the German people that this would never happen, yet fires were burning within full view of the city.

If the German people were disillusioned because British bombers could reach them, Hitler was outraged that the attacks had taken place. He met with Goering on August 30 and removed his earlier restrictions on bombing London. At this meeting he made it clear that he wanted revenge for the bombs dropped on Germany and that he specifically wanted that revenge to be exacted on the city of London. When Hitler spoke at the Berlin *Sportpalast* on September 4, he publicly threatened British cities and promised that the Germans would soon appear in the skies over London.

In outlining his strategy for the conduct of *Adlerangriff*, Goering had described a plan that included progressively bombing at ever-reduced ranges from London. The clear expectation was that the German fighters would have cleared the skies of British fighters, and the final assault on the city would be largely unopposed. As Goering outlined plans for the bombing to Sperrle and Kesselring, his *Luftflotten* commanders, their reactions were influenced by their differing views as to the best way to achieve the air superiority they had been tasked to deliver. Park had directed 11 Group air defense controllers to concentrate their efforts on bringing down bombers and not waste their precious assets engaging the Bf 109s. British fighters had stopped responding to free-chase provocations; they were reserving their efforts solely to attack bomber formations. In the meeting with Goering, Kesselring argued strongly that the best way to force British fighters into the air would be to attack a target that they would have to defend—London. Sperrle, convinced that the campaign against the fighters and airfields was having the desired effect, argued for continuing the attacks directed at the RAF. Kesselring responded that if the threat to their existence became too acute, British fighter squadrons would simply be redeployed to the north—out of the range of the Bf 109s. Sperrle's arguments were disregarded; the decision had really been made when Hitler met with Goering on August 30. On September 5, Hitler's supreme command directed the Luftwaffe to carry out harassing attacks by day and night on the inhabitants and air defenses of large British cities. London was the target.

By this time Fighter Command was near exhaustion on every conceivable scale. Far too many of the most experienced pilots and leaders had fallen; one-fifth of the squadron commanders and one-third of all flight leaders were by then dead or wounded (Hough and Richards 1990). Replacements were ill-trained and not ready for combat. The airfields had been bombed until they were marginally operational. Fighter Command was on the threshold of defeat.

September 7 began as all days had begun for the past two weeks. There were a number of engagements during the day, but everyone up to and including Dowding understood that a more demanding test was coming; it was simply a matter of when it would start. During the afternoon the Chain Home radar detected a large formation massing over the Pas de Calais, and fighters were launched to intercept and engage—the normal course of events. But on this day the size of the formation was uncommonly large. Goering had sent 348 bombers and 617 fighters in a giant formation toward the British coast. As the size of the attack became more apparent, Dowding fed more fighters into the fray until by 1630 (4:30 P.M.) Fighter Command had ordered 21 squadrons into the air. When the first British pilots finally caught sight of the formation, they saw 1000 aircraft stacked 1½ miles high and covering 800 square miles (Mason 1969).

Dowding and his staff only then fully comprehended that the battle had shifted to its next phase. A formation of the size encountered could mean only one thing—London. This marked the beginning of an ordeal that would not only test the Royal Air Force, but, in particular, would test the resilience of the British people. But as impressive and frightening as the German formation must have been, this day also marked the turning point in the Battle of Britain. The decision to turn toward London and abandon the attacks on Fighter Command and the airfields allowed the RAF to survive and recover when it appeared that the battle had been lost. The change gave Dowding a chance to fill his empty squadrons, build his reserves, and rest his exhausted crews enough to allow them to return to full combat effectiveness. Even though London and the civilian population of Britain would endure night bombing attacks for months—well into 1941—September 7 was, in retrospect, the point at which Germany lost the Battle of Britain.

35. September 15, 1940: Battle of Britain Day

As plotters moved their markers about the plotting table, Churchill and Park discussed the seriousness of the situation. With no reserves left, the outcome of the battle would depend on the pilots, crews, and aircraft of Fighter Command. On September 15, 1940, the Luftwaffe sent wave after wave of bombers to attack London. They were met by literally hundreds of fighters from 10, 11, and 12 Groups. At the end of the day, both the German and the British pilots and crews knew that the Battle of Britain had been decided. The strategic decision that ensured the RAF would prevail had been implemented on

Uxbridge, 15 September 1940. Churchill: "What reserves have we?" Park: "There are none."

September 7, when Hitler and Goering abandoned the attacks on the RAF and its airfields and turned to an all-out assault on the city of London. However, September 15 was the day when the tactical situation made it unequivocally certain that the Luftwaffe had lost the air campaign. This was the date later selected as Battle of Britain Day—the day when it became obvious to one and all that the German campaign had failed, that the Royal Air Force would prevail, and that Hitler would never defeat England.

Just 10 days earlier, Fighter Command—and 11 Group in particular—had been on the verge of collapse. Crews were exhausted, too many squadron commanders and flight leaders were either wounded or dead, and Dowding had reluctantly initiated his "stabilization scheme," the rotation policy aimed at balancing the experience levels among the squadrons that bore the brunt of the daily attacks. Daily attrition was beginning to have its inevitable effect as British losses mounted and exceeded German losses during early September. Defeat was at hand for Fighter Command.

Had the German attacks on the RAF continued, the question naturally occurs—faced with inevitable defeat in the south, would Dowding have reinforced 10 and 11 Groups with units from the northern 12 and 13 Groups? Or would he have redeployed the remaining units in 10 and 11 Groups northward to fields that were out of reach of the Bf 109s, which would have forced the Luftwaffe commanders to send unescorted bombers to achieve the air superiority they had originally set as a condition for invasion? One can never know the answer to a hypothetical question such as this, but it does make for interesting speculation. The authors favor the view that Dowding, who had moved heaven and earth to save his fighters when everyone from Churchill downward had favored sending additional aircraft to France, would once again have made the decision to save his forces to fight another day. He would have redeployed to the north and fought from the relative security of the areas assigned to 12 and 13 Groups.

When Hitler changed the fundamental objectives of the campaign to focus on punishing Britain rather than destroying the RAF, Fighter Command was saved. Even though the demands on the crews and aircraft were still heavy—they still had to intercept and engage the German bombers on a daily basis—the fact that they were no longer the primary targets took an immense amount of the daily pressure off of the fighter units. For the first time since July, pilots might enjoy a day when they were not scheduled to fly. Newly assigned pilots were given training to acquaint them with their aircraft and its weapons before being thrust into combat. By September 15, while hardly completely recovered from the exertions of the past two months, the air crews had caught their collective breaths, realized that they might, in fact, survive, and had time to once again consider tomorrow instead of wondering if they would live through the day. In effect, they were in better shape than they had been in weeks.

September 15 started normally, with a few German reconnaissance flights. By midday a large formation of Dornier Do 17s, escorted by still more fighters from the Pas de

Calais area, began the flight toward London. Observing this on the Chain Home radar, Fighter Command alerted several squadrons of Hurricanes and Spitfires, and positioned them to meet the bomber force. Fighters from 12 Group were also alerted and were ready when the German formation arrived in the areas south of London. A huge wing of over 40 fighters attacked the bombers head-on, and then turned to rip through the formation from its flank. Bombs were flung more or less randomly over the countryside as the Luftwaffe formation fought to maintain its cohesiveness as an airborne unit.

Winston Churchill had chosen this day to visit Park in his 11 Group operations center at Uxbridge. As the first wave of German bombers made its way across Kent, Churchill observed as Park fed first a few, then gradually all of his squadrons into the battle. Churchill is reported to have asked, "What reserves have we?" to which Park is reported to have responded, "There are none" (Hough and Richards 1990). It was truly an all-or-nothing roll of the dice—one that the British could not afford to lose.

This early attack was then followed by a second wave of bombers, consisting of over 150 bombers escorted by Bf 109s, which approached London in the early afternoon. Park had ordered his 11 Group aircraft to be quickly serviced and turned around in preparation for such an attack. Approximately 170 Hurricanes and Spitfires from 11 Group met the bombers as they penetrated British airspace. As they shuddered under the impact of this blow, they were met again by a five-squadron wing formation of fighters from 12 Group. It was more than the brave German crews could handle. Bombs were jettisoned and aircraft began turning short and heading for the European coast. The formation was broken, air discipline had collapsed, and the engagement took on the characteristics of a rout.

The German crews had been taking the attack to the British since July. They had done as their leaders had ordered, flying across the hated English Channel every day, and every day they had lost aircraft and friends. They had cynically observed the senior Nazi strategists and leaders as they first targeted shipping, then the fighters and airfields, and next London. They had been told too many times that the British fighters were down to their "last 50 aircraft." To then find themselves in a running air battle with hundreds of Hurricanes and Spitfires was too much. When they finally broke and dashed for home, both the pilots of the Luftwaffe and those of Fighter Command understood that this was it—the Battle of Britain had been decided—Britain had won. Later tallies indicated that the Luftwaffe had lost about 60 aircraft on the day, but far more important than aircraft, the Germans had lost their will to continue. From this point forward, it was simply a matter of time until the inevitable end of the German campaign in England. The bombing of London would continue well into 1941, and the civilian population of the country would hardly have agreed that the battle was over; however, to those in the RAF who were fighting the war in the air, the signs were unmistakable. No one knew it on that day, but Germany would soon call off its invasion plans, and, while the Luftwaffe might go on bombing for some time, the threat to Britain's existence as a sovereign state was no longer at issue. Britain would survive. September 15, 1940, was only later designated Battle of Britain Day in recognition of the valiant effort made by the Royal Air Force and Fighter Command and the fact that this day truly marked the beginning of the end of the German campaign to defeat Great Britain.

36. Group Tactics Differ

Fighter aircraft typically take off either as a single ship or in pairs. They then join into larger formations in the air, as shown in the illustration. The airborne join-up takes time, and the more aircraft involved, the longer the join-up requires. The issue of the time required to form a flight before it was ready to enter combat and the relative effectiveness of small formations in comparison to larger formations became the subject of an intense conflict between the commanders of 11 Group and 12 Group. In 11 Group, Park tended to favor relatively small groups; in his view, it took too long to form larger groups when the enemy bombers were just minutes away from the English coastline. In 12 Group, on the other hand, Leigh-Mallory had come to the conclusion that the mass firepower of larger formations was preferable as a tactic to confront German bombers. The employment by 12 Group of a huge fighter formation that became known as the *Big Wing,* or the *Duxford Wing* became the central issue in a growing conflict between Park and Leigh-Mallory. The differences between the two reflected not only the differing views of fighter pilots as to the best way to accomplish an assigned mission, but also the different circumstances that the two groups found themselves in as a result of the different geographical areas they were assigned to defend.

One of the strengths of Dowding's conduct of the tactical fighter war was the fact that he delegated responsibility for day-to-day decisions to the group commanders. As in most situations where highly competent and competitive officers are given broad responsibilities, the commanders of the four groups that made up Fighter Command did not hesitate to take the initiative in decisions regarding the tactics that their units would practice on daily missions. It should also come as no surprise that these were ambitious men who, on more than one occasion, sought to have their particular views accepted as the best practices and subsequently mandated as Fighter Command policy for all other groups to follow. No better example of this competitive approach to practice and policy exists than the controversy that arose between Park and Leigh-Mallory over the size of the flying units that would normally be employed in combat.

In considering this issue, it is important to recognize the advantages and limitations associated with both small and large formations of aircraft. In the first place, there is the fact that smaller units are inherently more flexible than are large formations. Consider a single aircraft in flight; the pilot of a single aircraft need not restrict his maneuvers at all. He can climb as fast as the airplane's performance permits; he can turn or dive as suddenly and as violently as he cares to, with no thought to whether a wingman is alert and able to follow. The single aircraft is, then, the ultimate in maneuverability in the air. But, as was amply demonstrated in World War I, the single aircraft is vulnerable in combat. It is often limited in firepower and enjoys none of the benefits of mutual protection that aircraft flying in formation offer.

The typical British squadron consisted of 12 aircraft. The general practice was to have the 12 divided into two flights of 6 airplanes, and each flight was further divided into two sections of 3 each. The three-ship vic formation was a natural outgrowth of this hierarchical subdivision. The background and gradual evolution from the vic to the finger four has been discussed earlier in this work. However, it is also important to recognize that the adoption of the finger four as standard practice did not occur until after the Battle of Britain was over. The typical squadron formation consisted of four vics of three aircraft each. In August 1939, Dowding wrote in a memorandum to the Air Ministry that, in his opinion, the squadron was the largest tactical unit that could be practically employed in combat. This was the common view throughout most of the Battle of Britain, and it was largely related to the view that a force any larger than the squadron would lack the necessary maneuverability to handle the demands of tactical warfare.

The fighting of August and September 1940 was often characterized by raids that originated over the area of the Pas de Calais, then proceeded along a route that took them over the area assigned to 11 Group for defense. As the formations were first detected, then analyzed for size, type, altitude, and route, fighter squadrons were alerted and ordered into the air to intercept. Even when tens, and sometimes hundreds, of British fighters were sent to intercept the German formations, which were composed of both bombers and escorting fighters, the individual squadron often found itself outnumbered 10 or more to 1. It is little wonder, then, that some units began to suggest that it might be well to send flying units consisting of more than single squadrons into the battle.

Park, commanding 11 Group, which had responsibility for the southeastern quadrant of England, faced a particularly vexing problem in dealing with approaching German attackers. The German formations were literally only 30 miles from the British coast when they departed the European mainland. For the British fighters to have sufficient time to climb to altitude and be prepared to intercept the Germans, they had to launch quickly, climb as rapidly as possible to altitude, and maneuver without hesitation to the point directed. Park wanted to intercept the Germans as far as possible from high-value targets such as his airfields, and later London. Forming several squadrons into a larger formation requires time as the aircraft maneuver to find each other and position themselves

Big wing theory. Fundamental shortcoming is that it takes too long into a large unit; moreover, once formed, the large unit cannot maneuver with the abruptness that the smaller squadron unit can. Park was, by the nature of the mission assigned him, compelled to think in terms of relatively small units as the preferred approach to intercept incoming German formations. In spite of his preference for operating in squadron units, Park acquiesced to the use of paired squadrons as the massive German formations became so large that a single squadron was hopelessly outnumbered.

To the north of 11 Group, however, the circumstances—and hence the conclusions—were different. The 12 Group defense area stretched essentially across the whole of England, beginning at a line that ran more or less east to west some 40 miles north of London. In general, squadrons in 12 Group had ample warning of inbound enemy

for a large number of fighters to launch and form up.

bomber formations. They were not confronted with the same time-to-climb pressures that 11 Group faced. In many instances, 12 Group squadrons were called upon to cover bases in the 11 Group area as the aircraft from those bases flew south and eastward to meet the German formations as far forward as possible. Much the same was true when the Germans began to bomb London on a regular basis. 11 Group would meet the German formations at the coast and attack as they proceeded toward the city, then 12 Group would hit them as they neared the urban area around London.

The commanding officer of 242 Squadron (12 Group), Douglas Bader, had come to the conclusion that attacking the German formations with larger fighter formations would be much more effective than having single squadrons, or sometimes paired squadrons, intercept them, as was the practice in 11 Group. Furthermore, the tactical organization and division of responsibilities by geographic area provided enough time for the squadrons in 12 Group to mass into larger units before moving to intercept attacking German formations. Bader became an active proponent of the concept that came to be known as the *Duxford Wing*—a large formation composed of several squadrons operating under the command of a single leader in the air. The Duxford Wing comprised up to 60 aircraft, and the idea was to have this massive fighter formation hit a German formation after it had already run the gauntlet of 11 Group fighters on its way inbound to London.

As is implied from the remarks made earlier in this section, problems arose when the conflicting opinions, egos, and ideas began to interfere with daily operations. Bader was anxious to lead the Duxford Wing into combat, convinced that the massive fighter formation would be much more lethal than incremental attacks by single squadrons. Park, on the other hand, insisted that the role for 12 Group in his area was to be limited to protecting 11 Group airbases when his squadrons flew off to meet the Germans. The differing views became an issue of considerable conflict and debate through the months of August and September. As August wore on, the debate continued, with Park insisting that the large wing took too long to form and that he needed squadrons ready to cover his bases when 11 Group interceptors vacated them. On more than one occasion, Park's requests for coverage went unanswered as 12 Group, employing the Duxford Wing formations, chose to engage enemy formations rather than fly high cover over Park's bases. Park became so agitated and the debate grew so rancorous that, in late August, Park instructed his controllers that they could call directly on 10 Group for support, but they must send any request for 12 Group support through Fighter Command, because, in his view, he could not depend on Leigh-Mallory's squadrons to react to a request from 11 Group. As the German attacks progressed further inland and the city of London increasingly became the focus of attacks, the 12 Group Duxford Wing was increasingly active, and the debate between Park and Leigh-Mallory had become an open spectacle, setting up a situation that would eventually have to be settled at higher levels.

37. RAF Tactics Meeting–October 1940

As the German raids progressed inland toward London, the Duxford Wing, led by Douglas Bader, became increasingly active in the air war. On several occasions Bader's wing claimed uncommonly good results, and, with this, Bader became increasingly adamant in his assertions that the big wing was tactically superior to smaller units. His confidence was accompanied by a growing frustration that he was not given more opportunities to exploit his tactical innovation; for this he held Air Vice Marshal Park responsible.

Air Vice Marshal Leigh-Mallory, 12 Group commander, was completely supportive of Bader's new tactics, and he also chafed at the relatively limited role that his group was forced to play due to the geographical location of its area of responsibility. It happened that Peter Macdonald, a member of Parliament, was also a pilot in 242 Squadron, Bader's unit. Sympathetic with Bader's frustrations, Macdonald brought the conflict and differences between 11 and 12 Groups to the attention of the undersecretary for air and the prime minister, both fellow members of Parliament. As often happens when politics and defense issues are mixed, pressure was subsequently brought on the air staff to investigate the situation and to resolve what, to the members, seemed a lack of decisiveness and leadership within Fighter Command. The chief of the air staff called a meeting to discuss day tactics in the fighter force. The meeting was held on October 17, 1940.

The picture shows Park conferring with Dowding. In no uncertain way, these two stood together—and on the defense—in the meeting called to resolve the growing differences between Park and Leigh-Mallory. Air Vice Marshal Sholto Douglas, the deputy chief of the air staff, chaired the meeting. Attending were Dowding, Park, Leigh-Mallory, Brand (10 Group commander), the newly designated chief of the air staff, several members of the air staff, and, surprisingly, Douglas Bader. There are many who have expressed surprise—even dismay—that officers of Dowding's and Park's status and position should

Park and Dowding reflecting on Leigh-Mallory's "Big Wing" argument, 17 Oct 40.

be called upon to defend themselves in a forum attended by the squadron commander whose actions had led to the meeting. The fact is that Bader was not in attendance during the substantive discussions. Leigh-Mallory had apparently brought Bader to the meeting, hoping that he would be permitted to make his case for the Big Wing tactics. However, he had not cleared Bader's attendance with Sholto Douglas, and the deputy chief refused to allow Bader into the closed-door session.

The records from this meeting are deliberately noncontroversial. In the minutes of high-level meetings, the emotions associated with the exchanges are rarely, if ever, recorded for posterity. Leigh-Mallory spoke of the need to challenge the German bomber formations with as much force as possible, calling attention to the results that Bader had achieved with the larger unit tactics. Park, in response, pointed out that his units had less time to position themselves for intercepts. He argued that, if he were going to meet the German formations as far forward as possible and before they reached high-value targets, he simply could not afford to delay while multiple squadrons joined together in the air. Park's position was that he only had time to form one or, at most, two squadrons before the opportunity to intercept forward of British airspace was lost. Sholto Douglas summarized the outcome, stating that, while he recognized the unique circumstances created by 11 Group's geographical location, he believed that there was opportunity to employ both tactical approaches—single or paired squadrons to intercept forward, and larger wings to bring more mass to the air battle. He called for cooperation from the two groups when that made operational sense. Dowding assured Sholto Douglas that cooperation could be achieved and that he would personally see to it that 12 Group large wings were given the opportunity to participate in suitable operations over 11 Group's territory.

As innocuous as the minutes of the meeting were, subsequent events gave some hint of the depth of the feelings that attended both the meeting itself and the events that precipitated it. Park sent typed remarks to the air staff, asking that they be incorporated into the minutes of the meeting. In these remarks he repeated the positions stated in the meeting, but went on to add a number of highly critical remarks concerning the Duxford Wing, the extent to which the formation was effective under difficult tactical conditions, and the confusion that the wing had brought by conducting uncoordinated incursions into 11 Group areas. The air staff secretariat declined to include them in the formal minutes of the meeting.

There are several issues that cloud and complicate the issue when one seeks to sort out which of the groups actually practiced the most effective tactics for the time and situation. It is clear that some of Park's own pilots thought they should be using larger formations, as he felt compelled to write a letter to his sector commanders explaining his reasons for not using units of three or more squadrons more often. Again, the time to intercept and maneuverability were at the core of his justification. It is also true that technology played a part in Park's decision. Even by October 1 only 16 squadrons had VHF radio; the rest still communicated in the air using HF, which was notoriously difficult to hear and often prone to fade in and out. Park felt that having very large numbers trying to coordinate their actions on the same HF radio frequency would have impeded, rather than improved, operational effectiveness. And there is also the fact, learned later, that the combat claims of the Big Wing far exaggerated the reality of the wing's performance. This is to be expected when up to 50 or 60 aircraft are jousting in relatively close proximity. One aircraft might shoot and hit an enemy bomber and judge it to be mortally damaged. Another might subsequently attack the same damaged bomber and damage it further, and both crews might claim a combat kill. While both claims might be made in good faith, the result was that the claims of the Duxford Wing were among the most exaggerated of all units in the battle. This was later confirmed by postwar research and analysis.

So who had it right—Park or Leigh-Mallory? The authors believe that attempting to resolve this question fails to allow for the possibility that they *both* had it right—each for the circumstances they encountered. Sholto Douglas had it right; Park was using the right tactics to intercept as far forward as possible, but there should have been better cooperation between 11 and 12 Groups to take better advantage of what they both could bring to the overall combat picture. What is certain is that Park and Leigh-Mallory should have been able to work the conflict out at their own level; or, conversely, Dowding should have demanded that they do so before the debate became politicized. As it was, the conflict between Park and Leigh-Mallory, compounded by friction between Dowding and the air staff, led shortly to Dowding's forced retirement. He left his position as the head of Fighter Command on November 25, 1940. Park was subsequently replaced as commander of 11 Group in December of the same year. No two officers had given more, nor were any more directly responsible for the Fighter Command's success during the Battle of Britain, than were these two leaders.

38. Last Large Daylight Raid

A Spitfire is shown attacking a flight of three Ju 88s—a sight that would not be seen again in the skies above Britain after September 30, 1940. By the end of September, the losses suffered by German bomber fleets had increased to levels that forced the Luftwaffe to cease conducting daylight raids. Not that the bombing stopped; in fact, the bombing would continue to be a daily feature of life in London until the following spring. Rather, September 30 marked the last time the Luftwaffe would send bombers over the English countryside during daylight hours.

As September ended, night bombing raids on British targets had been a regular fact of life in England for well over a month. The tactics conference called by Goering in mid-

30 Sept 1940. the last large daylight raid on London.

August had resulted in the relocation of most of Sperrle's fighters to the Pas de Calais under Kesselring's command. The purpose of this redeployment was to provide more aircraft for bomber escort missions and to locate the Bf 109s as close to the British coastline as possible in order to compensate for the lack of range of that aircraft. However, a side effect of the redeployment was that Sperrle's bombers were left without adequate fighter escort. Sperrle's response was to begin using his bombers at night. The immaturity of airborne intercept radar technology was such that the risk to bombers from intercepting British fighters was far less at night than during daylight hours. Of course, night operations raised other problems—maintaining formations and finding targets, to name just two.

As daylight bombing missions came under increasing pressure from the now rested and newly confident RAF, the incidence of night bombing grew substantially during the last few weeks in September. London and Liverpool were favored targets of the Luftwaffe, and casualties in the hundreds were a common occurrence in those unfortunate cities. Goering, in yet another turn in strategic direction, ordered his forces to once again make British fighters and aircraft factories a priority target—adding, no doubt, to the view among German crews that their top leadership had little or no idea of how to fight the air war. The bomber crews complied, however, and several attacks were conducted on Spitfire production facilities. However, by this time the British shadow factory system was fully operational and production was distributed among several plants, so these raids caused little or no disruption in production schedules. Toward the end of September, the Luftwaffe suffered several disastrous encounters with Fighter Command. The Luftwaffe began to use Me 110s as fast bombers in combination with Ju 88s, accompanied by massive formations of Bf 109s as escorts, but Park and Brand quickly adjusted to the new tactics and, in spite of heavier than normal losses, were able to render those attacks relatively ineffective as well.

On September 30, two large raids were mounted by the Luftwaffe, one consisting of about 200 aircraft in the morning and the other of approximately 300 aircraft in the afternoon. While a number of Spitfires and Hurricanes were damaged, relatively few were destroyed, but the Germans again suffered huge losses of both Bf 109s and bombers. The fighting in September had been disastrous for the Luftwaffe. Not only had Hitler made the drastic strategic error of switching targets from the airfields and fighters to attack London, but, in addition, the daylight raids during the last days of the month were marked by tactical and strategic confusion and lack of effectiveness in results. The Luftwaffe was now forced to change tactics once again. After September 30, the Germans would no longer send their bombers over England during daylight hours. The bombing would continue every night for months to come, and many British citizens would die, but the direction and trends were unmistakable—it was now only a matter of time before the Germans would finally admit the futility of their campaign and end the air war against Britain.

39. High-Altitude Nuisance Bombing

October brought the onset of the British autumn with its clouds and rain, which made the weather an integral part of daily tactical planning. October also saw the introduction of a new variant of Germany's Bf 109 fighter, the Bf 109E-7. As shown in the illustration, the aircraft carried a single 250-kg bomb that it dropped from high-altitude, level flight. While a fighter carrying a single bomb would seem to be a relatively insignificant threat, this aircraft produced substantial stresses on Fighter Command during the waning days of the Battle of Britain.

A single-seat fighter with no bombsight or navigator cannot drop a bomb from straight and level flight with any degree of precision. It was precisely this fact that led the Germans to embrace the dive-bomber with such enthusiasm as they searched for means to improve air support for ground operations. The Bf 109E-7 had a significantly improved engine that enabled it to approach the British coast at very high altitudes. Bombs were dropped from straight and level flight, meaning that they fell more or less indiscriminately on the areas targeted. Dropping their bombs on populated areas such as London, these aircraft were strategically insignificant, but they represented yet another challenge for Park and 11 Group to deal with.

The Bf 109E-7s typically approached the British coast at altitudes between 25,000 and 30,000 feet, making it extremely difficult for even the Spitfires—the only British fighters that could effectively engage another fighter at this height—to climb to altitude

Me 109E-7's bombing London from high altitude.

in time to intercept them. The net result was that the 109E-7s operated with a degree of impunity during the days when weather permitted fighter operations. The Luftwaffe sent these specially equipped aircraft one by one or in pairs toward England, often sending hundreds in a single day. On some days the numbers exceeded 1000 German fighter-bomber sorties. The pilots of 11 Group once again found themselves flying two, three, and even four sorties per day in the attempt to intercept them. A consistent concern expressed by the British pilots was that they invariably arrived at altitudes that were too low to effectively engage the high-flying 109s. Park, in an attempt to solve the problem, eventually had to resort to standing patrols. By orbiting Spitfires at 25,000 feet and Hurricanes at 20,000, he was able to improve the rate of interception, although at considerable expense to his air crews and machines.

As October wore on, the British weather gradually constrained tactical operations on both sides as the clouds, fog, rain, and shorter days became daily considerations. Park had increased the altitudes of his standing patrols, with Spitfires orbiting as high as 28,000 feet. On October 29, Park's fighters intercepted a rather large group of the high-flying Bf 109E-7s escorted by several squadrons of the more conventional 109Es. The 109E-7s were forced to jettison their bombs, and several of both variants were shot down. After this, although single-fighter raids continued as the weather permitted, the focus clearly became the night bombing campaign conducted against civilian population centers. It was essentially the end of the fighter-on-fighter engagements that had been so much a part of the battle.

The end of the fighter war brought no solace to the citizens of London and other population centers in Britain. Over 7000 tons of German bombs fell on British cities during October alone; over 13,000 people were killed and over 20,000 were wounded as a result (Hough and Richards 1990). The bombs would continue to fall nightly until May 1941, yet the Battle of Britain was essentially over. In the same sense that the date chosen to mark the beginning of the battle was more or less arbitrarily designated as July 10, 1940, so also was the date selected to mark the end—October 31, 1940. During these four intense months a force of just over 3000 young men had literally changed world history. No one has said it better than Winston Churchill, who in August 1940 spoke the famous line, "Never in the field of human conflict was so much owed by so many to so few."

CHAPTER 4

HOW AND WHY

WHY BRITAIN WON

Signs of the impending collapse of the German air campaign first appeared during the days leading up to mid-September 1940. On three separate days the Luftwaffe lost over 50 aircraft as the Germans pressed their assault on the city of London. Then on September 15, later designated as Battle of Britain Day, flight discipline broke and German forces retreated in chaos and disarray. On September 21, Britain lowered the state of alert from the highest level, "Cromwell," when it became apparent that the Germans were disassembling their barge and boat invasion fleet in northern Europe. By September 30 the Germans had decided to abandon daylight bombing and seek the cover of night to escape the relentless defenses that they encountered on a daily basis. Then, on October 12, although the British would not be aware of it until much later, Hitler postponed the invasion that had been held like a sword of Damocles over the British population since July. While insisting that the appearance of preparations be continued in order to maintain political and military pressure on the British and keeping open the possibility that the decision might be revisited in 1941, the facts were clear to anyone who might have read the decision: the campaign was over. The bombing of London would continue until spring of 1941, but its purpose would be punitive rather than with the expectation of conquering England. That hope had disappeared.

The Royal Air Force and Fighter Command in particular had survived continuous attack since the early days of July. It had started with attacks on British shipping in the English Channel intended to lure RAF fighters into the air so that Luftwaffe fighters, successful in every campaign dating back to the Spanish civil war in 1936, could destroy them. Fighter Command had barely survived the attacks directed specifically at the fight-

CHAPTER 4. HOW AND WHY

1940

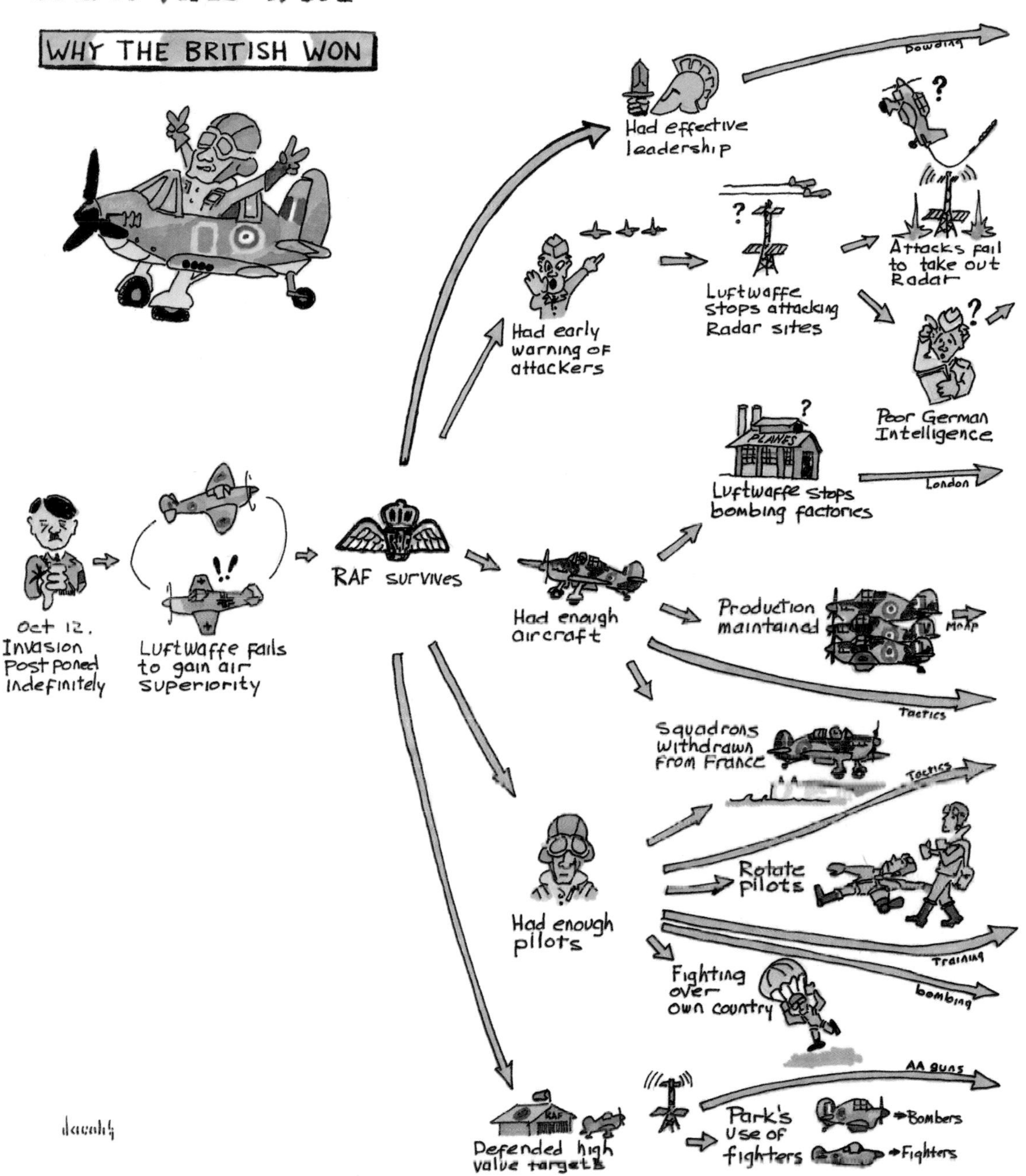

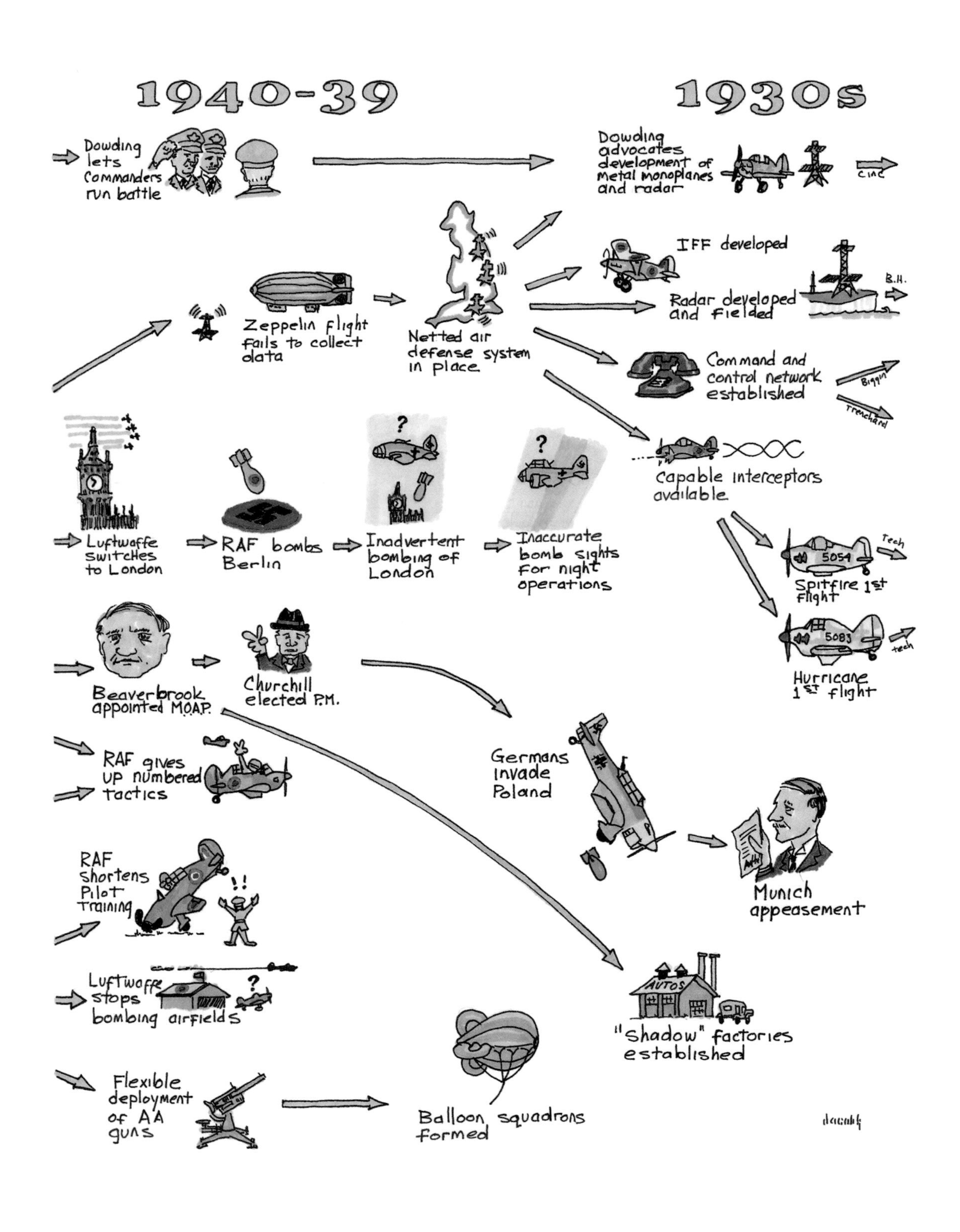

1940-39
1930s
Dowding lets commanders run battle
Dowding advocates development of metal monoplanes and radar
CinC
Zeppelin flight fails to collect data
Netted air defense system in place
IFF developed
Radar developed and fielded
B.H.
Command and control network established
Biggin
Trenchard
Capable interceptors available
Luftwaffe switches to London
RAF bombs Berlin
Inadvertent bombing of London
Inaccurate bomb sights for night operations
5054
Spitfire 1st flight
Tech
5083
Hurricane 1st flight
tech
Beaverbrook appointed M.O.A.P.
Churchill elected P.M.
RAF gives up numbered tactics
Germans invade Poland
Munich appeasement
RAF shortens pilot training
!!
Luftwaffe stops bombing airfields
?
AUTOS
"Shadow" factories established
Flexible deployment of AA guns
Balloon squadrons formed

1930s

Dowding becomes CinC of Fighter Command

Dowding appreciates value of technology as Air Member for R&D

Biggin Hill experiments

Daventry/Heyford bomber experiment

Tizzard committee to study RF applications
Baldwin
ships

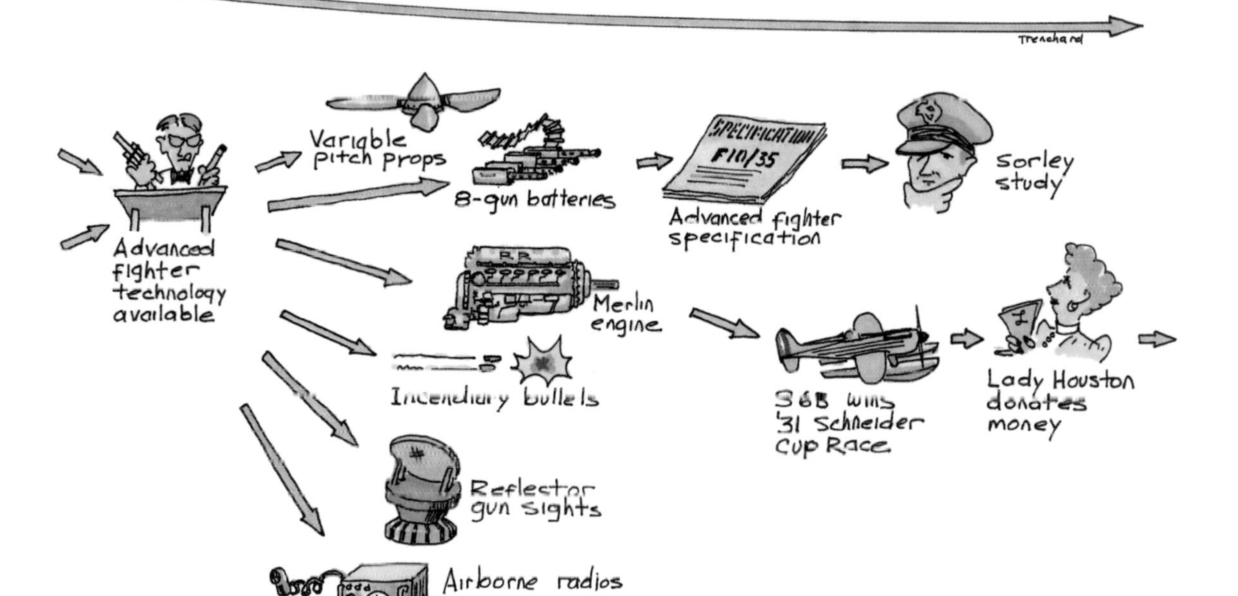

Trenchard
Advanced fighter technology available
Variable pitch props
8-gun batteries
SPECIFICATION F10/35
Advanced fighter specification
Sorley study
Merlin engine
Incendiary bullets
Reflector gun sights
Airborne radios
S6B wins '31 Schneider Cup Race
Lady Houston donates money

1920s

WWI

P.M. Baldwin's claim "the bomber...gets through"

Realization that passing ships interfere with radio signals

Trenchard keeps air defense elements in place during '20's

Primitive air defense system set up

Gothas and Zeppelins bomb London, 1916

Britain has success with Schneider races in '20's

WHY THE GERMANS LOST

1940

Invasion called off

Luftwaffe fails to gain air superiority

RAF survives

Poor leadership

understand

Poor German intelligence

Poor target selection

understand

key

Unacceptable aircraft losses

No surprise

key

Aircraft optimized for tactical support

Payloads

escort

Unacceptable air crew losses

EIN DEUTSCH SOLDAT

Channel forces air campaign

escort

aircrew

Deteriorating aircrew morale

channel

captured

Jacobs

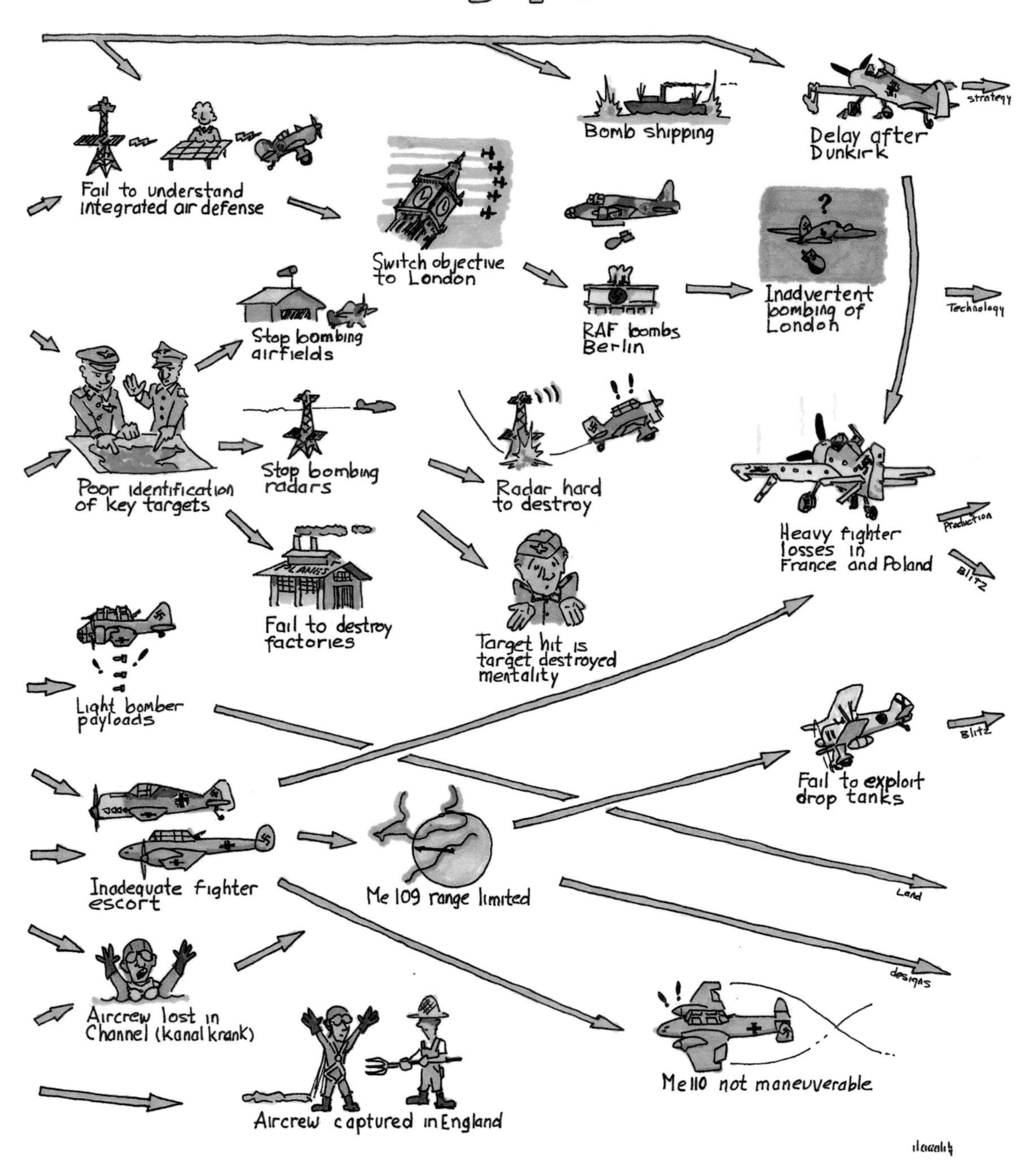
1940
strategy
Bomb shipping
Delay after Dunkirk
Fail to understand integrated air defense
Switch objective to London
RAF bombs Berlin
Inadvertent bombing of London
Technology
Stop bombing airfields
Poor identification of key targets
Stop bombing radars
Radar hard to destroy
Heavy fighter losses in France and Poland
Production
Blitz
PLANES
Fail to destroy factories
Target hit is target destroyed mentality
Light bomber payloads
Fail to exploit drop tanks
Blitz
Inadequate fighter escort
Me 109 range limited
Land
designs
Aircrew lost in Channel (kanal krank)
Me110 not maneuverable
Aircrew captured in England

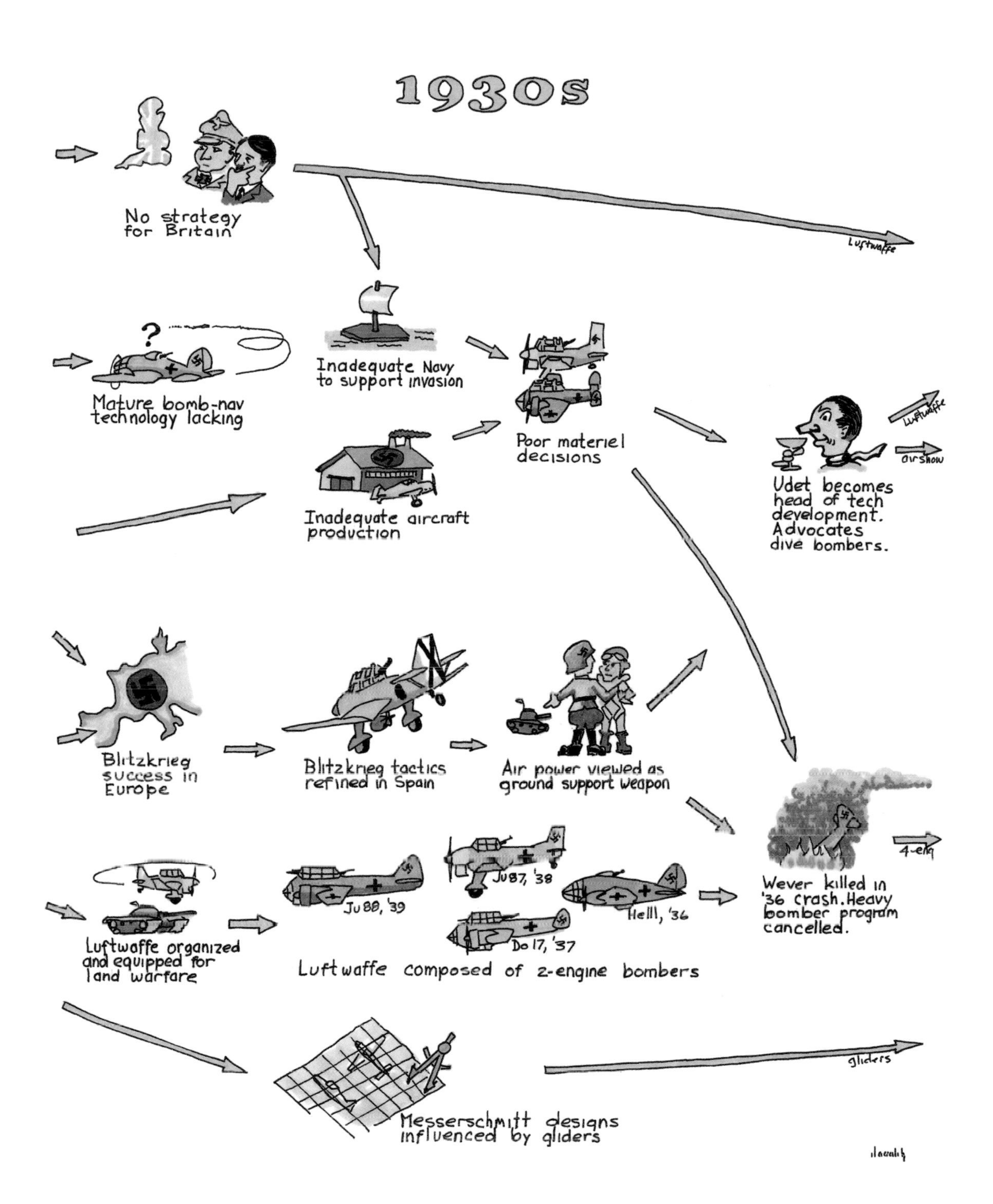

1930s
No strategy for Britain
Luftwaffe
Inadequate Navy to support invasion
Mature bomb-nav technology lacking
Poor materiel decisions
Inadequate aircraft production
Udet becomes head of tech development. Advocates dive bombers.
Luftwaffe
airshow
Blitzkrieg success in Europe
Blitzkrieg tactics refined in Spain
Air power viewed as ground support weapon
Wever killed in '36 crash. Heavy bomber program cancelled.
4-eng
Ju88, '39
Ju87, '38
Do17, '37
HeIII, '36
Luftwaffe organized and equipped for land warfare
Luftwaffe composed of 2-engine bombers
gliders
Messerschmitt designs influenced by gliders

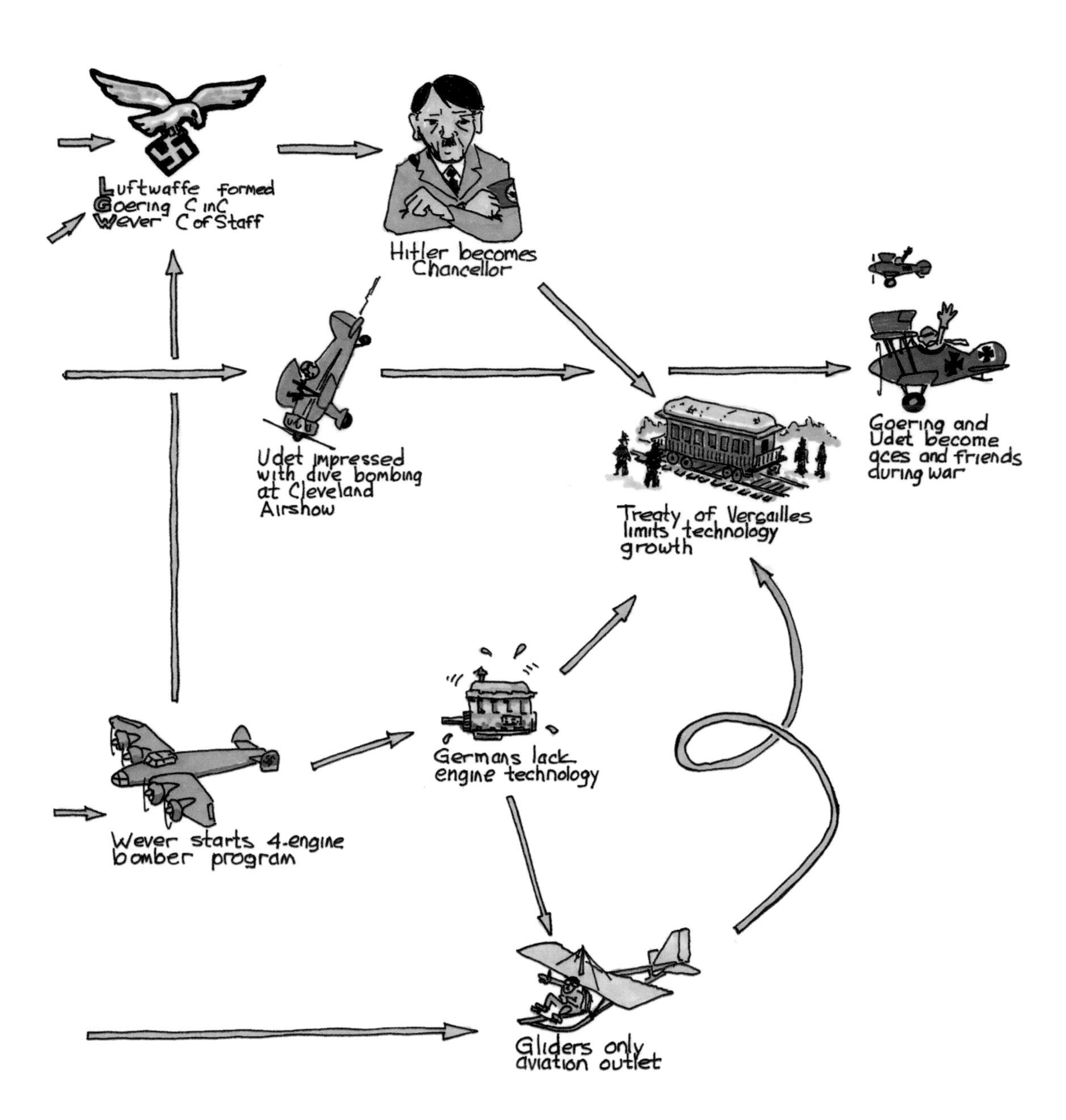

30s
20s
WWI
Luftwaffe formed
Goering CinC
Wever C of Staff
Hitler becomes Chancellor
Udet impressed with dive bombing at Cleveland Airshow
Treaty of Versailles limits technology growth
Goering and Udet become aces and friends during war
Germans lack engine technology
Wever starts 4-engine bomber program
Gliders only aviation outlet

ers and airfields, but the crews had more than held their own as the objective turned to London and the civilian population. Through it all, the Luftwaffe had been denied air superiority. The fact that the RAF survived was a victory in itself, given that the German objective was to literally eliminate this fighting force as a precondition for invasion. Their survival was attributable to five primary factors:

- Fighter Command benefited from superior leadership.
- Britain's integrated air defense network provided early warning of German attacks and enabled highly effective allocation and control of air assets.
- Britain produced sufficient numbers of highly capable fighters to weather the attacks mounted by the Luftwaffe.
- Fighter Command undertook extreme measures to put pilots into cockpits.
- The RAF provided flexible and innovative defenses for its high-value targets.

The following is the authors' analysis of the factors that led to the British victory. This is followed immediately by a similar analysis of the primary factors that led to the German defeat. As might be expected, a number of leaders on both sides of the conflict have provided comments and analyses of the battle and have expressed their opinions about how and why the outcomes developed as they did. Four ink drawings included with this chapter summarize some of the key views expressed by Churchill, Dowding, Galland, Goering, Hitler, Kesselring, and Milch regarding the battle and why it concluded as it did.

Superior Leadership

With strong and visionary leadership at the top, gifted and experienced leaders at the group level, and highly capable tactical leadership at the squadron level, Fighter Command was blessed with superior leadership at every level. At the very highest level, Dowding was a leader with vast experience in the operational, technical, and political aspects of command. He was an early advocate for the role that airpower could play in the nation's affairs, and he was committed to airpower as a natural and equal complement to ground and naval forces. Dowding had joined the Royal Flying Corps in 1913, had flown in World War I, and had subsequently joined the Royal Air Force. He became the air member for supply and research on the Air Council in 1930, and, as such, became one of the key forces in guiding the rearmament of the RAF during the early 1930s. Here his actions were nothing short of visionary as he presided over initial efforts to pursue modern fighter designs and develop radar for use in detection and tracking airborne targets. In 1936, Dowding took command of Fighter Command and set about preparing that fledgling organization for the task that lay ahead. The organization of Fighter Command, the transition to the Spitfire and Hurricane aircraft, and establishment of the operating procedures that would govern the defense of Britain were all accomplished on his watch. Dowding was an exceptionally strong leader, willing to take difficult positions when he believed it necessary. His willingness to oppose Churchill on the issue of sending aircraft to France toward the end of the battle for France and his insistence on rotating crews for rest even during the height of the battle are just two examples of his strength. Dowding was not a man given to seeking the easy way; his relationship with the RAF Air Council was often difficult. Although he was replaced as the commander of Fighter Command shortly after the Battle of Britain ended, his contributions to the preparation for and conduct of the battle were crucial to the British victory.

The group leaders in Fighter Command were all highly capable air commanders.

- Air Vice Marshal (AVM) Sir Keith Park, Commander of 11 Group, was a New Zealander and World War I fighter ace. He had Dowding's complete confidence, as evidenced by his assignment to 11 Group, the most critical of all the defense group assignments. An outspoken man of action, Park was known to fly into battle areas to get a firsthand feel for the tempo and conditions of the air war. He stood 6 feet 5 inches tall and was universally admired by those above and below his position.
- AVM Sir Quintin Brand, commander of 10 Group, was born in South Africa. He fought as a pilot in the Royal Flying Corps (RFC) during World War I and shot down a Gotha bomber in the last German raid on England. He held various posts in the RAF between the wars before assuming command of 10 Group in 1939. He and Park worked extremely well together, and Brand, like Park, was remarkably well thought of by his pilots and squadron commanders.
- AVM Sir Trafford Leigh-Mallory, an Englishman, began his air force career in 1916 as a pilot with the RFC. He won the Distinguished Service Order in France during World War I. He had joint service experience with the army and was known in the RAF as an expert in air-land operations. He took command of 12 Group in 1937, and he was a principal player in the Big Wing controversy. He succeeded Park as 11 Group Commander in 1942. He was killed in an air accident in 1944.
- AVM Sir Richard Saul, although not as well known as the other group commanders, was influential in shaping the organizational construct of Fighter Command during the 1930s. A World War I air combat veteran, he was the senior air staff officer at Fighter Command headquarters when he was tapped for command of 13 Group, which had responsibility for the northern regions of England.

Finally there was the leadership at the squadron level. The average pilot never came into contact with commanders at the group or Fighter Command level. To the pilot, leadership was personified in his squadron commander. The squadron commanders communicated policy, represented authority, set tactics, and generally organized and led the squadrons in their day-to-day operations. These were the leaders who kept the pilots focused, trained them to survive, rallied them to fight, and stood them down when they could not go on. Again, Britain was blessed with a number of truly outstanding individuals. It would be far too difficult and counterproductive to name them, but they included men of the stature of "Sailor" Malan, Peter Townsend, and Douglas Bader.

There are common threads in the hierarchy of Fighter Command leadership. At every level, these were experienced and battle-tested *air commanders.* They understood airpower. They knew the possibilities and the limitations of airpower; they understood the people who flew airplanes; they understood the machines and the problems associated with delivering the weapons they carried. In addition, the leaders in Fighter Command were the best officers available from all over the English-speaking world. They came from England, New Zealand, South Africa, and other countries. They were not bound by the constraints of a single culture or national tradition, nor did they bring a single point of view to the tactical and strategic problems they faced. They synthesized these differences into effective policies due in large part to Dowding's willingness to exercise command in a decentralized fashion, allowing his commanders the flexibility to set policy and tactics as needed to get the job done based on their own assessments of the combat situations they encountered.

Strong military leadership was essential to the British victory, but no discussion of leadership would be complete without acknowledging the civilian political leadership that provided the umbrella under which military leadership was exercised. As the 1920s dawned, the RAF had atrophied to a token force; there was considerable doubt that the young service would even survive as a separate service. The prime minister, Bonar Law, at the urging of the navy, had gone so far as to recommend that the RAF be absorbed back into the senior services. This proposal was rejected, and a committee was formed under Lord Salisbury to study the matter. In a report issued by the Salisbury Committee, the RAF was affirmed as an equal to the army and navy, and the committee further recommended that the RAF be significantly strengthened in order to provide adequate defenses for the nation. In addition, the Salisbury Committee recommended the establishment of a central Home Defense Force that would be the responsibility of the RAF. These actions—establishment of an independent air arm and the assignment of responsibility for centralized Home Defense to the RAF—were to have huge implications for the conduct of the Battle of Britain. A few years later, in 1933, facing a political atmosphere still dominated by pacifism, Winston Churchill rose to sound the warning that Germany was on a path to rearmament. This eventually resulted in the chain of events that provided the resources to develop the aircraft and technical systems that would become central to the defense of Britain. As prime minister, Churchill time and again rallied the British people

HOW AND WHY... CHURCHILL'S OPINION
IN HIS "HISTORY OF THE SECOND WORLD WAR," CHURCHILL GAVE SEVERAL REASONS WHY HE FELT THE BRITISH WON. HIS MAJOR REASONS...
CHANGING TARGET TO LONDON
FAILURE TO ATTACK COMMAND AND CONTROL NETWORK
CHANGING FROM CONCENTRATED TO DISPERSED ATTACKS
FIGHTING OVER AND ABOVE THE CHANNEL
BEAVERBROOK'S AIRCRAFT PRODUCTION, BEVIN'S MANAGEMENT OF LABORORS
DOWDING'S MANAGEMENT OF THE BATTLE
POOR GERMAN INTELLIGENCE
COMPARABLE FIGHTER PERFORMANCE
Jacobs

and the pilots of the RAF, uniting them in the bonds of a common struggle for survival. There can be no doubt that the political leadership in Britain contributed immeasurably to the victory that, in the end, was not only a victory for Fighter Command, but also a victory by and for the British people.

Integrated Air Defense System

It is pointless to argue that a single part of a system is more important than another, since, in a well-designed and balanced system, it is unlikely that the whole can function without any one of its key parts. What is certain is that the integrated air defense system that Britain fielded shortly before the Battle of Britain was the key to victory in the air battle. In considering the integrated air defense system (IADS), one must look at the detection and tracking, command and control, and interception aspects of the system. The military payoff from the system was obvious and of immense importance. It denied Germany the element of surprise. When the Luftwaffe sent its bombers, the British knew when, where, how many, and how high. This enabled them to employ a relatively small force of pilots and aircraft in the most efficient and effective way possible. In the absence of the IADS, Germany would surely have prevailed; the resources of Fighter Command would simply not have been sufficient to handle the force sent against them. This is the essence of a *force multiplier,* a term used commonly in modern defense parlance. A force multiplier enables a force of a determined size to perform tasks that would require a much larger force in the absence of the multiplier. That is what the IADS did for Britain.

The fundamental architecture of the integrated air defense system, the major components of the system, and the relationships among them were established as Britain faced German air attacks in World War I. In World War I, Generals Smuts and Ashmore developed the essential structure of a centrally controlled air defense system. The establishment of detection zones, communications linked to central control points, air interceptors operated under central command authority, the integration of antiaircraft artillery—all were initially conceived and implemented during the last year or so of World War I. The fundamental architecture was refined and improved, but the major functional elements remained the same when war with Germany again loomed in the late 1930s. The system still consisted of detection, tracking, communication, command and control, interception, and kill. The key to British victory was the way technology was exploited to accomplish these basic functions in the Battle of Britain.

The addition of the identification-friend-or-foe (IFF) transponder system in 1940 signaled the completion of the British integrated air defense system. The tragedy of the Battle of Barking Creek, where British Spitfires intercepted a squadron of British Hurricanes and shot down two of them, was the impetus for the addition of this vital capability. At that point the Chain Home radar system was operational, but there remained the vexing problem for controllers of how to differentiate friendly fighters from German aircraft in airborne fighter engagements. By July 1940, IFF capability was installed in all British fighters, thereby enabling positive control right up to the point at which the fighter had visual contact with the target. With this addition, the entire IADS was completed and the system that would be central to British victory in the air was in place.

The nature of an *integrated* air defense system requires that multiple capabilities and functions be accomplished in concert and in cooperation with each other. This demands a network of communications and data that enables the linking of the elements that make up this cooperative defense. The individual components are of marginal value in the absence of an integrated network; for example, there is little value in detecting targets if information cannot be distributed to the aircraft or artillery that must shoot at them. As noted, the early work toward a network had started in World War I, but the technology required to take full advantage of the new capabilities like radar was immensely more sophisticated. Development of the integrated network started at Bawdsey Manor, where the first three radar sites were linked. Bawdsey was also the site where subsequent changes and additions to the Chain Home system were first tested before integration into the fully networked system. Once the operational value of the system was demonstrated in experiments in 1937, the radar system was then integrated with the command and control elements—the filter room, plotting table, and command sections—at Biggin Hill. The linking of detection and tracking to command and control was fundamental to the development of both technical and operational procedures that would govern the functioning of the integrated system.

Central to the entire IADS was the Chain Home radar system. With rising concern about German rearmament and home defense centralized as an RAF responsibility, Dowding, as air member for supply and research, was an early proponent and sponsor of scientific investigations that led to the development of radar. Guided by Henry Tizard and Robert Watson-Watt, a series of experiments was conducted which conclusively established that radio frequency (RF) energy could be bounced off of airborne targets, allowing them to be detected and tracked using cathode-ray tubes. This demonstration of potential resulted in approval for construction of the first three Chain Home stations. When these were successfully linked, further technical work improved the detection range to 100 miles and operators were trained to correctly interpret the display information received. With this, further approval was received to expand the system. By February 1940, there were 29 operating Chain Home stations, each with two towers, covering essentially the entire southern and eastern approaches to the English coast.

One of the great failures—if not the greatest—of the German campaign was the failure to recognize the critical role that the Chain Home system played in the integrated defense of Britain. Poor intelligence was a hallmark of the German campaign, and the failure to understand and cripple the radar system stands as testimony to that fact. The

Germans failed to grasp the purpose of the system as it was being developed and constructed, each station standing on the coast with two towers, one 350 feet tall and the other 250 feet. In 1939 the Germans placed a Zeppelin dirigible off the coast of England to collect electronic data from the stations, but the operators failed to locate the frequencies on which the stations operated, leading to speculation that the stations were some sort of air traffic control communications link and that the British had no operational radar of any consequence. As the airfield phase of the battle began, the Germans seemed to have come to a conclusion that the sites were somehow instrumental in the success enjoyed by the RAF, and they designated them as primary targets. This effort was soon terminated, however, when the difficulties involved in the destruction of the lattice structures proved greater than originally envisioned. From start to finish, the German failure to understand and effectively target the Chain Home system was an important factor in the British victory.

Of course, the romance in the IADS was in the design, development, and operation of the aircraft that performed the vital intercept function. In so many ways, the story of the Battle of Britain is the story of the Hurricanes and Spitfires. From Sydney Camm and R. J. Mitchell, the designers, to those who flew and maintained them, the defense of England revolved around these great machines. They gave a human face to the integrated air defense system, and it was around them that the entire British nation rallied to join the mutual defense. As the last link in the chain—the element that took the information and direction provided by the other elements and actually intercepted enemy aircraft—this new generation of aircraft and their crews completed the kill chain that enabled the RAF to realize the full potential of the integrated air defense system.

Sufficient High-Performance Aircraft

Not only did Britain possess the high-performance airplanes needed to make integrated air defense against the German fleet a reality, but it also had these aircraft in sufficient numbers to withstand the losses incurred. This can be attributed to three factors: (1) the Germans turned the focus of their efforts from bombing the airfields and the factories that produced the fighters to bombing London; (2) production and repair rates were adequate to the needs of the defense; and (3) measures were taken as needed by Fighter Command to conserve scarce fighter resources.

The quixotic decision by Hitler to punish Britain for having bombed Berlin by focusing the bombing campaign on London and the civilian population simply guaranteed that the fighter fleet and facilities that produced them would survive and be adequate for the defense of Britain. Bomber Command responded to Churchill's order to attack Berlin following a disorganized and in all likelihood inadvertent attack by German bombers on London. At the time, Germany was executing its strategy of gradually moving its bombing targets inward toward London in ever-decreasing concentric circles. During the evening of August 24, the confusion created by the British defense and the fact that, at best, night bombing technologies were relatively imprecise and immature resulted in the Germans dropping bombs on various targets within the London metropolitan area. In response, Churchill directed attacks on Berlin, which in turn so infuriated Hitler that he abandoned the methodical campaign that was slowly but surely destroying Fighter Command in favor of attacks on London. That decision was the single most critical factor in the eventual collapse of the German campaign.

One of the miracles of the British effort was the production and fielding of ever-increasing quantities of high-performance fighters. The 1930s had been marked by a steady march to develop and incorporate advanced technologies into the designs of the Spitfire and Hurricane fighters. Britain's early successes in the Schneider Trophy races and the contribution by Lady Houston ensured the entry of the Supermarine S.6B in the 1931 races. The preparation and design efforts that preceded the races led to the engine and airframe technologies that would become the Merlin engine and the Spitfire airframe. Squadron Leader Ralph Sorley's study of air-to-air gunnery led to specification F.10/35, which required that British fighters have eight guns rather than the four previously believed adequate. Other technologies were added. Variable-pitch propellers were adopted for use in the new fighters. Incendiary bullets became the norm for air-to-air gunnery. The reflector gunsight was developed and initial installations begun, and air-to-air radio became a common addition to every aircraft in Fighter Command.

The first contract by the RAF for the Hurricane, designed and developed by Sydney Camm of Hawker Aviation, was written in 1936. Camm designed the Hurricane for ease of production, and high output rates were common from the start. By 1939, 18 squadrons had been equipped with the new fighter, and before the war ended over 14,000 were produced. The Spitfire, designed for performance over producibility, was also ordered in 1936, but this aircraft experienced considerable difficulty before reaching acceptable rates of production. Churchill, elected prime minister in 1939, named Lord W. M. A. Beaverbrook to the post of minister of aircraft production. Beaverbrook took immediate steps to intervene in the production strategy being followed by Supermarine in the building of the Spitfire. Finally, just in time for the Battle of Britain, production of this magnificent airplane was ramped up to rates that permitted ample fielding and operation during the battle.

Almost as important as original manufacturing of the aircraft were the policies that Beaverbrook instituted to repair and recover damaged aircraft. Using a system of quick evaluation and categorization of damage status, procedures for repair were implemented that enabled aircraft to be repaired on site at the airfields if the damage was sufficiently light or repaired at central depots if the damage was more severe. With German strategy turning to focus on London, weekly aircraft losses dropped to a level below the weekly rates of production, yielding a net positive rate of increase in RAF fighter inventories. So effective were the combined policies that governed production and repair that Dowding actually had more airplanes at his disposal at the end of the Battle of Britain than he had when the battle started. In fact, Dowding's problem was not the lack of aircraft; it was the lack of experienced and qualified pilots to fly them.

The final factor that ensured enough airplanes for the battle was the combination of measures and policies taken to conserve the fighter resources in possession of Fighter Command. Both Park and Dowding were quick to take action to avoid the unnecessary loss of aircraft. Park, on observing that too many of his fighters were falling victim to the free chase tactics employed by the Germans, issued a directive reminding controllers and pilots alike that the objective was to stop the bombers and ordered that fighters not be diverted from that task to pursue German fighters. Park also took measures to ensure that his fighters were appropriately matched against the adversaries they faced. He insisted that the Hurricanes should be primarily targeted against bombers and the Spitfire against the Bf 109s. This ensured that both aircraft were matched against enemy airplanes with which they could compete as equals or superiors. The policy was aimed at preventing the Hurricanes from needlessly being matched with the Bf 109, which had the performance advantage. Of course, nothing demonstrated a willingness to take extreme measures to conserve fighter assets more than did Dowding's "most famous letter," which challenged the government's intent to send additional fighters to the front in France at the end of the battle for France in 1939. This letter demonstrated a strength of resolve that is uncommon and a gift of foresight that was, without doubt, instrumental in ensuring that Britain was prepared for the initial onslaught by the Germans just weeks later.

Sufficient Pilots

The Germans had designated the fighters of Fighter Command and their airfields and factories as the primary objectives of the phase of the bombing campaign that the Germans called *Adlerangriff,* which began on August 13. By the first week in September, exhaustion and attrition had so decimated the pilots of Fighter Command that defeat stared Dowding squarely in the face. In the two weeks leading up to September 7, Fighter Command lost 103 pilots, and 128 more were wounded. By that time, 20 percent of the squadron commanders and one-third of the flight commanders had been killed or wounded since the battle began (Hough and Richards 1990). Those remaining flew up to four combat sorties every day. Although Dowding and the group commanders of Fighter Command took extraordinary measures to man the aircraft and avoid total capitulation, salvation actually came from the most unlikely of sources—Hitler himself. When he made the decision to target London and to abandon the campaign against the airfields, Fighter Command was literally days away from defeat. Hitler provided the respite that enabled the young men who flew the fighters to rest and regroup before they plunged back into the battle.

Dowding also took a number of innovative steps to bring more experienced pilots into Fighter Command and to transition pilots to fighters more quickly than had been the norm. He allowed Polish and Czech pilots to join the command, at first as members of existing units, then later he agreed to form special squadrons for foreign nationals. During the Battle of Britain, 154 Polish pilots flew for the RAF, and 30 lost their lives in doing so. Josef František, a Czech pilot with 17 confirmed kills, was the leading ace among all who flew in Fighter Command during the battle. Dowding also requested that pilots from other commands be transferred into Fighter Command. Twenty volunteers from light bombers and twelve from Lysander units were transferred and, after a brief period of upgrade training, joined Fighter Command as active pilots by the end of August. By mid-August, attrition in Fighter Command had gradually reached the point that there were fewer than 1000 pilots, at a time when approved manning was between 1300 and 1400 (Mason 1969). To transition pilots into operational units more rapidly, radical steps were taken to shorten the training afforded them before they were sent to their squadrons. Transition training, the training that upgraded new pilots from trainers to Spitfires, was reduced from a period of months to two weeks—and later to seven days. This had the effect of placing raw trainees into combat cockpits, but it was a step considered necessary to manning the fleet. Needless to say, placing untrained pilots into combat situations exposed them to extreme risks at the hands of the far more experienced German fighter pilots.

Dowding also implemented policies to slow the loss rates among the pilots assigned to the command. Dowding realized from the outset—as early as the Channel phase—that manning his fighters would be the crucial variable in the battle. A number of the steps already described in the earlier discussion of measures taken to preserve aircraft were directed equally at avoiding unnecessary losses of pilots. These included steps to avoid engagements with free-chase German fighters and to ensure that the British fighters were not directed into engagements in which they were at an avoidable disadvantage in performance. In addition, directions were issued to controllers that they should attempt to time intercepts such that they would occur as near land as possible so that pilots could either coast to land if their aircraft were disabled or bail out over British soil if forced to leave their aircraft.

Dowding initiated his rest-rotation policy to allow whole squadrons to rotate into less intense areas to rest before returning to the battle. In spite of these efforts, manpower became an ever-increasing concern for Dowding, and he was forced to begin rotating experienced crewmembers from squadron to squadron just to maintain a balance of experience across the force. By early September, losses had mounted to the point that many squadrons were without effective leadership or experience. The single force that seemed to

HOW AND WHY... DOWDING'S OPINION
IN A SERIES OF ARTICLES PUBLISHED IN THE "LONDON GAZETTE" IN 1946, DOWDING GAVE SEVERAL REASONS WHY HE FELT THE RAF WON. AMONG THOSE...
DEFIANTS USED ONLY AT NIGHT
POOR TARGET CHOICES BY GERMANS
RECALLING HURRICANES FROM FRANCE
H. C. T. Dowding
USE OF RADAR
FLEXIBILITY OF ANTI-AIRCRAFT COMMAND AND EFFECT OF AA GUNS
BEAVERBROOK AND SHADOW FACTORIES
THE IMPROVISED AIRCRAFT REPAIR SYSTEM
FIGHTING OVER ENGLAND AND THE CHANNEL FAVORED RAF AIRCREWS
BARRAGE BALLOONS PROTECTED VITAL TARGETS
USE OF POLISH AND CZECH PILOTS

keep the pilots going was the recognition that they were fighting over their own homeland and the belief that the very survival of Britain depended on them. Dowding finally introduced his stabilization scheme, a policy in which he categorized squadrons as Category A, operationally ready; Category B, partially ready; or Category C, not ready. All Category A squadrons were assigned to 11 Group and the Duxford and Middle Wallop areas of 12 Group. All Category B squadrons were placed in 10 Group and 12 Group. The plan was to bring them up to strength, then deploy them as Category A squadrons. Category C squadrons were generally deployed only in the least likely combat areas, primarily in 13 Group. It had become clear that desperation was driving Fighter Command to the edge of chaos. Had Hitler not changed the strategic objective on September 7, there is little doubt that Fighter Command would have ceased to be an effective fighting force before the month was out.

Flexible Defences of High-Value Targets

The final factor leading to the British victory was the flexible and innovative defense established to protect high-value targets that were the objectives of the Luftwaffe's bombing campaign. Again recalling that the Germans planned to mount an invasion once they had gained control of the air, this strategy implied that the Royal Air Force itself, specifically Fighter Command, was a primary target. The measures taken to preserve and defend the aircraft and personnel of Fighter Command have been discussed at length in foregoing paragraphs.

Integral to the fabric of the integrated air defense system were the antiaircraft artillery elements. Again, the concept of artillery to complement air assets had descended from the air defense system developed during World War I. Artillery positions were integrated into the defense network, much as were the radar sites and airfields. They received information through the network and, likewise, reported sightings of aircraft inbound toward targets inside Britain. As radar and height-finding technologies were gradually integrated into artillery capabilities, they became even more effective as defenses against the German attackers.

Barrage balloons, deployed to protect a variety of cities, ports, and key airfields, were also an important contributor to the defense of high-value targets. Balloons had first been used in World War I, and they proved at least as effective in World War II. Trailing cables below them, they forced German bombers to approach targets at altitudes significantly higher than they would otherwise have done. The balloons were often floating at altitudes up to 5000 feet, and aircraft forced to fly above them could not deliver their bombs with the accuracy they might have achieved at lower altitudes. Furthermore, the higher altitudes exposed the bombers to antiaircraft artillery that was preset and positioned to defend the altitudes above the balloons. The balloons were sufficiently important to the defense that they were organized and deployed under a separate Balloon Command.

The airfields were major targets of the German attack. Not only did the RAF provide air defenses of the airfields, but in addition various innovations were introduced to defend and preserve these assets. The airfields came under significant threat when the Germans discovered that they could penetrate British airspace without detection by flying very low as they crossed the Channel. These attacks would appear as if from nowhere, and there was little or no time to react and defend using conventional means. To combat these attacks the RAF began to use parachute-and-cable rockets to snarl the low-flying aircraft and force them to fly at higher altitudes—similar to the barrage balloon defensive tactics. When the defenses were breached and the fields damaged in bombing raids, every effort was made to repair them as quickly as possible. Churchill took notice of the importance of airfield repair, suggesting that civilian repair teams be formed to relieve the aircraft maintenance crews from repair duties so they could concentrate on getting the airplanes back into the air. He even went so far as to suggest that repairs be camouflaged so the fields would appear to be out of service when viewed by German reconnaissance aircraft.

When the primary objective became the city of London, all of these elements—barrage balloons, artillery, and air assets—were employed to defend and save the city. However, if there is a symbol of the resilience and determination of Londoners to survive and even prevail, it has to have been the firefighters and repair crews that night after night fought to save the city. They were sufficiently successful that the Germans began to target them as well as the city itself, first using incendiary bombs to draw them into the streets and then dropping conventional bombs with their blast and shrapnel to incapacitate the crews.

These, then, were the factors that were primarily responsible for the British victory. They can be summarized as *leadership, technical innovation,* and *human perseverance.* In no uncertain way, the British victory was more than a military triumph; it was a triumph of the human spirit. This was a battle in which the military forces of the country were joined with the civilian population in a common struggle that is impossible to achieve in wars fought on foreign soil. It was England's victory and it was also a victory that changed the course of western history. In scale, it may not have been one of the great battles of history, but in importance to western civilization it would be difficult to overestimate the importance of the events that took place in Britain during that summer of 1940.

WHY GERMANY LOST

Hitler's order to postpone the planned invasion of Britain, issued on October 12, 1940, signaled the failure of the air campaign that had been waged against Great Britain since July of that year. The decline in Germany's fortunes during the battle had accelerated with the decision to change strategic objectives to focus on London and the civilian population. By the end of September, the leaders of the Luftwaffe had concluded that daylight bombing was not sustainable, and they retreated into the cover of night for the conduct of bombing raids that would continue through the long winter and into the spring of 1941.

From the outset of the battle, air superiority had been established as a necessary condition for invasion. Only by controlling the air could the German navy hope to transport ground forces across the Channel and land them on the beaches of England. In the absence of air superiority, Germany understood that their forces would come under intense air attack throughout the Channel crossing and landing, and furthermore, the British army, weakened though it was after Dunkirk, would defend the beaches with all resources at their command. To successfully invade would require that German fighters own the skies above Britain. The recognition of these facts was evident in Hitler's Directives 16 and 17. In July, Directive 16 stated that, because of Britain's unwillingness to negotiate an acceptable nonaggression pact, German armed forces were to prepare for an invasion. This was followed in August by Directive 17, which established the fundamental requirement for air superiority—control of the air—as a condition necessary for invasion.

The reasons for Germany's failure to gain air superiority, and hence the failure of the invasion plan, can be attributed to three primary factors:

- Poor senior leadership that failed to provide the vision and strategic direction needed to win the air battle
- Poor intelligence resulting in a failure to properly target British assets
- Unacceptable losses created by the fact that the Luftwaffe was not organized and equipped to conduct the battle it was directed to prosecute

Poor Senior Leadership

Among factors contributing to the German defeat was the failure of Germany's most senior leadership to provide the strategic vision needed to guide both the preparation and the employment of the Luftwaffe in its assigned missions. The failure in strategic vision was largely due to (1) the blurring of political and military roles that resulted in confused command relationships and a lack of strategic clarity and (2) the fact that leadership positions at the top of the Nazi hierarchy were awarded primarily to those who had personal and political links rather than to the best qualified.

Adolf Hitler, having served as an enlisted soldier during World War I, joined the political party that would come to be known as the Nazi Party in 1919. By 1921 he had become the chairman of the party, and his gifts of oratory and persuasion soon attracted growing attention and party membership. Hitler actively recruited Hermann Goering, aristocrat and war hero, as a member of the party and appointed him leader of the Brownshirts in 1922. Using his powers to focus the outrage of his followers on the humiliation and injustices of the Treaty of Versailles, Hitler attempted to overthrow the Bavarian government in 1923—an attempt that failed. As a result of the failed coup, the Nazi Party was banned from German politics, Hitler was jailed, and Goering, who had been wounded, fled the country.

The ban against the Nazi Party was lifted in 1925, and Hitler soon returned to assume leadership of the party once again. Goering returned to Germany in 1927, where he also rejoined the Nazis. With Hitler's support, he was elected as a deputy to the Reichstag one year later. By 1930, Hitler had attracted almost 20 percent of the electorate. The professional military leaders in Germany had allied themselves with the Nazis as a result of Hitler's apparent willingness to openly oppose the humiliations of the Treaty of Versailles and his increasingly successful attempts to unify Germany in his cause. Hitler decided to run for the presidency in 1932, but von Hindenburg defeated him in a runoff election. Due to the nature of the coalition government, von Hindenburg was persuaded to name Hitler as Reich chancellor in 1933. Hitler soon named his old friend and political ally, Goering, as minister of aviation. When von Hindenburg died in 1934, Hitler assumed the combined roles of chancellor and president; he became führer. By 1935 Hitler had renounced the Versailles treaty and had declared his intent to rebuild the armed forces of Germany. Goering was named commander in chief of the Luftwaffe in 1935.

This sequence of events provides certain insights into the nature of Germany's most senior leadership. Fundamental to the German hierarchy was the blurred line that separated political from military leadership. Hitler, unquestionably a brilliant political tactician, imagined himself equally brilliant as a military strategist and reserved to himself much of the definition of the military strategy that would govern the war. Naming Goering, a political ally, as commander in chief of the Luftwaffe further blurred the separation. Goering's World War I service hardly qualified him to lead the Luftwaffe; his appointment clearly signaled the politicization of senior military leadership. The other key insight gained is Hitler's apparent conviction that political loyalty was more important than competence at the top of the military chain of command. The professional military was allied with the party, but the senior military officers were never really fully accepted, trusted, or integrated into the topmost ranks. These positions, whether or not they were military commands, were dominated by Hitler's political allies. Goering had never shown any facility for strategy—had never led a military organization larger than a squadron—yet he was thrust into a position in which he would be responsible for crafting the strategy to defeat England in just a few years.

Goering's selection effectively bypassed other, better-qualified candidates for the position of commander in chief of the Luftwaffe. Among them was Hugo Sperrle, who had commanded the Condor Legion in Spain, was a graduate of the training program at

Lipetsk, and later commanded *Luftflotte 3* during the Battle of Britain. Equally well qualified was Erhard Milch, the brilliant organizer and manager of the state airline, Lufthansa. Under Milch's leadership Lufthansa rose to prominence in the 1930s. Milch understood large organizations and the multiple concerns—research and development of aircraft, training, and logistics, to name some of them—that attend the development and management of a large air operation.

Although Milch was selected to be Goering's deputy and eventually rose to the rank of *Feldmarschall* in the German politico-military hierarchy, his methodical approach to building the Luftwaffe soon frustrated both Hitler and Goering. Milch envisioned the Luftwaffe as a balanced and powerful air arm and estimated 8 to 10 years would be required to build it; Hitler wanted a war machine immediately. Inevitably, conflict rose between Milch and Goering, who was committed to Hitler's agenda. Milch was gradually marginalized, and decisions involving the Luftwaffe were increasingly made based on Hitler's political and military ambitions in Europe and Russia.

In similar fashion, Goering chose his own staff based on personal relationships rather than on careful selection of those best qualified for the position. He chose his old friend and fellow fighter pilot, Ernst Udet, to be the director of technical development in the Air Ministry. Udet, the fighter pilot's fighter pilot, the barnstorming demonstration pilot enamored of tactical operations, was an advocate of the dive-bomber, and his efforts eventually led to the Ju 87 Stuka. Udet's infectious enthusiasm won Goering's admiration and support and helped establish the mind-set among German leaders at every level that the fundamental purpose of the Luftwaffe was to support ground tactical operations. These rather limited views of the role of airpower left little room for the consideration of independent air operations or the role that heavy bombers might play in such operations. This would turn out to be a significant shortcoming once the Luftwaffe was confronted with the task of conducting independent operations in the Battle of Britain.

The blurred lines between political and military leadership tended to confuse and weaken the role of the professional military in formulating both strategy and tactics for military operations. To illustrate, Generalfeldmarschall Keitel was the military commander of Oberkommando der Wehrmacht (OKW)—equivalent to the chairman of the Joint Chiefs of Staff in the military structure of the United States—the highest military post in the German hierarchy. Under OKW was the Luftwaffe, headed by Hermann Goering, Hitler's personal friend and designated second in the Nazi Party line of succession. For Keitel to be openly critical of Goering would have been difficult to impossible.

The result was reflected in the fact that military leaders were, at several critical points, required to delay initiating military operations as they waited for clear directions from the political leadership. An example of the paralysis caused by the overlapping political and military roles was evident in Germany's failure to adequately prepare for the battle with Britain. Hitler presumed that the British would realize the hopelessness of their

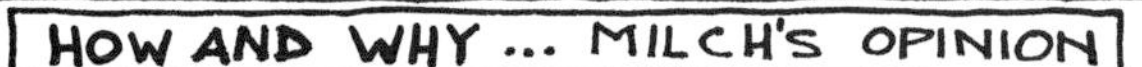
HOW AND WHY ... MILCH'S OPINION

AFTER THE WAR, ERHARD MILCH, GOERING'S DEPUTY, COMMENTED ON OPINIONS BY HISTORIANS REGARDING THE WAR'S OUTCOME. AMONG HIS REASONS WHY THE GERMANS LOST THE BATTLE OF BRITAIN WERE ...

IN LATE '39, LUFTWAFFE WAS STILL DEVELOPING. NOT READY FOR LARGE MILITARY OPERATIONS.

THE LUFTWAFFE WAS INCORRECTLY ORGANIZED AT THE TOP LEVEL.

CANCELLATION OF THE FOUR-ENGINE BOMBER PROGRAMS.

POOR DROP TANK DESIGN (?). MEANT LACK OF USE AND INADEQUATE FIGHTER ESCORT RANGE.

BOMBING TACTICS AND TECHNOLOGY
DAYLIGHT BOMBING DOCTRINE PLUS INADEQUATE FIGHTER ESCORT COMPELLED SWITCH TO NIGHT OPERATIONS.

NIGHT BOMBING DOCTRINE NEVER EVOLVED.

COULDN'T HIT PINPOINT TARGETS. AREA TARGETS AND LIGHT PAYLOADS RENDERED BOMBING INEFFECTIVE.

situation and would either surrender or agree to a nonaggression pact that would permit him to turn his attention eastward toward Russia. So confident was he in this assumption that he did not demand that his general staff prepare adequately for an invasion. As a result, the Luftwaffe did not have a clear strategy for the Battle of Britain, nor had the German navy adequately prepared for invasion.

This lack of clarity and definition left the Luftwaffe commanders, Sperrle and Kesselring, at something of a loss regarding what was expected of them as they deployed their forces along the northern European coast opposite England. Lacking a clear strategy for the campaign, they initially began the battle in early July by bombing British shipping in the Channel as a means of luring the fighters of the Royal Air Force into the skies. A clearly defined strategy—*Adlerangriff*—was not finally enunciated until early in August. The same sort of delay and confusion had been in evidence in the battle in northern France when Guderian had the British army trapped between his forces and the Channel and then delayed the final assault for 48 hours, thus permitting the British to escape at Dunkirk. And finally, the ultimate in strategic confusion occurred on September 7, 1940, when Hitler, in an emotional response to an otherwise token bombing of Berlin by the RAF, intervened in the bombing campaign, which was on the verge of success, and ordered that the objectives be changed to bomb London rather than the fighters and their airfields.

The lack of strategic vision and clarity at the highest levels of command also led to a failure to develop adequate matériel strategies. A prime example was the failure to build an adequate naval infrastructure to support an invasion of Britain. It is clear that the navy began to consider the matériel requirements to support an invasion well before 1940, yet little or nothing was done to prepare for the operation. Whether Hitler believed that an invasion would never actually be necessary, or whether OKW believed that crossing the Channel was equivalent to a "wide river crossing," the fact remains that the gathering of barges and boats betrayed the reality that the German navy was essentially unprepared to conduct an invasion.

The failure of Germany to produce enough aircraft to replace combat losses was another example of the failed matériel policies. Although Germany produced a huge armada of bombers and fighters, it did not take the necessary steps to ensure that follow-on production would be adequate. The result was that, as campaign followed campaign, the availability of combat aircraft gradually became one of the limitations and constraints that German operational commanders had to compensate for in their planning. Inexplicably, the Germans failed to exploit their own development of drop tanks for fighter aircraft. The limited range of the Bf 109 became a significant factor in the bombing campaign, and the use of drop tanks might have largely remedied that shortcoming. Range limitations forced the consolidation of Luftwaffe fighter assets in the Pas de Calais, lessening the ability of the Germans to disguise their movements; it limited the time that fighters could linger over Britain; and it lessened the ability of the fighters to provide adequate bomber escort.

In summary, the lack of strategic vision among the topmost German leadership was a severe handicap throughout the Battle of Britain. Not only were strategic objectives often unclear, but the communications within the politico-military hierarchy were also confused. However, as debilitating as these factors were during the conduct of operations, the more critical failures were those made well before the battle started. These were the decisions that shaped the force structure and defined the capability of the German war machine. The lack of strategic vision among German leaders was reflected in the limited view of aerial warfare that never contemplated fighting a battle in which the blitzkrieg tactics used in every country since 1936 would not be available to them. The certainty that the same formula could be applied to every situation was finally exposed as unjustified when circumstances forced the Luftwaffe to conduct a campaign for which it was essentially unprepared.

Poor Intelligence

Intelligence is linked directly to leadership in the conduct of military operations. Good leadership depends on good intelligence, and good intelligence results from the policies and support of insightful and informed leadership. Poor intelligence leads to poor decisions, and hence the quality of leadership is inextricably tied to the quality of the intelligence available. It is a circular relationship. The quality of the intelligence available to the leaders of the Luftwaffe was consistently lacking, and the weakness of German intelligence resulted in a pattern of repeated failures in target selection. The Luftwaffe and German senior leadership in general did not have accurate information about either the state of the RAF or its airfields, failing to identify and effectively target the factories that produced the fighters they were trying so hard to destroy; and they utterly failed to either comprehend or give adequate priority to disrupting the Chain Home radar sites that were critical to the operation of the British integrated air defense system (IADS).

The decision that turned the tide of the battle was Hitler's insistence that the Luftwaffe abandon the *Adlerangriff* strategy that primarily targeted the fighters and airfields of the Royal Air Force and turn instead to focus exclusively on bombing the cities of England. Hitler's decision was an emotional reaction to a relatively unimportant series of strikes on Berlin by Bomber Command.

The *Adlerangriff* strategy called for a gradual tightening of the noose around London, bombing targets in ever smaller circles around the city. On the night of August 24, the difficulties of night bombing, compounded by the lack of mature navigational and bombing technologies, resulted in the accidental bombing of London proper. Churchill immediately responded by requesting that Bomber Command attack Berlin in kind. Even though relatively insignificant in terms of the destructive results achieved, the mere fact that Berlin had been struck both embarrassed and outraged Hitler. With little further thought, he launched the London phase of the campaign. With victory in its grasp, the Luftwaffe ceased its relentless campaign against the RAF, giving the British the respite that enabled them to regroup and eventually prevail.

Nothing is more evident in this critical decision than the fact that Germany had no idea how close Fighter Command was to total exhaustion and defeat. This failure to fully understand the status of the forces against which they fought was evident throughout the airfield phase of the battle. The Germans had never fully grasped which airfields were critical to the defense; key airfields were often ignored while fields of only marginal importance were, at times, repeatedly attacked. German poststrike bomb damage assessment was also lacking; leaders seldom seemed to fully understand the extent to which an attack had been successful or not and, in far too many cases, appeared to take the attitude that a target, once struck and damaged, was a target destroyed. They never fully understood that targets like airfields had to be struck again and again to keep them out of action. Another intelligence failure was showcased on August 15, when *Luftflotte 5* struck the northern areas of England on the premise that the fighter forces normally located in those areas had been dispatched to reinforce the southern groups. This was a sheer guess—a supposition based on the stubborn defenses mounted by the southern groups; it had no basis in hard intelligence. The result was a debacle for the German forces and the end of *Luftflotte 5* as a contributor to Goering's battle plans.

The German leaders also often failed to appreciate the effectiveness of their own tactics. German free-chase tactics were extremely effective against the fighters of the RAF. Dowding had recognized the danger that they represented to his forces and had directed that controllers not deliberately engage the free-chase fighters, but, rather, focus on stopping the bombers. Yet Goering, in his mid-August tactics conference, insisted that his fighter pilots escort the bombers at close range, thereby foregoing one of his most productive tactics.

As August ended and September began, desperation had descended upon Fighter Command. Training for new pilots had been reduced to days, pilots were being shipped from squadron to squadron to rebalance and compensate for the growing losses, yet the Germans seemed oblivious to the grim situation that faced their foe. With no more than days left before its collapse, Hitler saved Fighter Command with his redirection of the bombing campaign. Nothing in the Battle of Britain stands as starker testimony to the lack of German intelligence than does this decision, nor is any single action more indicative of the interdependence that exists between good leadership and good intelligence.

The intelligence failures that afflicted the campaign against the airfields and fighters were also evident in the bombing directed at Britain's aircraft industry. Here again, the Germans seemed to have had difficulty understanding which factories were critical to the

HOW AND WHY ... GERMAN OPINIONS
AT VARIOUS TIMES, GALLAND, GOERING, KESSELRING, AND HITLER GAVE OPINIONS WHY THE LUFTWAFFE LOST THE BATTLE. AMONG THE REASONS FROM THE GERMAN PERSPECTIVE...
POOR LEADERSHIP, CHANGING ORDERS, CONSTANT CRITICISM. -GALLAND
LACK OF AGGRESSIVENESS WITH COMMANDERS AND PILOTS. -GOERING
RADAR AND FIGHTER CONTROL SURPRISED GERMANS. -GALLAND
LOSS OF PILOTS OVER CHANNEL. -GALLAND
INADEQUATE RANGE FOR FIGHTER ESCORTS. -GALLAND
KESSELRING AGREED THAT FIGHTING OVER THE CHANNEL WAS DECISIVE.
POOR TARGET SELECTION. -GALLAND
STUKAS WERE ILL-SUITED FOR THE BATTLE. -GALLAND
INADEQUATE BOMBERS FOR THE MISSION. -GALLAND
OCT 4. HITLER TELLS MUSSOLINI HE ONLY NEEDED 5 CONSECUTIVE DAYS OF GOOD WEATHER TO CONDUCT INVASION.

production of war matériel, especially fighter aircraft. For example, one would have thought that the Spitfire production facility at Woolston would have been targeted and hit again and again from the outset. It should have been a primary target, yet attacks were relatively sporadic and appear to have suffered from the same "a target hit is a target destroyed" mentality that afflicted the campaign against the airfields. By the time attacks appeared to be fully organized and the factories effectively targeted, Lord Beaverbrook's shadow-factory policies had sufficiently distributed aircraft production that damage to a single factory did not significantly alter total production output. Factory production, plus the additional aircraft that were repaired and returned to the fight, ensured that the number of fighters available to Fighter Command was more than adequate during most of the duration of the battle.

Another critical lapse in German intelligence was evident in the extent to which it failed to understand the integrated air defense system and, in particular, the critical role that the Chain Home radar system played in making that system the force multiplier that it was. The Chain towers were no secret. Each site had two giant towers that could be seen for miles; before the battle began there were 29 sites arrayed along the southern and eastern approaches to England. In 1939, the Germans, suspecting that the towers might be associated with air defense, sent a Zeppelin dirigible to hover off the coast and collect electronic intelligence. The electronic collection mission was inconclusive, and the Germans apparently assumed that the Chain Home stations were somehow related to air traffic control, dismissing the likelihood that the British possessed a significant radar capability. The Germans had a radar facility, *Freya* by name, located in the Pas de Calais, that they used for detecting shipping in the Channel. It appears that they were again victim to their predisposition to think of airpower as an adjunct to surface warfare. Whatever the reason, they did not understand the British IADS as a key to independent air operations and failed to attach the appropriate urgency to the destruction of the radar sites. Once *Adlerangriff* was initiated in mid-August, the Luftwaffe did in fact target the radar stations. However, they found them difficult to destroy due to the steel lattice construction of the towers, and Goering soon directed that the effort to destroy them be abandoned.

The failure of German intelligence is surprising. We tend to think of the Germans as having been very thorough and methodical in their preparations for war, and that reputation often tends to obscure the fact that their ability to conduct military intelligence operations was clearly lacking. The Germans performed an intelligence estimate in 1940 (*Studie Blau*) that they never updated as additional data became available. They accepted the conclusions of *Studie Blau* and either rejected any information that conflicted with them or, for reasons that are patently unfathomable, believed that updates were unnecessary. They clearly did not develop adequate sources of intelligence inside Britain that might have informed them on key issues such as airfield usage and status, factories, and radar; nor did they make adequate use of intelligence that might have been gleaned from their own airborne crews. The Germans had no intelligence apparatus within the operational units of the Luftwaffe. Without current and accurate information, the Luftwaffe was, on several occasions, reduced to guessing about critical issues such as the disposition and status of Fighter Command resources. In the absence of this kind of information, the targets selected and the decisions made about strategy and tactics were too often wrong—and those errors lead again to the conclusion that the top leaders at the Hitler and Goering levels were inadequate to the task they set for themselves and their air force.

Unacceptable Losses

The Luftwaffe, for a variety of reasons, was structured and equipped to perform as a complement to the German Wehrmacht in the conduct of ground operations. It was an integral part of the blitzkrieg attacks launched on the countries of continental Europe, and it was a force optimized to support highly mobile ground operations. When German forces arrived on the coast of northern France and looked northward to Britain as their next, and final, western target, they probably did not immediately consider that the English Channel would force them to fight a completely different kind of battle than any they had fought up to that point. So different would this next campaign be, in fact, that Germany would find its vaunted Luftwaffe ill structured and ill equipped to conduct the independent air war that the Channel would force it to fight. The price would be losses that exceeded expectations and a level of combat effectiveness that failed to meet either expectations or projections. It exposed the weaknesses of German leadership and finally resulted in the deterioration of aircrew morale that finally culminated in the defeat that first became evident to the aircrews on both sides on September 15—Battle of Britain Day.

Events and history conspired to make the German Luftwaffe a formidable air force, but very much a force optimized for tactical operations. The Treaty of Versailles denied Germany access to aeronautical technologies and, in particular, propulsion technology, with the result that the country turned to gliders as the primary outlet for innovation, design, and development related to aircraft. Willy Messerschmitt learned aircraft design in this environment, and the lessons he took from this experience—the techniques required to keep airframes as light and adaptable as possible—would show up in his later designs that led to the Bf 109, Germany's primary fighter throughout World War II.

Generalmajor Walther Wever, the first chief of the air staff, had envisioned airpower as the cornerstone for a German force that would have supported a strategic view of Germany as the single central European power. Wever's vision called for Germany to have strategic air forces that could reach all corners of Europe with sufficient firepower to exert its will as necessary to control the continent. Key to his vision was development and deployment of a heavy, four-engine bomber that would have long striking range and could carry huge bomb loads. In 1934, Wever initiated a program to build such a bomber. Wever died in an aircraft accident in 1936, and was succeeded by General Kesselring, later *Luftflotte 2* commander during the Battle of Britain. Kesselring had been an army officer, and his view of airpower was limited to its tactical applications in support of ground forces. With Goering's approval, Kesselring canceled the four-engine bomber project in favor of building more two-engine bombers, which he viewed as adequate to the tasks he envisioned for the Luftwaffe. This decision ensured that the German bomber fleet, consisting of the He 111, Do 17, and the Ju 88, would be a force of light, two-engine bombers with limited bomb load capacity.

In the late 1930s, Germany entered the Spanish civil war and used that experience to both test its new equipment and refine the tactics and procedures that would be used to fight a few years later in Europe. The Condor Legion was a combined-arms force—air and ground units that were joined in highly coordinated and cooperative operations. Here the elements of blitzkrieg were developed and refined. The new systems—the Ju 87 Stuka dive-bomber and the Bf 109—performed magnificently, confirming the judgment made earlier to design the Luftwaffe to support ground tactical operations.

Using the lessons learned in Spain, Germany moved to consolidate its position in Europe. Blitzkrieg tactics were employed with stunning success again and again, as Germany defeated Austria, Czechoslovakia, the Netherlands, Belgium, and then France. In every case, victory was overwhelmingly decisive; nothing occurred that would have caused Germany to doubt the wisdom of the decisions made earlier that had shaped the structure and fundamental tactics of its air forces—and then the Battle of Britain started.

The reality of the English Channel meant that the campaign launched to defeat Great Britain would be completely different from the earlier battles. The usual combined-arms campaign tactics would be impossible. Here there would be no artillery to soften targets, no highly mobile ground assault to consolidate gains, and no surprise air raids that appeared as if from nowhere and lasted only a few minutes before the airplanes once again disappeared over the horizon. No, this would be different, and Germany would soon find that it was poorly prepared to fight the battle it had chosen.

The Luftwaffe soon found that the fighter designs that had served so well in its continental campaigns were not well suited to the missions they were assigned. The design of the Bf 109, which combined the structural lightness evolved from glider designs with the largest engine possible, had produced a fighter that was extremely capable in terms of speed and maneuverability, but which was also extremely limited in combat radius. The airplane that was to destroy the RAF and achieve air superiority could linger over British territory for only a few minutes before it was forced to return to Europe to refuel. To compound matters, the Me 110, designed as a multirole fighter-escort, lacked the maneuverability to compete with the new generation of British fighters, forcing the Bf 109s to assume ever-increased bomber escort responsibilities, a mission for which they were ill suited due to the range limitations they suffered.

The shortsightedness that left Germany without a heavy bomber now began to become apparent as well. The relatively light loads that the tactical bomber fleet carried necessitated formations of huge numbers of aircraft to produce the levels of destruction desired. Once their limited range had forced the escorts to turn back, the bombers were vulnerable to attacking British fighters. Losses mounted, forcing unusual—and sometimes unwise—tactical policy shifts. For example, Goering insisted that no aircraft have more than a single officer on board, and he further demanded that the Bf 109 give up free-chasing, one of the truly successful tactics, in order to provide additional close escort for the bombers.

Nor was the German navy structured adequately for the tasks it faced in the battle. The German general staff had dismissed the challenge of crossing the Channel as no more than a "wide river crossing," an assessment that, in retrospect, is breathtaking in its naïveté. Even at its narrowest point, the Channel is 30 miles wide. Thirty miles of open water is hardly insignificant—recall the efforts that the Allies expended in their preparations to cross the Channel in 1944. As a result of this rather cavalier dismissal of the scale of the problem, adequate force structure to accomplish the crossing was never provided. Had the German navy actually been called upon to transport ground forces to the English coast, there is every probability that the result would have been disastrous.

Blitzkrieg depended on surprise as an essential element to succeed, and in its campaigns on the Continent, surprise had played a major role in the rapidity with which the defenders had collapsed in face of the German onslaught. In the campaign against Britain, there was no surprise. The integrated air defense system, with its Chain Home radar detection system, ensured that Fighter Command was aware of every major attack. Defenders could watch the bombers and fighters forming up over the Pas de Calais and see them as they headed toward England. They knew how many aircraft were in the formation, at what altitudes they were flying, and where they appeared to be headed. This information enabled them to prioritize among attacks, deciding where to concentrate the defenses and to choose where they would intercept. The ability to efficiently allocate fighter resources to maximum advantage was the key to Britain's ability to defend effectively with a relatively small fighter force.

The German leaders had consistently underestimated the size of the defending British fighter forces. Failing to appreciate the combined effects of the IADS and Britain's fighter production and repair capabilities, they had repeatedly told their aircrews that the British were running out of fighters. The crews knew only that the fighters rose to meet them every time they entered British airspace. The crews began to observe cynically that every day they faced the "last 50" Spitfires. This loss of faith in leadership had the twin effects of destroying the morale of the flying crews and lessening their confidence that their leaders knew what they were doing in terms of tactics and strategy. The morale of German aircrews was further lowered by the fact that, when they were forced to bail out, they knew that no good could result. If they bailed out over British soil, they became prisoners of war; if they bailed out over water—and they spent considerable time over water—the best they could hope for was rescue after the discomfort and terror of being downed in the Channel or the North Sea. In fact, a common condition among German crewmen was known as *Kanalkrank,* which described a growing obsession with being downed at sea and manifested itself as lowered morale and, sometimes, outright fear of flying over the water.

As the morale of the Luftwaffe crews steadily eroded and losses mounted, the daily demand that they form up and fly into the face of an enemy they seemed unable to eradicate finally began to impact their fighting ability. On September 15, 1940, the combat losses, the inability to destroy the RAF, and the seemingly endless futility of it all combined in a way that somehow impacted the entire German attacking force. Flight discipline broke, and the formations disintegrated in disarray and chaos. On that day, the aircrews on both the English and the German sides knew in their hearts that the Battle of Britain had been won by the stubborn pilots of Fighter Command. The leadership did not, at first, recognize the inevitable, and the bombing was to continue into the spring of 1941, but, for all the reasons discussed here, the outcome had been determined. The Battle of Britain, which had officially begun on July 10, 1940, was later declared to have officially ended on October 31 of the same year. It is forever remembered as a battle fought entirely in the air and as the battle that turned the tide of western history in World War II.

REFERENCES AND BIBLIOGRAPHY

Air University. 2001. E. Milch. "The Combined Bomber Offensive." Translated by W. Geffen. Maxwell Air Force Base, AL: www.au.af.mil.

Batchelor, J., and B. Cooper. 1973. *Fighter: A History of Fighter Aircraft.* New York: Charles Scribner's Sons.

Brickhill, P. 1964. *Reach for the Sky.* London: Collins Clear Type Press.

Brown, M. 2000. *Spitfire Summer.* London: Carlton Books.

Chant, C. 1979. *Pictorial History of Air Warfare.* London: Octopus Books.

Churchill, W. C. 1949. *The Second World War.* Vol. 2, *Their Finest Hour.* Boston: Houghton Mifflin.

Cooling, B. F. (ed.). 1994. *Case Studies in the Achievement of Air Superiority.* Washington, DC: Center for Air Force History.

Coombs, L. F. E. 1997. *The Lion Has Wings.* Shrewsbury, England: Airlife Publishing.

Daniels, J. (ed.). 1988. *Aircraft, World Wars I and II.* Stamford, CT: Longmeadow Press.

Deere, A. C. 1959. *Nine Lives.* London: Hodder and Stoughton.

Deighton, L. 1979. *Fighter: The True Story of the Battle of Britain.* London: Pluriform Publishing.

Deighton, L., and M. Hastings. 1980. *Battle of Britain.* Hertfordshire, England: Wordsworth Editions.

Dowding, H. 1946. "Official Dispatch," *The London Gazette.* September.

Flack, J. 1985. *Spitfire: A Living Legend.* London: Osprey Publishing.

Freeman, R. 1993. *The Royal Air Force of World War II in Color.* Stillwater, MN: Specialty Press.

Goss, C. 1994. *Brothers in Arms.* Bodwin, England: Hartnolls.

Green, W., and G. Swanborough. 1981. *Flying Colors.* Carrollton, TX: Squadron/Signal Publications.

Haining, P. 1985. *The Spitfire Log.* London: Souvenir Press.

Hallion, R. P. 1998. "The Battle of Britain in American Context and Perspective" Air Force History and Museum Programs, Bolling Air Force Base.

Heaton, C. D. 1997. "Luftwaffe General Adolf Galland." www.thehistorynet.com.

Hough, R., and D. Richards. 1990. *The Battle of Britain: The Greatest Air Battle of World War II.* New York: W. W. Norton.

Ishoven, A. van. 1977. *Messerschmitt Bf 109 at War.* New York: Charles Scribner's Sons.

Jackson, R. 1976. *Aerial Combat.* London: Wiedenfeld and Nicolson.

Kaplan, P., and R. Collier. 1989. *Their Finest Hour: The Battle of Britain Remembered.* New York: Abbeville Press.

Mason, F. K. 1969. *Battle Over Britain.* London: McWhirter Twins.

Moll, N. 1990. "The Battle of Britain," *Flying*, September, p. 40.

Mosley, L. 1977. *The Battle of Britain.* Alexandria, VA: Time-Life Books.

Murray, W. 1983. *Strategy for Defeat: The Luftwaffe 1933–1945.* Maxwell Air Force Base, AL: Air University Press.

National Aviation Hall of Fame. 1998. "Raoul Gervais Lufbery," *Enshrinee Home Page.* www.nationalaviation.org/enshrinee/lufbery.html.

Ovary, R. 1999. *The Battle of Britain: The Myth and the Reality.* New York: W. W. Norton.

Parkinson, R. 1977. *Summer, 1940.* New York: David McKay.

Pimlott, J. 1998. *Die Luftwaffe—Die Geschichte der deutschen Luftwaffe im Zweiten Weltkrieg.* London: Brown Packaging Books.

Price, A. 1982. *The Spitfire Story* London: Jane's Publishing.

Public Broadcasting System. 2001. *Decoding Nazi Secrets.* September.

Quill, J. 1986. *Birth of a Legend: The Spitfire.* Washington, DC: Smithsonian Institution Press.

Rimell, R. 1985. *Spitfire, Supermarine Mark V.* Vista, CA: Aerolus Publishing.

Rimell, R. 1990. *Battle of Britain Aircraft.* Hertfordshire, England: Argus Books.

Robinson, A. (ed.). 1979. *In the Cockpit: Flying the World's Great Aircraft.* New York: Ziff-Davis.

Sarkar, D. 1999. *Battle of Britain: The Photographic Kaleidoscope.* Worcester, England: Ramrod Publications.

Scutts, J. 1980. *Spitfire in Action.* Carrollton, TX: Squadron/Signal Publications.

Scutts, J. 1986. *Hurricane in Action.* Carrollton, TX: Squadron/Signal Publications.

Townsend, P. 1970. *Duel of Eagles.* New York: Simon & Schuster.

INDEX

ABOUT THE AUTHORS

PAUL JACOBS is an aerospace engineer in Alexandria, Virginia, specializing in stealth technology and flight testing. A career Air Force officer, he graduated from West Point in 1971. After a tour with the U.S. Army and the 82nd Airborne Division, he earned a master's degree in aerospace engineering from Penn State University. After graduate school, he returned to active duty with the U.S. Air Force. He was stationed at the Air Force Flight Test Center, Edwards Air Force Base, where he graduated from the Air Force Test Pilot School. He participated in the initial flight testing of the B-1, B-2, KC-135R, and X-29 aircraft, as well as several classified flight test programs. He accumulated over 500 hours of flying time in 14 different aircraft as a flight test engineer. He completed his Air Force career with a tour as a professor at the Defense Systems Management College. He taught systems engineering and served as the deputy chairman of that department. He currently teaches and consults in the area of stealth technology and flight test engineering with Modern Technology Solutions, Inc. He is also a widely recognized aviation artist. He has published aviation art prints and has held one-artist shows of his work. He works part-time as a commission artist and illustrator. He is married to a special education teacher, has three children, and is a passionate fly fisherman.

ROBERT LIGHTSEY teaches at the Defense Acquisition University, near Washington, D.C. A career Air Force officer, he graduated from the U.S. Air Force Academy in 1962. He became a pilot upon graduation, and he has over 4000 hours of flying time in various types of aircraft, including trainers, cargo planes, and fighters. He flew as a forward air controller in Vietnam, where he won the Distinguished Flying Cross. Upon his return from Southeast Asia, he earned a master's degree in aeronautical engineering at the University of Illinois. His career then turned from flying and operational assignments to assignments in engineering and project management associated with the research and development of Air Force systems. He has worked on aircraft, simulators, and weapons at various times in his career in R&D. His assignments have spanned various levels of responsibility, from project office to the headquarters level in the Pentagon. He retired from the Air Force as a colonel in 1986. Upon his retirement, he worked for five years as a consultant in technologies related to national defense with a small company in the Washington area. He then took a position as a professor at the Defense Acquisition University, where he currently chairs the technology and engineering department. He earned his doctorate in engineering management from the George Washington University in 1997. He is married with two children, both of whom are teachers.

Aviation Week Books is an imprint of McGraw-Hill Professional Book Group in conjunction with the *Aviation Week* division of The McGraw-Hill Companies. With nearly 50 products and services and a core audience of some 1 million professionals and enthusiasts, *Aviation Week* is the world's largest multimedia information and service provider to the global aviation and aerospace market. For more information, use www.aviationnow.com

Editorial Director: Stanley Kandebo, Assistant Managing Editor, *Aviation Week & Space Technology*